Hanford's Battle with Nuclear Waste Tank SY-101

Bubbles, Toils, and Troubles

Chuck Stewart

Battelle Press

Columbus • Richland

Library of Congress Cataloging-in-Publication Data

Stewart, Chuck, 1947– .
    Hanford's battle with nuclear waste tank SY-101 : bubbles, toils,
and troubles / by Chuck Stewart.
        p. cm.
    Includes bibliographical references and index.
    ISBN 1-57477-155-8 (pbk.)
        1. Radioactive waste sites—Cleanup—Washington (State)—
Hanford site—Case studies. 2. Radioactive wastes—Storage—
Washington (State)—Hanford site—Case studies. 3. Nuclear accidents—
Washington (State)—Hanford site—Prevention—Case studies.
4. Underground storage tanks—Washington (State)—Hanford site—
Case studies. I. Title.

    TD898.12.W2S74 2006
    363.72'8909797—dc22

                                                        2005057065

Printed in the United States of America

Copyright © 2006 Battelle Memorial Institute. All rights reserved.
This document, or parts thereof, may not be reproduced in any form
without the written permission of Battelle Memorial Institute.

Battelle Press
505 King Avenue
Columbus, Ohio 43201-2693, USA
614-424-6393 or 1-800-451-3543
Fax: 614-424-3819
E-mail: press@battelle.org
Website: www.battelle.org/bookstore

The PNNL Information Release number assigned to this Book/Conference Proceedings
release is: PNNL-SA-43778.

To

Kaylene, Robin, and Whitney

Contents

Preface vii
Acknowledgments xiii
Foreword by Rick Raymond xvii

Beginnings: The Hanford Works, 1943–1980 1
    A Short History of Hanford 3
    The Workings of the Hanford Works 9
    The Tanks 19
    Tank Troubles Begin 31

Part One: Flammable Gas 35
    Storm Warnings: 1977–1989 37
    Landfall: October 1989–March 1990 55
    Blowing in the Wind: March–October 1990 63
    Open Windows: 1990–1992 97

Part Two: The Mixer Pump Era 141
    Deciding to Mix: 1991–1992 143
    Preparing the Pump: 1992 163
    A Window for Mitigation: 1992–1993 199
    The Pump Goes In: March–July 1993 235
    Mitigation Happens: 1993–1995 261

vi | Contents

Part Three: Surface Level Rise 289
The Eye of the Storm: 1995–1996 291
The Approaching Eye Wall: 1995–1997 301
Richland, We Have a Problem: 1997–1998 311
A Crusty Christmas: 1998–1999 341
Ready to Remediate: 1999 369
Remediation Happens: December 1999–April 2000 391
Are We There Yet? 2000–2001 409
Epilogue: 2001–2003 417

Chronologies 425

Glossary 437

Bibliography 445

Index 451

Preface

This book tells the story of a tank—not just any tank, but Tank 241-SY-101, one of 177 big underground radioactive waste storage tanks on the Hanford Nuclear Reservation in the south-central Washington State desert. So why does this tank deserve to have book written about it? SY-101 was the only nuclear waste tank at the Hanford Site and maybe one of only two or three in the United States that was demonstrably dangerous and a huge amount of money and energy was expended to deal with the danger. Besides, it's a good story.

It was an explosive subject. The chemical reactions and radioactivity in all radioactive wastes makes bubbles of gas, mainly hydrogen along with a little methane, nitrous oxide, and ammonia. This gas mixture is not only toxic, but also flammable! All the Hanford tanks generate these gases, but SY-101 was the most potent gas producer of all. Countless bubbles built up in its million gallons of thick, muddy waste until it couldn't hold any more. Then, every few months, thousands of cubic feet of gas suddenly rushed up in great roiling "burps." The largest burps actually made the air space at the top of the tank flammable for a few hours. The fear that this flammable gas might explode, blow the top off the tank, and spew radioactivity into the environment became a national issue.

Unfortunately, these fears were not entirely groundless. In 1957, a chemical explosion occurred in a Soviet radioactive waste tank at the Kyshtym site in the Ural Mountains. This accident shaped the public perception of how bad a flammable gas accident in SY-101 might be. The Kyshtym tank held mainly nitrate and acetate compounds with about ten times as much radioactivity as in Hanford Tank SY-101, but the Kyshtym waste was packed into a smaller volume. The high radiation load made the Soviet tank so hot that it had to be water-cooled. Apparently, the cooling process failed, the waste heated to over 600 degrees Fahrenheit, the nitrate-acetate salts dried out, and the tank exploded. About 10 percent of the radioactive material in the tank rose in a mile-high dust cloud that drifted with the wind contaminating 400 square miles with radioactivity measured at levels above what were considered allowable limits. Traces of strontium-90 from the accident were detected over another 7,300 square miles.

A Kyshtym-type chemical explosion could not have happened in SY-101. We now know that its waste was too wet, too cool, and had too little organic fuel to burn, let alone explode. Nevertheless, a hydrogen gas burn in the tank dome space was then at least a theoretical possibility, and it threatened similar kinds of consequences. The public, the media, and the government were concerned such an accident might actually happen and we didn't understand the tank well enough to prove it wouldn't. Ignoring the technical details, it was not hard for people to picture a cloud of radioactive contamination from Hanford floating over Spokane, Seattle, or Portland—like the ash eruptions from Mt. St. Helens, the worldwide plume from Chernobyl, or the mushroom cloud from an atomic bomb! After all, they made those bombs at Hanford, didn't they?

This is why SY-101 and the safety issues surrounding it demanded top priority at the Department of Energy (DOE), dominated waste management politics, and obstructed Hanford cleanup for the last decade of the 20th century. The tank seemed to almost have a personality—acting with violence and apparent malice, hiding information about itself, deceiving us with false indications, and sometimes lulling us into complacency only to attack in a new way. The wayward bubbles in SY-101 demanded the toil of scientists, managers, and officials from the time it was filled in 1980 until it was finally declared safe in January 2001. From 1990 through 1993, the tank's flammable gas troubles were acknowledged as the highest priority safety issue in the entire DOE complex. Uncontrolled waste level rise demanded another high-priority remedial effort from 1998 through April 2000.

A succession of Secretaries of Energy, Environmental Protection Agency administrators, the governors of Washington and Oregon and their own Departments of Ecology gave the tank their attention. Congress passed several laws attempting to control it. It gained national notoriety and grabbed headlines. Even the *New York Times* assigned a reporter to cover it. It appeared in at least one documentary on PBS television.

SY-101 Time Capsule

Tank 241-SY-101 is one of three 1.16-million-gallon underground radioactive waste storage tanks in the SY tank farm in the 200 West Area of the Hanford Site in south-central Washington State. Built between 1974 and 1976, SY-101 is one of the newer double-shell tanks with an outer steel shell wrapped around the inner primary steel tank to contain potential leaks. The inner tank is 75 feet in diameter and about 50 feet high at the center of the curved dome, which lays under 6 to 8 feet of gravelly soil. The waste in the tank is 35 feet deep when filled to its capacity of 1.16 million gallons.

SY-101 was filled in stages from 1977 to 1980 with a thick slurry of sodium salts laced with radioactive isotopes that settled into a sediment layer 16 to 18 feet thick under 14 feet of syrupy liquid topped by a 3-foot layer of crust. Radioactivity, mainly from decay of cesium-137 and strontium-90, heated the sediment layer to 135–150 degrees Fahrenheit and generated roughly 120 cubic feet of gas per day, including 40 cubic feet of hydrogen. From 1980 to 1993, the waste accumulated gas and suddenly released up to 10,000 cubic feet of it every 100 to 150 days in what were called "burps." The largest burps created a hydrogen concentration of over 5 percent in the air space above the waste. Four percent hydrogen is flammable.

The flammable gas hazard in SY-101 gained national attention in 1990. The hazard was defined in 1991 and a mixing pump was chosen as the method to stop the burps by stirring the sediment layer to prevent it from accumulating gas. The mixer pump was built in 1992, installed in July 1993, and mixed the tank by December 1993. The last burp occurred June 26, 1993.

But the troubles weren't over. The absence of periodic burps and the stirring action of the mixer pump caused the floating x | Preface

> crust layer to accumulate gas and the waste level to rise. Level rise became noticeable in 1996 and accelerated into an emergency in 1998. In early 2000, about half the waste in SY-101 was pumped out and water added back to dissolve much of the soluble solids and prevent gas accumulation. The mixer pump was permanently shut down in April 2000, and the tank was formally declared safe in January 2001. The tank received its first transfer of waste as a working tank in November 2002. A detailed chronology of events pertinent to SY-101 is given in the back of this book.

The direct cost of the bubbles, toils, and troubles was high. Overall, the price of dealing with the real and imagined hazards in SY-101 may have reached $250 million. Spending all this money fighting SY-101's safety issues only stirred the radioactive waste and moved it around–it accomplished no cleanup whatsoever. In fact, SY-101's flammable gas problem spawned suspicions of a much wider danger that impeded and complicated cleanup in the other 176 waste tanks for a decade.

The real tank waste cleanup job has yet to begin. SY-101 can be studied as a time capsule of all the broad scientific, managerial, political, and personal decisions that will have to be made to accomplish it. Thousands of similar time capsules will be created before the work is done, so we need to learn what SY-101 has to teach. The most important lesson, especially where radioactive waste is concerned, is that no action is risk-free. But doing nothing also carries an unacceptable risk! Uncertainty increases the risk, and there will always be uncertainty. Unless government agencies and their contractors accept reasonable risk, uncertainties overwhelm science and political forces stop progress dead. Uncertainty must be accepted—we just don't know enough to prescribe everything. But it also must be managed to keep the risk acceptable—we must learn as we go. Mechanical and managerial systems must be designed to adapt to changing knowledge and conditions, and even to encourage change—a little innovation can make the work much easier, quicker, and cheaper.

Although these big tasks are important, the story of SY-101 is not all about past safety hazards or future cleanup. At the core, it's the collective experience of people, from pervasive misconception to grand insight, near miss to sweeping success, meddling interference to close teamwork, all on an uncommonly large scale. It was a necessary catharsis that transformed the entire Hanford culture from a closed defense production operation to an open environmental cleanup project. Its tight project discipline and close teamwork became the Hanford standard. The final remediation of SY-101 placed second in an international project-of-the-year competition. Many consider SY-101 work the peak of their careers and measure all other experience by it. SY-101 defines some of the worst and the best of Hanford history. This book attempts to narrate and explain the whole human chronicle.

There is a vast amount of history in SY-101's story, but I did not write it as a history. By my own inclination and the advice of many of the participants I interviewed, it is a straightforward narrative without footnotes or formal reference citations. Most of the raw material for this book comes from recollections of those who lived it, handed down to me through interviews. I also had the use of a decade of notes in the personal notebooks of Jerry Johnson and Carl Hanson who were leaders in the struggle. Carl's notebooks also contain a rare collection of e-mail messages and informal memos that reveal what people knew, when they knew it, and what they thought and did about it. I also relied on my own experiences on the projects to focus the story on what I believe are important events and issues.

The story generally follows a chronological thread, with an occasional break to parallel narrative when too much was happening at the same time. I chose to illustrate the chronology with periodic plots of SY-101's waste depth that graphically show each burp, when the burps stopped, and how the level growth emergency developed. I have placed a brief list of important events at the head of each "book" and a large selection of lists and tables in the back to supply the details. These include a glossary of acronyms and uncommon words, a multi-page chronology of important events in SY-101's history, tables listing all the Hanford reactors, separation plants, and tank farms, and the government and contractor organizations that ran them.

I find, as most historians undoubtedly do, that there are often as many versions of an event as there are participants and observers describing it. For example, at least five people are claimed to have originated the buoyant gob theory for explaining SY-101's burps.

Unfortunately, we can't replay events to see which is true. When several versions of an event differed substantially, I had to either chose one, tell all of them, or concoct a composite that would accommodate as many as possible.

This book focuses on Tank SY-101, with only a few side trips to give a background or to add interest. This leaves out a lot of hard work and interesting research that was done in the broader Flammable Gas Project that Jerry Johnson led in parallel with Jack Lentsch's Hydrogen Mitigation Project. The tight focus neglects whole areas like the gas generation studies of Sam Bryan, Jim Person, and Albert Hu; most of the gas retention research by Phil Gauglitz and Scot Rassat; much of the theoretical work of Perry Meyer, Paul Whitney, and Bill Kubic; and the chemistry studies of Larry Pederson and Dan Herting. Some future book will need to tell those stories.

The focus of the book is also heavily influenced by my own viewpoint and Hanford experience. My role was to try to understand this tank's behavior and to help guide efforts to change the behavior. Thus I probably pay less attention than I should to upper management, local public perception, and the down-and-dirty field work out in the farms. I hope the information I gained from interviews makes up for at least part of the neglect.

I have tried to make this story simple and clear, so no one needs an advanced degree to understand and follow it. At the same time, I wanted to give enough technical details to interest the scientists and engineers. Above all, I emphasized the human side of the story wherever I could. I have set off interesting stories of personal incidents, passions, successes, and failures in a different typeface to provide occasional relief.

It was a rare privilege to play a part in the last ten years of the struggle with SY-101. The greatest privilege, rather honor, was in working with and gaining the respect of so many really good people. It was teamwork at its best that may never be equaled. I sincerely hope this book does it justice!

CHUCK STEWART
Kennewick, Washington
January 2005

Acknowledgments

Hundreds of people gave their best efforts to the conquest of SY-101 and this book is their story. Without the kind and gracious help and support from a large group of them, writing it would have been impossible. "In a multitude of counselors there is safety" *Proverbs 11:14.* I want to begin with my sincere thanks to Jerry Johnson and Carl Hanson for trusting me with stacks of their personal notebooks from the "glory days" of SY-101 mitigation and remediation. These treasuries of experience taught me many things I never realized had happened and gave me a broader view of others I already knew about. I am also indebted to Tony Benegas who loaned me literally boxes of color photographs from the days of building and installing the mixer pump.

The core of the book is built on the experiences of the many people who granted me interviews. In chronological order of their interview, grateful thanks to Cal Delegard, who also loaned me his copy of the "Blush report" and his collection of newspaper clippings; Nick Kirch, a valued co-worker, for his excellent chronology of the early years; Jack Lentsch, who personified the SY-101 mitigation project more than anyone, for insights on the political issues and management perspective in the mitigation era; Kemal Pasamehmetoglu, who gave me time between his son's soccer matches in Albuquerque; Scott Slezak, who gave me valuable details on the evolution of the Tank Advisory Panel and minutes of early

| xiii

xiv | **Acknowledgments**

meetings; Bob White, for instructing me on early Los Alamos participation; Jerry Johnson, who, besides his notebooks, also loaned me his file of newspaper clippings and magazine articles, and his extensive SY-101 files including what may be the only complete set of "Window" documents.

Thanks to Rick Raymond, for the privilege of serving on his remediation team, for access to his collection of digital project photos, for answering numerous questions about the startup of the remediation project, and for providing his foreword to the book; Rudy Allemann, my SY-101 Mitigation Project mentor, who gave me two boxes of his notes, historic documents, and calculations from the mitigation project; Bob Marusich, for describing the frantic press of early safety analyses, and for providing several historical reports; Carl Hanson, for his personal notebooks full of historical gems and photos depicting the mixer pump era; Craig Groendyke, for giving me the Department of Energy's perspective and describing what people "downtown" were thinking; Perry Meyer, my valued co-inquirer in understanding flammable gas; and Harry Harmon, for his *very* thorough review of the manuscript, and for copies of his collection of newspaper clippings and magazine articles on mixer pump installation.

I appreciate the Dan Reynolds' help, for imparting a little of his vast knowledge of the tanks, including his experience in the evaporator campaign that made SY-101 what it was, for the two "scrolls" of SY-101 waste level history plots annotated with important events, and for the three file drawers of his accumulated historic papers and documents; Paul D'Entremont, who told me how the initial safety issue ignited; Don Trent, for his collection window evaluation memos and unique anecdotes about his duty at the old Hanford missile sites; Joe Brothers, consummate "Mr. Nice Guy," for people photos and valuable tidbits of information; Tom Michener, for the loan of tapes of television news interviews on TEMPEST modeling; Max Kreiter for the perspective of the first Pacific Northwest National Laboratory involvement in the Mitigation Project; Ben Johnson, who hired me in 1973; Tony Benegas, for innumerable anecdotal "nuggets" and his photo collection; and JR "Hans" Biggs, for his insights on experience in the tank farm and how engineers and scientists should behave.

Thank you, Doug Larsen, for your narrative that gave life to running the mixer pump, and for numerous consultations on event chronology; Fred Schmorde, for providing the operations viewpoint and the heritage of the nuclear navy; Dan Neibuhr, truly a "walking history book" on SY-101; and to Steve Marchetti, about whom I heard more wild stories than anyone else, and who verified that at least some of them were true. Thanks to John Wagoner who kindly described some of the frustrations and satisfactions of being DOE Richland's point man during both periods of SY-101's misbehavior.

I also need to recognize Del Lessor for his guidance on nuclear physics, the excellent resources of the DOE Public Reading Room, the "PictureThis" website of historic photos organized by Pacific Northwest National Laboratory, and the huge collection of declassified Hanford photographs and documents on the DOE Hanford Historic Site website. I am very grateful to Laura Cooper for gathering up a list of archival photographs from the files of the *Hanford Reach*.

A very special thanks to Sadie Haff who did the first full edit of the manuscript and educated me on the fine points of grammatical construction, punctuation, and the limits of author's license.

I am grateful to Hanford authors Roy Gephart and Michelle Gerber, and to Joe Sheldrick of Battelle Press for giving me the confidence that writing a book like this was actually possible. At the other end of the chronology, I am very grateful to late Battelle Vice President Walt Apley for the final management review and for weaving all the administrative, financial, and political threads together to reach the final decision to "publish it!" Special thanks to Kristin Manke, Sheila Bennett, and Barbara Wilson for doing the final polishing edit. It was a pleasure working with Susan Vianna of Fishergate, Inc., who constructed the layout and with Karen Buxton of PNNL who created the index.

Most of all, I am greatly blessed by my wonderful wife, Claudia, for tolerating the pacing, muttering, and interminable keyboard clatter for the 18 months it took to produce the manuscript. She also gets credit for many of the constructs and concepts I used to illustrate and add interest to the book. Her "big picture" insights on the flow of the story were extremely helpful.

But I must credit my Lord and God for crafting the conjunction of all the people, experiences, circumstances, resources, and abilities that allowed this to happen. This book is hard evidence of His promise that "I can do all things through Christ, who strengthens me" *Philippians 4:13.*

Foreword

I was pleased for several reasons when Chuck asked me to write the foreword to this book. First, it gives me the opportunity to discuss Chuck's role in the history of Tank SY-101 in a way that he could not do himself without appearing to be self serving.

I was introduced to Chuck Stewart by Joe Brothers when I was organizing the team to work on the Surface Level Rise Remediation Project. Admiral H. G. Rickover once told me to "Trust your experts, or find experts you can trust." As a result I asked Joe to name the person he considered to be "the world's expert in scientific understanding of tank SY-101 waste behavior." Joe named Chuck. Joe was right.

We frequently asked Chuck to predict tank waste behavior. Our requests went something like this: We would have a team brainstorming session and come up with ideas on what to do next. Then, we would ask Chuck to predict how the waste would behave if we did as suggested. He was not always able to give a precise prediction; he was ALWAYS able to give a bounding analysis. This was essential to the success of the project for the following reason.

We were working with a high hazard situation: over a million gallons of waste that was chemically hazardous, highly radioactive, contained gasses that were highly toxic, AND contained gasses that were flammable (we never liked it when others called them explosive). To make matters worse, we did not have an option to "do nothing" since the trapping of flammable gases was continuing and

xviii | **Foreword**

the waste level was rising! It was a slow-motion emergency with the condition getting worse, at an increasing rate, every day!

We had to be able to predict whether our actions could be done safely and effectively. Chuck, with help from those he gives well-deserved and appropriate credit in this book, was in charge of predictions. He provided the scientific understanding that was the safety basis for all of our actions. Without this basis, we would have been at dead stop. We could take action only after we could prove that it could be done safely. Chuck provided the foundation for that proof with his bounding predictions. Without Chuck and his tireless efforts in support of the team, the history of SY-101 might have been very different.

My second comment is on safety and quality versus productivity. In a project like this, where the goal was to improve the safety of a worsening condition, it might have been tempting to play loose with safety or quality with an "end justifies the means" type of thinking. The team described in this book was committed to the principle that the fastest and cheapest way to meet the goal was also the safest one with high quality. Aldous Huxley says it this way, "The end cannot justify the means for the simple and obvious reason that the means employed determine the nature of the ends produced." The team behaviors resulting from this commitment to safety and quality were the key: a constant self assessment of the safety and quality of every action, a willingness to stop in the face of uncertainty or criticism (from either inside or outside the team), constantly questioning each other and watching out for, and identifying and helping to correct, any and all mistakes in an open and constructive manner.

The project statistics resulting from this behavior speak for themselves: the project baseline was for a four-year, $55 million project to stop the waste level rise within SY-101. The project finished in two years for $35 million. As for safety and quality, the project ended with one first aid injury and an otherwise perfect safety record, all quality issues were self detected and corrected before being deployed or applied in the field. As for technical accomplishments, the goal was to stop the waste level growth in the tank; the results were to reduce the safety hazard sufficiently that the tank could be returned to beneficial service. It is now used for routine service in the Hanford tank farm cleanup mission. Several other technical and operational innovations came out of this team that have gone on to wider use to improve safety, quality, and reduce costs of other projects.

Lastly, this foreword gives me the opportunity to discuss the team, including Chuck, which I had the great fortune to work with on this project. There is something wonderfully powerful in a team of talented, experienced people with a shared vision of a worthwhile goal. We often read stories in the newspaper of safety problems, or management problems, or union problems, or engineering mistakes, or government mis-management of projects. The team at Hanford, including scientists, engineers, union personnel, government managers, regulators, and, yes, Hanford contractor managers, worked together, as a team, using their talents to best effect.

I can say from personal experience that in this case the shared vision of all of these groups to perform work, with the highest standards for worker and public safety, to effectively solve a problem was wonderfully powerful. Almost nothing was impossible for this team! We would watch each other in a positive and supportive way.

No single person deserves the credit for the success of the work on Tank SY-101. The team was the real success story. Pliny may have said it best, "True glory consists in doing what deserves to be written; in writing what deserves to be read; and in so living as to make the world happier and better for our living in it." Thank you, Chuck, for doing what deserves to be written and writing what deserves to be read about a team that has made the world safer and better.

RICHARD RAYMOND
Richland, Washington
2004

BEGINNINGS

The Hanford Works
1943–1980

August 1942 – Manhattan Project established to build an atomic bomb

March 1943 – Corps of Engineers begins building the Hanford Works

September 1944 – B Reactor, Hanford's first, starts up

July 1945 – Trinity Test used Hanford plutonium for the first atomic bomb

1952-57 – Uranium recovery process in U Plant adds ferrocyanide compounds to some waste tanks

1959 – Old single-shell tanks begin leaking

1968-85 – Strontium recovery process in B Plant adds organic complexants to some waste tanks

1968 – Double-shell tank construction begins with the AY farm

1974-77 – SY double-shell tank farm constructed

April 1977 – SY-101 begins receiving waste

November 1980 – SY-101 filled

A Short History of Hanford

About 200 years ago, President Jefferson sent Captains Meriwether Lewis and William Clark to lead the Corps of Discovery up the Missouri River to search out a route to the Pacific Ocean. On their way, they explored the southern edge of the Columbia River Basin just below the present Hanford area. The expedition accomplished one of the first systematic scientific studies on this continent. The Corps of Discovery carefully recorded what they observed in nature as they passed, and left few physical signs except their journals, maps, letters, and some preserved specimens.

About 60 years ago, President Roosevelt authorized General Leslie Groves and Colonel Frank Matthias to lead the Army Corps of Engineers in building the Hanford Engineer Works along the Columbia River in Washington State to make plutonium for the atomic bomb. The Hanford Works was part of the Manhattan Project, a super-sized industrial development program to scale up atomic theory and laboratory experiments to full production. Unlike the Corps of Discovery, the Manhattan Project and its successors did not just observe and record nature—they fundamentally changed nature, creating elements in their reactors that had never existed before. The footsteps of Lewis and Clark and the Corps of Discovery

| 3

have long since vanished. The Corps of Engineers and the Hanford Works, on the other hand, left a lot more evidence. The nuclear waste in Tank SY-101 was one of their artifacts.

President Roosevelt hoped the Manhattan Project would shorten World War II, and it did. Bombs built with plutonium from Hanford and enriched uranium from the Oak Ridge Site in Tennessee ended the war with Japan without the immeasurably greater loss of life that an invasion of Japan would have cost. Hanford affected the Cold War too. Keeping up with our plutonium production and nuclear weapons development exhausted the Soviet Union's economy, leaving the United States the sole world superpower. Ironically, by fencing off the Hanford area for half a century the government created a unique natural preserve much larger than the area actually occupied by the production plant, still undisturbed by development.

These strategic gains were costly. The Manhattan Project and its successor, the Atomic Energy Commission, altered the landscape with millions of tons of steel and concrete, poured contaminants into the soil, and emitted radioactive particles into the air that affected populations a hundred miles downwind. For a decade, they even controlled the Columbia River's flow for official purposes. A by-product of the primary plutonium production mission was the millions of gallons of dangerous, and even deadly, radioactive waste that must be isolated from the public for the foreseeable future.

Managing this waste was relatively easy through the 1960s, when nuclear weapons work had top priority. There was usually money available to develop or improve processes to create less waste or to build new tanks when the old ones filled or started to leak. But after national plutonium stockpiles filled and production dropped off in the 1970s, Hanford's priority plummeted and funding stagnated. More and more of the oldest tanks began leaking, and their waste had to be pumped out, but only a few new tanks were available to accept it. The new Hanford waste vitrification plant was supposed to provide some relief in the mid-1990s by turning the waste into radioactive glass. But the glass plant was delayed and is only now under construction. The one viable option left was to cram more and more waste into the few new double-shell tanks by concentrating it into a thicker and thicker soup. Unfortunately, it was this well-intentioned process that made the waste in SY-101 go bad.

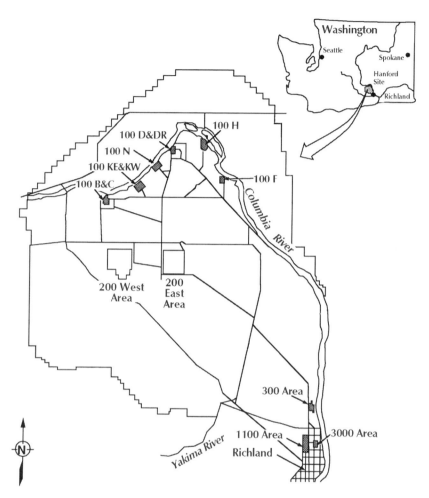

Hanford Area in mid-1960s showing reactors (100 Areas), production areas (200 Areas), and fuel fabrication site (300 Area) in relation to the administrative center (1100 Area), research labs (3000 Area) and the town of Richland and Washington State.
Courtesy DOE-RL Declassified Document Retrieval System.

WHY HANFORD?

When the United States entered World War II, physicists informed President Roosevelt about the potential power of an atomic bomb, and the possibility of building one. It would be disastrous if Nazi Germany or Japan did it first. After the attack on Pearl Harbor, Roosevelt authorized a top priority, secret project to turn the might

6 | Beginnings: The Hanford Works, 1943–1980

of U.S. industry to building atomic bombs. He hoped that having such unimaginably powerful weapons would help end the war, and strengthen our position in post-war negotiations with the Soviet Union. The Army Corps of Engineers would manage the work. The Corps organized the effort as the Manhattan Project and put General Leslie Groves, who had just finished building the Pentagon, in charge.

The Manhattan Project developed two different kinds of bombs in hopes that at least one would be successful. Both were nuclear fission bombs that liberated an enormous, instantaneous blaze of energy by suddenly bringing together a critical mass of fissionable material to ignite a runaway chain reaction. One design created the critical mass with uranium-235, an easily fissionable isotope of natural uranium-238, by literally shooting one piece of it into another with a "gun." Success of the "gun bomb" itself was almost a sure thing, but separating enough uranium-235 from chemically identical but infinitesimally heavier uranium-238 was almost impossibly difficult.

The other bomb design used the much more powerful plutonium-239, manmade in a reactor by forcing natural uranium-238 to swallow a neutron. But plutonium tended to fission earlier than uranium, and the gun method could not assemble a critical mass fast enough. The first fissions would blow the pieces apart before the main reaction got going. Instead a chemical explosion was used to almost instantaneously compress or "implode" a hollow sphere of plutonium into a tightly packed critical mass. Creating a sufficiently fast and uniform implosion was the real challenge of the plutonium bomb. However, because plutonium is chemically different from uranium, it is much easier to obtain than uranium-235, though the process is a lot messier.

Making a bomb was so important that General Groves pursued both the gun and implosion concepts at full industrial scale. There were work sites all across the country. Project headquarters were in Washington, D.C.; the bombs themselves were designed and built at Los Alamos, New Mexico; and the slow and difficult job of separating uranium-235 from natural uranium for the gun bomb was done at the big Clinton Engineer Works, now Oak Ridge National Laboratory, near Knoxville, Tennessee. But the nuclear reactors to create plutonium and the chemical plants to separate and purify it for the implosion bomb were too dangerous to put at the Clinton Works. An accident might stop work on both bombs and could put a large number of people in the densely populated eastern United States at risk.

Accordingly, Groves went looking for a new site. It had to be large, remote from towns and settled areas, and relatively flat. It had to have both a large supply of water for cooling the nuclear reactors and a dependable electric power supply. The Hanford area, bordered by the Columbia River and crossed by a high voltage power line from Grand Coulee Dam, was an obvious choice. The Secretary of War began proceedings to acquire the land on February 8, 1943, and the thousand families residing in and around Hanford, White Bluffs, and Richland, got their eviction notices March 9. Almost all the structures were razed, and the orchards were cut down. Only the steel vault from the White Bluffs bank, the shell of the old Hanford High School, street patterns, traces of irrigation ditches, and clumps of trees marking old homesteads are visible today.

Horace Burton Stewart was born in early 1867, son of a western Pennsylvania blacksmith. In the fall of 1868 he immigrated with his parents to a farm in southwestern Iowa where he grew up. He went to college in Nebraska in the 1890s and became a teacher. While teaching he met and married his wife Emma in 1904.

Full of enthusiasm for a simpler life closer to the land, the couple turned from teaching to farming. They homesteaded first at Cody, Wyoming, where their four children were born. In 1914, Horace and Emma responded to publicity of farming opportunities in northern Montana and joined thousands of other "honyockers" taking a dry half-section near the town of Malta. They worked hard, but the farm failed after five successive drought years. In 1919, the Stewarts said goodbye to "Old Dry" and moved to their final homestead, 160 acres of second growth timber and poor soil near Clayton, 25 miles north of Spokane, Washington, a small town named for the clay deposits discovered there in the 1880s.

Thirteen years of hardscrabble farming evidently satisfied his every desire to get closer to the land, and Horace turned back to teaching. He took the job of Superintendent of Schools at Hanford, Washington, and held it into the mid-1920s. During this time the family spent the summers at the Clayton farm and returned to Hanford for the school term.

Failing health forced Horace to retire from Hanford, and the family again had to live off the farm. The Depression made this difficult, and their only son, Burton, had to forego college to put up hay and milk cows. Horace died in 1940. Burton was drafted and served as a radio technician in the Army Air Corps. Meanwhile, the Manhattan Project conscripted the town of Hanford to house construction workers and built the huge plant that took its name.

After the war, Burton and his wife Linda came back to the Clayton farm, where he became a rural mail carrier. I am Burton's son and Horace's grandson. I joined Pacific Northwest Laboratory at Richland in 1973 and started working on SY-101's flammable gas releases in 1993. My Hanford experiences of the next 10 years inspired this book.

Ruins of the Hanford High School c2000.
Courtesy PictureThis.

The Workings of Hanford Works

From March 1943 to February 1945, the Corps of Engineers transformed the Hanford area from open desert and small irrigated farms into a closed military-industrial complex with guards at the gates. The very possibility of nuclear reactors and bombs was new and untested when the Manhattan Project set out to build them. Nuclear fission was only discovered in 1939, and the first nuclear reactor at the University of Chicago had just started up in December 1942. The bomb itself was only a theoretical concept. Given this primitive state of knowledge, it is amazing that in just two years, three full-scale nuclear reactors were operating, and two huge plants were separating plutonium at Hanford. Less than a year after production began, Hanford plutonium imploded at the Trinity test in July 1945 and in the "Fat Man" bomb that destroyed Nagasaki in August 1945. The "Little Boy" gun-type bomb dropped on Hiroshima a week earlier used uranium-235 from Oak Ridge.

Plutonium is not natural. It has to be manmade. The Hanford Works made plutonium by exposing uranium to neutrons created inside huge nuclear fission reactors. A fission reactor contains and controls a self-sustaining fission reaction where atoms of uranium-235 split apart, or fission, after their nucleus absorbs a neutron. The two or more fission products— fragments of uranium-235 atoms, like cesium-137, iodine-131, and strontium-90—are the primary radioactive byproducts of the reactors. Each fission event releases one or more neutrons and a lot of heat. Some of the liberated neutrons hit other uranium-235 nuclei and cause more fission events that sustain the reaction. Other atoms that do not fission absorb neutrons become radioactive activation products. And a very small number of neutrons are absorbed in nuclei of natural uranium-238, which does not itself fission, but becomes imminently fissionable plutonium-239. One ton of uranium makes about 1.4 pounds of plutonium.

Believed to be the most dangerous part of the process, the reactors were built farthest from the downriver communities in what was called the 100 Area along the upper bend of the Columbia River in what is today called the Hanford Reach. Six self-contained reactor complexes were planned, identified from west to east as 100-A through 100-F. Three reactors—B, D, and F—were built by the Manhattan Project. Six more were built by the Atomic Energy Commission from the late 1940s to the late 1950s.

The early reactor cores were a literal pile of 1,200 tons of precisely machined graphite bricks about 4 inches square and 4 feet long. The bricks were fitted together into a core 28 feet wide, 26 feet long, and 36 feet high. Holes drilled in the graphite blocks housed 2,004 steel process tubes that held 200 tons of 8-inch long by 1.5-inch-diameter uranium fuel slugs. Fabrication of the fuel slugs was done in the 300 Area, considered a sufficiently safe process to site close to the town of Richland. Each of the three original reactors

100 F Area in 1962. F Reactor is at far right, White Bluffs of the Columbia River in the background.
Courtesy DOE-RL Declassified Document Retrieval System.

Hanford Production Reactors

Hanford Reactors	B	D	F	DR	H	C	KW	KE	N
Construction start	8/43	11/43	12/43	12/47	3/48	6/51	11/52	1/53	5/59
Start up	9/44	12/44	2/45	10/50	10/49	11/52	12/54	2/55	3/64
Shut down	2/68	1/67	6/65	12/64	4/65	4/69	2/70	1/71	1/87
Years of operation	22	12	20	14	15	17	15	16	22
Design power (MW)	250	250	250	250	250	750	1,800	1,800	4,000
Peak power (MW)	1,940	2,005	1,935	1,925	1,955	2,310	4,400	4,400	3,950
Total plutonium created (tons)	6.1	6.0	4.9	4.2	4.5	7.1	14.1	14.8	12.1

produced about 250 megawatts of heat while gulping 30,000 gallons per minute of treated river water through the process tubes to carry the heat away. The buildings housing the systems for filtering, treating, pumping, and dumping the water actually took many times more space than the reactors themselves.

Cooking the fuel slugs in the reactor about 30 days allowed the optimum number of uranium-238 atoms to capture a neutron and turn into plutonium-239. Wait longer and too much of the plutonium-239 would also swallow neutrons to become plutonium-240 or -241, which was undesirable for bomb making. At the proper time, the reactor shut down and now plutonium-rich uranium slugs were dumped out the rear face of the reactor into pools of water 20 feet deep that shielded the workers from the lethal radiation from the fission products in the hot fuel. In fact, the radiation dose was even too high for processing, and the hot slugs had to stay in the pool for another month or two, until the short-lived radioisotopes like iodine-131 decayed away.

The plutonium separation plants were built in the 200 Areas, high on a plateau in the middle of the site about 10 miles south of the reactors. The basalt ridge of Gable Butte and Gable Mountain formed a natural safety barrier between the reactors and the separation plants should anything go wrong at either. Two plants, T and U, each with their own underground tank farms for radioactive waste, were planned for the 200 West Area. B and C Plants were slated for the 200 East Area some 4 miles away. Because the B and T

Plants were so successful, C Plant was never built, and U Plant was held as a spare and did not run during the war. These old plants were colossal concrete boxes, 800 feet long, 85 feet wide, and 100 feet high, 20 feet of which was below the surface, or below grade. The buildings were called canyons because their deep narrow excavations had that appearance during construction.

A dedicated tank farm—groups of underground tanks connected by piping—served each of the plants and more were added as the original tanks filled up and new plants were built. As I describe the separation processes, and other operations designed to concentrate the waste, I'll point out which waste streams were destined for or affected SY-101.

The first step in processing the irradiated fuel from the reactors was dissolving the aluminum cladding, or coating, off the slugs with boiling sodium hydroxide. All of the dissolved cladding went directly into the waste tanks. Next, the naked uranium slugs were dissolved in hot nitric acid to form the liquid feed stream for the actual plutonium separation process. Besides plutonium and uranium, the slugs contained the radioactive fission fragments and activation products from the fission reaction. These contaminants went into the waste tanks after the valuable plutonium atoms were coaxed out.

Dissolving the uranium slugs in the canyon buildings made a vicious brown fog of nitrous gases, radioactive particles, and toxic vapors that belched out of high smoke stacks and drifted downwind over the towns, farms, and ranches of central and eastern Washington. The stacks did not have filters until 1948, and the first ones were not very efficient. Quite a bit of airborne contamination escaped in the 1940s.

The notorious Green Run of early December 1949 deliberately processed a ton of fuel only 16 days out of the reactor. This was supposed to compare the T Plant plume with Soviet plants suspected of processing "green" fuel that had not cooled long enough for the iodine-131 to decay away. Although emissions from the Green Run were small compared to total airborne releases of this era, it intentionally violated safety limits, even to the point of disabling the T Plant stack filters. Knowledge of this deliberate release later triggered class action lawsuits by people who lived downwind of the iodine-laden plume, or "down winders," claiming health damage from iodine exposure.

The old B and T Plants ran the bismuth phosphate process. Adding bismuth nitrate and phosphoric acid to the dissolved fuel solution made solid bismuth phosphate fall out of, or precipitate from, the liquid, taking plutonium with it. The fission products as well as highly valuable uranium stayed in the liquid, and went to tanks as waste. This process was slow and wasteful, but it served during the rush of World War II and the early days of the Cold War until better methods were ready. The bismuth phosphate process ran in B Plant from April 1945 until late 1956, and in T Plant from December 1944 to March 1956.

Hanford production expanded during the Cold War to stay ahead of the Soviet Union. Five more production reactors—C, DR, H, K East, and K West—were built between 1949 and 1955 and the dual-purpose N Reactor that made both plutonium and steam for electric power generation in 1963. The eight older reactors were shut down by 1971 while N Reactor survived until January 1987. Altogether, the 9 reactors made 74 tons of plutonium, over 65 percent of the total U.S. supply.

200 West Area in 1944 with U Plant under construction in foreground, coal-fired power plant with twin stacks and T Plant in operation in background.
Courtesy PictureThis.

> The Columbia River had plenty of water to cool the three original reactors. But by the early 1960s, 8 reactors were operating and producing a heat load over 30 times that of the original Manhattan Project. This was enough to raise the river temperature downstream by several degrees Fahrenheit. The Atomic Energy Commission feared that warmer water would harm the salmon. It also made it difficult to run the downstream reactors at maximum power during the summer. With the Cold War at its height, plutonium production could not be slowed. The only solution was to lower the river temperature by spilling large volumes of frigid water from the bottom of Grand Coulee Dam! The Corps of Engineers cooled the river this way during the summer months from 1959 through 1965.

Two huge new plants with improved separation processes were built during the Cold War to handle the fuel coming out of the new reactors. S Plant, or REDOX, was built in the 200 West Area in 1952, and the 200 East Area got the A Plant, or PUREX, in 1956. In 1952 the unused U Plant was converted to recover uranium that had been discarded in B and T Plant waste. All these plants used versions of the solvent extraction process where the dissolved uranium and plutonium were coaxed out of the nitric acid feed stream into an immiscible oil, leaving the radioactive fission products in the acid solution that went to the tanks as waste. Because oil and water don't mix, solvent extraction was much easier, faster, and "cleaner" than bismuth phosphate precipitation. A large fraction of the waste that ended up in SY-101 was generated in the U Plant and REDOX.

The uranium recovery process in the U Plant deserves special mention. Sluicing the uranium-bearing waste out of the tanks into the plant took a lot of water that was returned to the tanks. This made about twice as much new waste as was already in the tanks. The Atomic Energy Commission solved this problem by adding a chemical called nickel ferrocyanide to some of the tanks to make radioactive cesium and strontium fall out as a solid. The leftover liquid was clean enough to be poured into the ground. But the ferrocyanide left in some of the tanks could theoretically explode if it were concentrated, dried, and heated. Public concern over the explosion hazard in tanks with ferrocyanide in them arose in the

The Workings of Hanford Works | 15

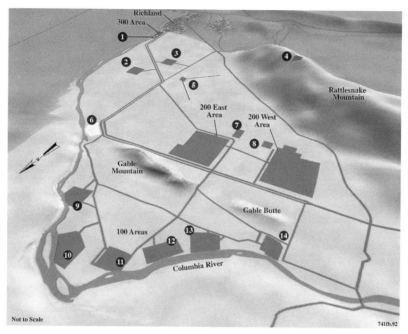

1. 300 Area Liquid Effluent Treatment Facility
2. Commercial Operating Nuclear Power Plant
3. Fast Flux Test Facility
4. Observatory
5. Laser Interferometer Gravitational Wave Observatory (LIGO)
6. Old Hanford Townsite
7. U.S. Ecology Commercial Solid Waste Site
8. Environmental Restoration Disposal Facility (ERDF)
9. F Reactor
10. H Reactor
11. D and DR Reactors
12. N Reactor
13. KE and KW Reactors; Cold Vacuum Drying Facility
14. B and C Reactors

Plutonium Separation Production

Process	Fuel Processed (tons)	Waste Produced (million gallons)	Average Waste Rate (gallons/ton fuel)
Bismuth phosphate	8,900 (8%)	98	10,000
REDOX	24,600 (23%)	42	1,600
PUREX	73,100 (69%)	39	500
Total	106,600	179	1,700

mid-1980s. This was the first in a series of tank-related safety issues that culminated in the alarm over SY-101 in 1990.

By the mid-1960s Hanford production was in high gear, and the lack of tank space was again becoming a critical problem. If the radioactive cesium and strontium could be removed from the liquid

Top: REDOX or S Plant in 1953 with 222-S Lab right. Bottom: PUREX or A Plant c1960.
Courtesy DOE-RL Declassified Document Retrieval System.

waste, it could be dumped into the ground and would not have to be stored in the tanks. To help justify the expense, the Atomic Energy Commission proposed using purified cesium and strontium extracted from the waste as a resource. Capsules of the material could be used in space power systems, food preservation, medical sterilization, chemical manufacturing, and even powering runway lights at backcountry airports. In 1968 the retired B Plant was refurbished to extract and encapsulate cesium and strontium from PUREX waste and from waste already in the tanks. The purified strontium and cesium were finally pressed into small screaming hot capsules and stored in the Waste Encapsulation and Storage Facility. Not only did the cesium and strontium capsules never realize their imagined value, but the process also created the waste most responsible for making SY-101 dangerous.

The process used to remove strontium was the culprit. It was a solvent extraction process that pulled strontium out of an acid solution into a stream of immiscible oil. But aluminum, which was supposed to stay in the waste, tended to follow the strontium. To solve this problem, soluble organic "complexants" had to be added to make aluminum stay in the acid solution. An organic chemist's nightmare, these complexants were glycolate, ethylenediamine-tetra-acetate (EDTA), and hydroxyethylethylenediaminetriacetate (HEDTA). They all ended up in B Plant waste. When this "complexed" waste was concentrated and the complexants began to decompose from gamma radiation, the cascade of resulting chemical reactions generated hydrogen gas at a prodigious rate. This is what made SY-101 burp so often.

Hanford Processing Plants

Plant	Function	Process	Start Up	Shut Down
T Plant	Pu separation	Bismuth phosphate precipitation	Dec. 1944	Mar. 1956
	Equipment decontamination, repair, and storage		1957	1983
B Plant	Pu separation	Bismuth phosphate precipitation	Apr. 1945	Nov. 1956
	Cs, Sr recovery	Cs: precipitation & ion exchange Sr: solvent extraction TBP, complexants)	1968	Cs - Sept. 1983 Sr - Feb. 1985
U Plant	U recovery	Solvent extraction (TBP + kerosene), nickel ferrocyanide to precipitate cesium	Jul. 1952	Mar. 1957
S Plant	Pu, U, Np separation	REDOX solvent extraction (hexone)	Jan. 1952	Jul. 1967
A Plant	Pu, U, Np	PUREX solvent extraction	Jan. 1956	June 1972

Cs	cesium	REDOX	reduction-oxidation process	
Hexone	a commercial organic solvent like kerosene	Sr	strontium	
Np	neptunium	TBP	tributyl phosphate, an organic solvent	
Pu	plutonium	U	uranium	
PUREX	plutonium-uranium extraction			

The Tanks

Making plutonium created great volumes of hot caustic waste liquid full of radioactivity. Subsequent processing stirred this waste around and added new chemicals, but it was still highly radioactive and had to be kept isolated in large underground tanks. SY-101 was one of these tanks, but it was not built until the mid-1970s. First came the single-shell tanks. As the name implies, the single-shell tanks contained waste in one thin cup-shaped steel shell encased with reinforced concrete and buried under 6 to 7 feet of gravelly soil. The first tanks were up to 75 feet in diameter designed to hold 530,000 gallons of waste. Later designs increased the capacity to 758,000 gallons, and finally, to 1,000,000 gallons. Sixteen smaller tanks were also built, 20 feet in diameter and 26 feet tall, holding 55,000 gallons of waste.

A total of 149 single-shell tanks were built from 1943 to 1964. The B, C, T, and U Tank Farms were built in 1943 and 1944 for the four planned separations plants, even though only the B and T Plants operated. Each of these farms had twelve 530,000-gallon and four 55,000-gallon tanks. The bismuth phosphate process created so much waste that more tanks were quickly needed. B Plant waste capacity expanded with the 12-tank BX farm built in 1946 and 1947 and 12 more tanks in the BY farm built in 1948 and 1949. Likewise,

T Plant got a whopping 18 new tanks in the TX farm in 1947 and 1948 and 6 more in TY in 1951 and 1952. The larger BY, TX, and TY tanks held 758,000 gallons each.

The new REDOX, or S, Plant was built with the 12-tank S farm in 1950 and 1951. The SX farm added 15 tanks for REDOX waste in 1953 and 1954. Likewise, PUREX, or A, Plant got the six-tank A farm in 1954 and 1955. The PUREX waste volume was so much less than its predecessors that the new tanks were not needed for 9 years. The AX farm, with only four tanks, was finally placed in 1963 and 1964. The S farm used 758,000-gallon tanks, while the SX, A, and AX tanks held 1 million gallons. AX was the last single-shell tank farm. It brought the total waste storage capacity to just over 126 million gallons.

LEAKING OLD TANKS AND A FEW NEW TANKS

The single-shell tanks were designed to last only 10 to 20 years and began to crack and leak radioactive waste as early as 1959. Thirteen tanks had leaked or were suspected to have leaked by 1970 and 35 more tanks were classified as "leakers" between 1970 and 1980, when all single-shell tanks were pulled from operation. There were 67 known or suspected leakers in January 2004.

The leaking tanks were a severe problem that had to be corrected. In the late 1960s, the Atomic Energy Commission began building new double-shell tank farms. The tanks used an improved design with a second steel shell wrapped around the primary tank to contain any leaks. All the double-shell tanks are 75 feet in diameter with a 1,116,000-gallon capacity. Twenty-eight double-shell tanks were built from 1968 to 1986 with a total capacity of almost 94 million gallons.

Most of the double-shell tank farms were sited in the 200 East Area near the giant PUREX Plant, the only separation plant still running during double-shell tank construction. The AY farm, built from 1968 to 1970, had the first two double-shell tanks, followed by two more in the adjacent AZ farm built from 1971 to 1977. The six-tank AW farm was built from 1974 to 1976. Seven tanks were added in the AN farm in 1980 and 1981, and the eight tanks in the AP farm were built from 1983 to 1986. The three tanks in the SY farm, the only double-shell tanks in the 200 West Area, were built between 1974 and 1976. This book focuses on the first of these tanks, SY-101.

The Tanks | 21

Single-shell tanks in S farm under construction in 1951, REDOX in background.
Courtesy DOE-RL Declassified Document Retrieval System.

Double-shell tanks in AW farm under construction in 200 East Area about 1975, 242A evaporator in background.
Courtesy PictureThis.

MAKING MORE TANK SPACE

To stop leaks in single-shell tanks, a big urgent program began in the late 1970s to pump excess liquid out before it could leak out. Long screens called "saltwells" were sunk into single-shell tanks from which jet pumps could pull liquid that drained from the waste. This liquid went into the new double-shell tanks. After almost 30 years, saltwell pumping was finally finished in April 2004. The single-shell tanks now have little liquid left to leak.

MAKING MORE TANK SPACE

No matter how many tanks were built, there was never enough space and inventive methods to reduce waste volume were always welcome. Any liquid waste reasonably clean of long-lived fission products was poured into the ground through various drains such as perforated pipes (also called "cribs") or open trenches. Most of the liquid from the later, "cleaner", cycles of the bismuth phosphate process in B and T Plants was drained into the soil directly from the plant or cascaded to into cribs after settling radioactivity-laden solids in the smaller waste tanks. Liquid waste from uranium recovery in U Plant was also cribbed after precipitating out cesium sludge with nickel ferrocyanide. Ferrocyanide was also added directly to some tanks so that the liquid could be cribbed. All told, some 120 million gallons of waste, almost the total capacity of all 177 Hanford tanks, was poured into the soil through cribs and trenches during Hanford production.

Boiling off extra water also saved a lot of space. All waste is warmed by radioactive decay, now mainly from cesium-137 and strontium-90. The higher the concentration of these isotopes, the hotter the waste. REDOX and PUREX waste generated so much decay heat that some A, S, and SX tanks actually got hot enough to boil. These were labeled self-boiling or "self-concentrating" and were actually designed to do so. To take advantage of this process with waste that was not hot enough to boil by itself, steam and electric heating coils were installed to boil the waste in three BY tanks.

The hot waste from REDOX and PUREX caused some dramatic events called "steam bumps." A muddy, wet sediment layer settled out in the tanks full of hot strontium-90 and cesium-137. Convection kept the liquid layer cool, but the sediment just got

The Tanks | 23

hotter and hotter. Finally, the water in the mud flashed into steam and burst towards the surface. It was like heating a can of condensed soup on a hot plate without stirring it.

Some of the steam condensed suddenly in the cool liquid with a distinct "bump," but the rising blob of superheated mud boiled up creating a powerful eruption of steam that occasionally blew out the ventilation filters, called tank breather filters, and forced radioactive mist out the ventilation stack. Rumor has it that tank farm workers knew to start running when they felt the ground shaking. Air bubble-powered waste circulators were installed in some tanks to keep the sediment mixed and prevent steam bumps.

The heat caused other interesting problems. In A-105, extremely hot sediment apparently boiled water that had leaked in between the steel tank bottom and the concrete base. A flash of steam caused about three quarters of the steel floor to rip away from the side wall and the tank bottom to bulge upward 5 feet! In C-106, the waste was so hot that a few thousand gallons of water had to be added every month or so to keep it from boiling dry! If it had boiled dry, the high temperatures might have cracked the concrete shell and cause the tank to collapse.

Evaporators concentrated waste by boiling it outside the tanks in steam-heated pots to avoid messy things like steam bumps that happened when waste boiled in the tanks. The first evaporator, 242-B, boiled dilute liquids from bismuth phosphate and uranium recovery wastes in 200 East Area tanks from 1951 through 1954. The 242-T evaporator did the same in the 200 West Area from 1951 to 1955, and after upgrades, from 1965 through 1976.

Because only 28 double-shell tanks were built, the volume of waste from the single-shell tanks had to be cut down still more to fit the available space. New evaporators were built to boil off more water so the waste would fit in fewer tanks. The newer 242-A and 242-S evaporators applied a vacuum to boil the waste at a lower temperature and create a thicker slurry. The 242-A evaporator began concentrating liquids from tanks in the 200 East Area in 1976 and is still operational. The 242-S evaporator, sited adjacent to the SY farm in the 200 West Area, had a special role because it prepared the concentrated waste that filled SY-101. It concentrated liquids in 200 West from 1972 through 1980.

Altogether the boiling tanks and evaporators removed about 300 million gallons of water up through 1988. This is equivalent to more than twice the total capacity of all 177 Hanford waste tanks! But the very effectiveness of these processes in concentrating the waste and making tank space is what created the hazard in SY-101.

Underground Waste Storage Tanks

Tank Farm	Number of Tanks	Capacity (1000 gallons)	Years Built	Note
Single-Shell Tanks				
A	6	1,000	1954–55	Self boiling
AX	4	1,000	1963–64	Self boiling
B	4	55	1943–44	
	12	530		
BX	12	530	1946–47	
BY	12	750	1948–49	In-tank solidification
C	4	55	1943–44	
	12	530		
S	12	750	1950–51	Self boiling, DSSF
SX	15	1,000	1953–54	Self boiling, DSSF
T	4	55	1943–44	
	12	530		
TX	18	750	1947–48	
TY	6	750	1951–52	
U	4	55	1943–44	DSSF
	12	530		
ALL SST	149	93,680	1943–64	
Double-Shell Tanks				
AN	7	1,160	1980–81	DSS, DSSF
AP	8	1,160	1983–86	
AW	6	1,160	1974–76	DSSF
AY	2	1,160	1968–70	High temperature
AZ	2	1,160	1971–77	High temperature
SY	3	1,160	1974–77	DSS
ALL DST	28	34,280	1968–86	
HANFORD	177	126,160	1943–86	

DSS Double-shell slurry
DSSF Double-shell slurry feed
DST Double-shell tank
SST Single-shell tank

WHAT'S IN THE TANKS?

The waste in the tanks came from a multitude of sources. It has been treated, transferred, and evaporated so that the current waste volume is only about 10% of what was generated. The records of what kind and how much waste was pumped into and out of each tank over the last 55 years are voluminous and intricate, but are neither exact nor entirely complete. As a result, tracing the waste in a specific tank back to its origin, or determining the composition of the waste from its origins, is almost impossible. This has led to accusations to the effect that "you don't know what's in your tanks!" While this may be true for the precise concentrations of all compounds present and their exact physical nature, the overall character and the main components of the waste are pretty clear in terms of what we need to know to store the waste safely. Let's see what's there and what it's like.

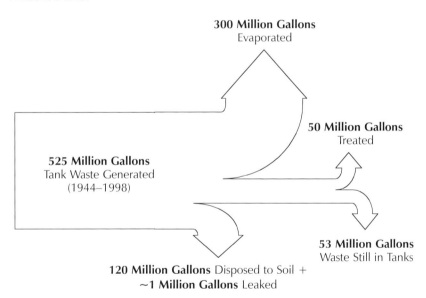

Even after 50 years of evaporation, the most abundant component of the waste is still plain old water. After water comes sodium nitrate, an extremely soluble colorless salt, sometimes used as a substitute for table salt. Sodium nitrate makes up the largest part of the waste in almost every tank. It originated from adding sodium hydroxide, called caustic, to neutralize the nitric acid waste from the separation plants so it wouldn't corrode the steel tank walls. Most waste also has a lot of aluminum generated by 25 years of dissolving the aluminum alloy cladding from the uranium slugs. In the same way, zirconium, a tin-like metal, entered the tanks from dissolving the zirconium alloy cladding on N and K Reactor fuel processed in the PUREX Plant.

The ancient bismuth phosphate process generated sodium phosphate, bismuth phosphate, plus some chromium, iron, and sodium sulfate. Some of the waste uranium—uranium that was not completely recovered in U Plant—remains in some tanks as part of the bismuth phosphate legacy. The chemicals used in the REDOX and PUREX processes made more iron oxide and sodium sulfate. Some iron oxide also slowly accumulates in the waste by slow corrosion of the steel tank itself. Some tanks have ferrocyanide compounds left over from cleaning up the liquid waste from the uranium recovery process in U Plant.

The oil stream from the solvent extraction processes in U Plant and PUREX as well as strontium separation in B Plant was washed with carbonate before recycling. This generated a background concentration of sodium carbonate in the tanks. A little sodium carbonate also accumulated in the tanks as the caustic waste absorbed carbon dioxide from the air. Only minute traces of the solvent oil itself entered the tanks. The little that did floated on the waste surface and evaporated away. On the other hand, the soluble organic complexants from the strontium recovery process in B Plant did not evaporate; they stayed in the waste. Radioactive and chemical destruction of these organic compounds in the presence of high concentrations of sodium nitrate has generated a large inventory of ammonia, a gas that dissolves eagerly in water.

Besides chemicals traceable directly to the separation and recovery processes, some other things were dumped in, dropped in, or even fell into some of the tanks. Sixty-three tons of portland cement was dumped into BY-105 in 1972 as an experimental "stop leak." The cement never set up in the caustic waste and that method was abandoned. Tons of diatomaceous earth, almost entirely silicon oxide, were sprinkled into four tanks from 1970 to 1972 in another stop leak experiment. Several casks of experimental enriched uranium fuel slugs, a collection of "poison" balls for emergency reactor shutdowns, and some cobalt-60 slugs were disposed of in Tank U-101.

The contaminated steel tapes used for waste level measurement were typically clipped off and snaked into the tank when they were replaced. Likewise, broken pipes, failed pumps, and other contaminated hardware were often cut off and allowed to fall into the tanks. There are stories of gloves, hard hats, lunch boxes, even well preserved dead bugs and birds, meeting a similar fate. Pictures of such debris in the tanks have incited a variety of reactions, such as a shaking of heads and clicking of tongues. At the time, however, it was believed wiser to avoid the huge cost of shielding and decontaminating an item covered with or full of radioactive waste by simply leaving it in the tank.

All the stirring, pouring, pumping, and boiling that the waste endured over its long history created two distinct kinds of waste: sludge and saltcake. Sludge is a slick, dark brown, black, or gray mud made of the less soluble compounds like metal oxides and hydroxides. It has high concentrations of sodium hydroxide, aluminum hydroxide, iron oxide, chromium oxide, and sometimes bismuth phosphate and uranium. The liquid portion of sludge waste is very watery, dilute enough that the ubiquitous sodium nitrate and other soluble salts stay dissolved.

Saltcake forms when water evaporates from liquid waste and drives the sodium salts out of solution as crystalline solids. Dry saltcake is a hard, gray to whitish solid. Wet saltcake is like gray to white wet sand or a coarser "snow cone." The rough, pond-pocked saltpans of Death Valley in California look a lot like the waste in a saltcake tank. Saltcake waste is mostly sodium nitrate with some sodium phosphate, sulfate, and carbonate thrown in. The waste in SY-101 was mostly saltcake.

Because of the fear of leaks, all the liquid possible has been pumped out of the single-shell tanks. The drained sludge tanks have a surface pattern of polygonal cracks, like a sun-baked mud flat. Pumped saltcake tanks have a dry and crusty surface. Any remaining liquid in these tanks tends to migrate away from the cool walls to the warmer center, creating a "jelly donut" configuration that's soft and creamy in the middle and dry and crunchy on the outside.

On the other hand, the double-shell tanks all have a lot of liquid, even those that have been concentrated. Whether the waste is sludge or saltcake, the solids in these tanks settle into a sediment layer on the bottom with a deep layer of supernatant liquid (from the Latin supernatare, "to float") on top. The viscous liquid ranges in color from almost clear to translucent yellow or green to opaque brown or gray-brown. The appearance of cylindrical core samples of the sediment that have been extruded onto a tray in a hot cell is of a thick dark gray or gray-brown mud.

Highly concentrated saltcake tanks also tend to evolve a foamy crust layer, where rising gas bubbles carry particles to the surface where they float and dry. Runaway crust growth turned into an emergency in SY-101, as later chapters will tell.

What would it be like if you happened to fall into a typical waste tank? In one of the hotter single-shell tanks, the first thing you'd notice would be a sinus-searing stench of ammonia. Breathing would be totally intolerable. Even at the concentrations of 100 parts per million found in many tanks, your respiratory agony would be so great you would not notice the pitch darkness, the humid heat, or the crunch of moist salt chunks underfoot. You'd soon be dead if the ammonia concentration topped 500 parts per million as it is known to do.

If you fell into a similarly hot double-shell tank, the ammonia smell might only be extremely unpleasant, not because double-shell tank waste has less ammonia, but because their domespace is ventilated to keep ammonia down to 20 to 30 parts per million. Much worse, you would be splashing around in a hot liquid so intensely caustic that it would already be dissolving your skin. You might be encouraged to find that you are floating with your shoulders and chest out of the super-dense liquid and in no danger of sinking. In several of the tanks, there would be a disagreeably crunchy crust, probably not strong enough to support your weight, but strong enough to bruise and abrade your thrashing body pretty badly.

What about the radiation? In an old single-shell tank, you would have to stay in over an hour before even noticing any ill effects. But double-shell Tank SY-101, before being diluted, was much hotter. After about 10 minutes, you would start feeling nausea, maybe muscle weakness. In half an hour, nausea would advance to vomiting, and weakness would be debilitating. Death would probably come in a few hours in SY-101, but radiation would preserve your remains from biological decay.

Though these scenarios are grim indeed, you don't need to worry about falling into a tank. Every opening large enough to admit a person is sealed with a heavy concrete slab or similar barrier. The real issue is the long-term environmental and human health risk of this concentrated waste getting out and contaminating large areas of soil and ground water.

WHO'S MINDING THE (TANK) FARM?

Hanford has been introduced as a geographic area, an industrial complex, and a pre-war agricultural community. Like representing a nation by its capitol city, "Hanford" also represents the people and organizations that made the decisions and ran the complex. Hanford, the organization, has three parts: the contractor doing the work, the local government office monitoring the work, and the national government agency directing the work.

DuPont was the contractor that built and ran the plant for General Groves and the Manhattan Project during the war. General Electric succeeded DuPont in 1946, and oversaw the Cold War expansion of the 1950s. In 1967 the Hanford contract expanded to seven companies, with Atlantic Richfield Hanford Company as prime, or lead, contractor. They got the job of shutting plutonium production down. Rockwell Hanford Operations became prime contractor after 1977, when people began to fret about the leaking tanks. In 1987 Hanford work was partially re-consolidated under Westinghouse Hanford Company until 1996, when it was farmed out to a group of six companies under the Project Hanford Management Contract with the Fluor Daniel Hanford Company, now Fluor Hanford, Inc., as the prime contractor. Lockheed Martin Hanford Company was initially responsible for the tank farms under Project Hanford but was succeeded by CH2M HILL Hanford Group in 1999. To keep acronyms under control, I will call contractors by their primary company name most of the time. Thus Rockwell Hanford Operations is "Rockwell," Lockheed Martin Hanford Company is "Lockheed," and so on.

Another important, but somewhat peripheral, piece of the Hanford organization is Pacific Northwest National Laboratory operated for the U.S. Department of Energy, DOE, by Battelle (not Bechtel the big construction company, nor Mattel the toy company) of Columbus, Ohio. Set up in 1965 as Pacific Northwest Laboratory (PNL) in a cluster of light brown buildings in north Richland affectionately called the "sand castles," it became one of DOE's multiprogram research laboratories in 1986, though DOE did not officially add "National" to its name until 1995. As Pacific Northwest *National* Laboratory, PNNL pursues a broad research agenda, and Hanford support, including environmental monitoring for DOE plus other work for DOE subcontractors as needed, is a relatively small fraction of its overall effort. However, during the SY-101 mitigation and flammable gas era from about 1991 through about 1995, PNNL had one of the largest and most successful Hanford programs in its history. I use "PNNL" throughout to indicate this laboratory.

On the government side, the Atomic Energy Commission ruled Hanford and the rest of the U.S. nuclear weapons complex from shortly after World War II until 1975, when Congress split the commission into the Nuclear Regulatory Commission and the Energy Research and Development Administration. The energy administration ran Hanford only 2 years before becoming the Department of Energy in 1977. The DOE is still in power today. The Office of Environmental Restoration and Waste Management has direct oversight of Hanford. In 1989 the Tri-Party Agreement on Hanford cleanup among DOE, Washington State Department of Ecology, and the U.S. Environmental Protection Agency suddenly complicated Hanford oversight. Now DOE could not act unilaterally, and the public had much more influence.

The local DOE monitor of the Hanford Site is the DOE Richland office. This office has mirrored its parent agency, whether Atomic Energy Commission, Energy Research and Development Administration, or DOE. Under DOE, the local office is known as "RL," short for Richland Office. Any DOE group that is not part of DOE Headquarters in Washington, D.C., is considered part of an operations or field office. To help focus the effort on its Tri-Party Agreement obligations, the Office of River Protection split from DOE Richland Operations in 1998 to handle the tank waste cleanup projects. The Office of River Protection is responsible for the new vitrification plant and for tank farm operations. I use "DOE Richland" to refer to the local DOE office, usually without distinguishing it from the Office of River Protection. I call the national entity "DOE Headquarters" or "Headquarters."

Tank Troubles Begin

From 1977 through 1980, Rockwell ran the "partial neutralization" campaign in the 242-S evaporator, where nitric acid and potassium permanganate were added to precipitate more sodium nitrate from the slurry when it cooled in the receiving tank. Some of partial neutralization waste was "complexed concentrate." It not only had an above-average fraction of organic complexants that originated in the strontium recovery process in B Plant, but was also concentrated by a trip through the evaporator. The evaporator sent this complexed waste to tanks SX-106 and U-111 and some other tanks, eventually ending up in SY-101.

Also in the late 1970s Rockwell began the "double-shell slurry" technology program. Its aim was to solve the tank space problem by running the 242-S evaporator flat out to wring as much water out of the waste as possible. The plan was to feed the process with waste that had already made one trip through the evaporator in the partial neutralization campaign. The product would be as thick as possible, just barely pumpable. SY-101 and next-door SY-103 would receive this new waste.

The first double-shell slurry test run was made in April 1977, pumping 239,000 gallons of the thick salty soup into SY-101. In

31

November that year, SY-101 got 365,000 gallons of complexed concentrate from the 242-S evaporator with complexed feed from next-door SY-102. In June 1978, 133,000 gallons of complexed concentrate arrived from SX-106 and in August another 60,000 gallons came in from U-111. In November 1980, SY-101 got its final load of double-shell slurry. This was a production run, not a test, and pushed the machinery to the limit. The pumps were shaking and the evaporator groaned to make 231,000 gallons of slurry, almost twice the density of water! SY-101 now held the 1,065,000 gallons of waste that should have filled it to 387 inches. But the waste depth measured 413 inches! The mysterious 26 inches of "slurry growth" would soon make the tank famous!

The waste formed three separate layers. About 18 feet of muddy sediment on the bottom lay beneath 13 to 14 feet of heavy liquid with about 3 feet of a foamy floating crust on top. The hydrostatic pressure at the bottom of the tank was 2.6 atmospheres, 30 percent higher than the same depth of water. The surface of the floating crust looked mostly gray with lighter patches of dry nitrate salts and black-looking pools of wet salt. Core samples taken in 1991 and 1998 showed that SY-101's waste was about 1/3 water, 1/5 sodium nitrate, 1/5 sodium nitrite, 1/8 sodium aluminate, 4 percent sodium hydroxide, and 1.5 percent organic complexants, by weight. Smaller fractions of sodium phosphate, sulfate, carbonate, and chloride were also present, along with traces of chromium, potassium, iron, calcium, nickel, and zinc. This composition was about the same everywhere in the tank, from top to bottom, crust to sediment. The liquid drained from core samples appeared dark greenish gray, not quite opaque. Extruded core samples from the sediment layer looked like dark gray clay.

About 90 percent of the radioactivity in SY-101 came from cesium-137 and 10 percent from strontium-90. By 1990 the total decay heat load was 13,000 watts, comparable to 130 hundred-watt incandescent heat lamps. This heat held the average temperature of the supernatant liquid at around 118 degrees Fahrenheit and the sediment about 130 degrees. It was much hotter in the early days with temperatures approaching 150 degrees in the sediment and 130 degrees in the liquid.

SY-101 was the only tank fed the potent concoction of thick double-shell slurry combined with solvent-laced complexed concentrate. Indigestion soon set in, and the tank came to life. The radioactive decay chain of cesium-137 into stable barium-137, or strontium-90 into zirconium-90, gives off gamma rays that break up water molecules to make hydrogen gas. Gamma radiation also breaks down complex organic compounds setting off a chain of chemical reactions that make more hydrogen along with other gases. Concentrating the waste speeds up hydrogen production from both these processes. At the same time, it congeals the slurry into a deep muddy sediment or sludge layer that traps the gas bubbles. The growing bubbles make the waste "rise" like bread dough. This shows up as a steady increase in the waste surface level. This phenomenon, first seen in Hanford tanks after evaporator runs in the 1970s, was called "slurry growth." It was the source of the 26 extra inches measured in SY-101 after it was filled.

The trouble was that the waste level didn't just rise. In 1980, a few months after the tank was filled, the level suddenly dropped! Then it rose and dropped again in a cycle that recurred about every 2 to 4 months. Because the level rise or slurry growth was attributed to gas buildup, the level drops must be gas releases or "burps."

You can do your own experiment to demonstrate "slurry growth" at home in the bathtub! Place a strip of masking tape up one side of the tub and have a soft pencil handy. Fill the tub enough to submerge your torso and chest a few inches when lying flat on the bottom (head out, of course!). Now, while lying flat with chest submerged, exhale fully and hold it for a few seconds. Make a mark on the tape at the waterline with the pencil. Now inhale deeply, hold it, and make another mark at the higher waterline. This might also work in a hot tub, but you'll need several people breathing in unison to change the water level enough to measure.

The difference between the two water levels is a measure of the volume of air you drew into your lungs, just as the waste level rise in a tank is a measure of the gas accumulation. If you knew the surface area of the tub (less the area of neck and legs sticking out of the water), you could calculate the volume exactly as the change in level times the area. Of course, you'd have to allow for the slightly higher pressure of the air in your lungs, just as we have to allow for the higher pressure of gas deep under the waste surface when calculating gas volumes in tanks.

This same method of measurement works just as well for gas release. The level drop times the area is equal to the volume of gas released under the pressure at which it had accumulated. But this method can get complicated because the level often doesn't drop all at once, but it subsides over several days as gas continues to exit. Sometimes the waste level rises first, as a big gob of waste rises to the surface and the gas expands in the lower pressure before it actually releases. Nevertheless, one can confidently conclude that a big level change, either up or down, means a big change in gas volume. These calculations were used continually to feel the three or four beats-per-year pulse of SY-101's gas release events.

According to the level drops, some of these burps were pretty big. When in-tank video became available, it showed a burp to be a series of huge violent upwellings from different parts of the tank with hydraulic forces strong enough to bend 3-inch pipes. The bad part was that the gas released was flammable. Occasionally a large burp made the whole tank domespace flammable for a few minutes or hours. If a stray spark ignited this flammable gas mixture, people feared it could blow up the tank and spew radioactive waste over a big area, maybe even beyond the Hanford boundary.

This possibility was horrific, but Rockwell and Westinghouse engineers believed for years that the real risk was low because a spark was unlikely; especially considering the short time the air space was actually flammable. But in late 1989 an outside reviewer reminded them that nitrous oxide, a gas generated along with hydrogen, could make the gas mixture burn without oxygen. The gas trapped down in the waste theoretically did not have to be released into the domespace to get enough oxygen to burn. This, the idea that the gas resided in a big bubble under the crust layer, and the raft of adverse revelations about other DOE weapons sites about this time, elevated the perceived risk of a tank explosion to a much higher level. On March 23, 1990, Mike Lawrence, the manager of DOE Richland, announced the possible danger in a press conference. The public reaction to this revelation energized DOE to start the frenzy of work and worry that is the main subject of this book.

PART ONE
Flammable Gas

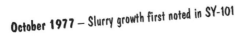

October 1977 — Slurry growth first noted in SY-101

January 1981 — First "burp" observed in SY-101

July 1983 — First dome pressurization alarm from a large burp in SY-101

July 1985 — Flammable hydrogen concentration first measured in SY-101 dome space

February 6, 1990 — "White paper" reviewer Paul d'Entremont advises "I'd watch out for SY-101"

March 14, 1990 — Mike Lawrence announces flammable gas hazard in SY-101

October 24, 1990 — First named burp, Event A

November 1990 — "Wyden amendment" establishes the tank safety "watch lists"

May 17, 1991 — Event C, first core samples of SY-101 waste

December 4, 1991 — Event E, largest burp ever, bends in-tank hardware, first in-tank video of a burp

1977–1989

Storm Warnings

Now let's narrow the focus and look at how SY-101 began to shake and reshape the U.S. Department of Energy, or DOE for short, and Hanford cultures. The progress of events might be compared to what happens when a hurricane approaches the coast. Today, meteorologists study radar and satellite images to make predictions, and airplanes fly out to measure the winds and pressures. Based on this wealth of data and much prior experience, officials issue public warnings as needed. But if nobody had been through a hurricane before, there would be no predictions or useful warnings until bands of dark clouds appeared on the horizon. Even then, most would not be aware of any danger, and those who sensed a threat could only watch things develop and speculate about what might happen. The actual storm would far exceed their speculations.

From 1977 to early 1990, the growing perception of the danger of SY-101's burps was like the approach of a hurricane. But, like the residents of Galveston before the great 1900 hurricane, most didn't recognize the signs of what was coming and went about their business as usual. The recently inaugurated National Weather Service knew relatively little about what made a hurricane go and did not actually forecast storms; it only tracked them and recorded data as they passed. Likewise, Hanford had some idea of what SY-101 was

| 37

doing but didn't know why or what the real hazards were. A few people got concerned, but the threat did not seem serious, and little could be done but puzzle over the sparse tank monitoring data and speculate about what might happen next. Like the Galveston hurricane, experience would prove much worse than the speculation.

When a hurricane actually arrives, the terrifying winds, deluges of rain, and flooding storm surge make it impossible even to communicate, let alone repair damage. One can only try to survive. This worst stage of SY-101's "perfect storm" began with DOE Richland Manager Mike Lawrence's news conference in March 1990, when he publicly announced the flammable gas hazard. Soon after this first gust, the whole world suddenly came down on Hanford, pointing fingers, making accusations, and demanding quick action. The raging storm of criticism and contention made it almost impossible to communicate, let alone find and fix the cause of SY-101's problem. The combined Hanford organizations could only hang on and try to survive, hoping the tempest would eventually abate.

When a hurricane is at maximum intensity, the wind abruptly subsides, rain ceases, and the blue sky of the eye appears. People leave their shelters, see the sun and survey the damage. Eventually, the burps were stopped and the storm around SY-101 calmed like the hurricane's eye. To the uninitiated, the eye might seem like the end of the storm, just as we thought the mixer pump would permanently make the tank safe. But the eye really signals the second half of the storm, maybe not quite as violent as the first but still incredibly disruptive. The second part of the SY-101 storm began with the runaway crust growth in 1998. The 2-year flurry of effort to reverse crust growth was not as bad as the 3-year agony leading up to the mixer pump, but it was still a stressful and urgent experience.

After a hurricane, casualties are counted, the damage estimated, insurance claims are filed, and rebuilding begins. The experience may make the victims better able to survive the next hurricane but certainly won't prevent it. For SY-101, the confidence built on knowledge gained in the first half of the storm and in the relatively wide "eye" provided by the mixer pump enabled us to begin rebuilding during, not after, the second half of the SY-101 storm. The claims were settled more quickly than expected. Not only that, but the work in the last part of the storm has actually prevented future storms in this tank and maybe in other Hanford tanks as well. Time will tell.

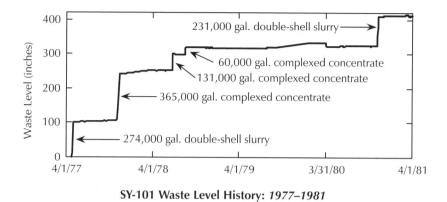

SY-101 Waste Level History: *1977–1981*

SLURRY GROWTH

With this introduction, it's time to start the story. The best place to start is when Dan Reynolds opened the valve on the 242-S evaporator and topped off the tank with double-shell slurry. Half the waste was double-shell slurry and half was complexed concentrate. The 1,061,000 gallons of waste that had been pumped in to that point accounted for only 387 inches of the 413-inch waste depth. The other 26 inches was "slurry growth" that built up since the first waste arrived. In fact, Rockwell engineer Jim Honeyman, who ran the first double-shell slurry into SY-101 as a "process test," reported "the double-shell slurry product increased in volume by 19,000 gallons from April to October 1977." This is equivalent to 7 inches of level growth in just 6 months!

It was clearly a problem, but only an operational one. Nobody thought it was a safety hazard then. A lot of time and money was already invested in double-shell slurry as a way to make tank space. The 242-S evaporator had been preparing double-shell slurry feed, the not-quite-so-concentrated precursor to the final product, through the last half of the 1970s and storing it in tanks all around the 200 West Area. The 242-A evaporator was doing the same thing in the 200 East Area. But slurry growth was eating into the space savings. It first emerged in 1976, and by late 1977, a tank farm memo noted 7 "growing" tanks in the 200 West Area alone. SY-101 was one of them. In 1981 the list had grown to 10 tanks. If double-shell slurry was to live up to its expectations, the cause and implications of slurry growth had to be found. A series of studies began to do just that.

From the start, slurry growth was correctly attributed to the buildup of gas in the waste. For a couple of years, the gas was thought to be hydrogen generated from radiolysis, the radioactive breakdown of water into hydrogen and oxygen. Then, in a 1978 experiment, Rockwell scientist Tom Lane found enough hydrogen in a closed pot of simulated waste to be flammable. This simulant had the same chemical composition as double-shell slurry but without the radioactivity. In January 1979, Lane measured hydrogen concentrations of 12 percent along with 21 percent nitrous oxide. So hydrogen came from chemical reactions as well as radiolysis of water, and flammable mixtures were possible. He warned that hydrogen could burn in a nitrous oxide atmosphere without oxygen present. Somebody should have heeded his advice!

After a quarter-century of hard work helping put the various waste tank safety issues to rest, Dan Reynolds is the heart, soul, and conscience of the Hanford tank farms. He has more Hanford history in his file cabinets and in his head than a DOE public reading room. He knows each tank by name and speaks their language. At the same time, he has to live with the fact that he filled SY-101 with double-shell slurry and thus is ultimately responsible for its notoriety—and this book!

Tall and taciturn, Dan is an authentic "graybeard" in all senses of the word. He is never arrogant, though he has earned the right, because he knows the limitations of the tank data and appreciates the incredible complexity of the chemistry and physics that rule them. With this wisdom, Dan can respect others who understand things differently, knowing they just might be right.

At meetings, Dan stays quiet most of the time. But, after the youngsters have confused things sufficiently, Dan speaks! When Dan stands up and ambles to the white board, listen carefully! He will interpret tank data and give you important historical facts, clearly laid out with simple logic, that answer the important question, even if it hasn't been asked. It has been an honor and privilege to work with Dan and have his great knowledge available through the years.

These results prompted Rockwell to start sampling the gas in the domespace of several of the growing tanks. But no high concentration of flammable gas was found. The gas was not generated very fast, and even the slow natural ventilation from thermal convection currents in the tank domespace was enough to dilute the hydrogen to barely detectable concentrations. This was a reassuring result. Nobody realized yet that if the gas making the slurry grow were released suddenly, the hydrogen concentration would be much higher, even flammable. But catching a tank in the act would take several more years.

> Jack Lentsch, who managed installation and testing of the mixer pump in SY-101 in 1992–1994, told me about his experience with hydrogen-nitrous oxide mixtures. "In the '60s when I was a chemist, we used a device called an atomic absorption spectrometer. It used a mixture of nitrous oxide and hydrogen to vaporize water samples. If you just happened to chance upon a stochiometric mixture [theoretically ideal proportions for chemical reaction], it could blacken an entire room. I've seen entire rooms covered with shards of glass. It happened about once a month. It was quite a violent reaction. I was shocked to find that it was a possibility at Hanford."

In 1980, Rockwell chemist Cal Delegard did more extensive studies on slurry growth using a nonradioactive double-shell slurry simulant with the same composition as in SY-101, including the load of organic complexants that were added in the complexed concentrate. He found that chemical degradation of one of the complexants generated not only hydrogen but also nitrous oxide and nitrogen. The gas generation rate was found to increase exponentially with temperature in a special way that implied that the same chemical reactions were happening over a wide range of temperatures. Even more striking, the growth rate of the waste measured in SY-101, to that time, was consistent with the rate predicted from the experiments. Larry Gale continued nonradioactive slurry growth tests that supported Delegard's results. In 1983, Gale found that one of the complexants produced gas, but some others did not (or at least produced much less). He admitted, "The reaction is more complicated than originally assumed." Gale also found that slurries without the organic complexants grew, but at less than half the rate of those with organics.

These early tests confirmed that slurry growth was a symptom of gas building up in the waste, that the gas contained a large fraction of highly flammable hydrogen along with nitrous oxide that could theoretically allow it to burn without oxygen present. The work also indicated that organic complexants caused, or at least greatly accelerated, gas generation. So, slurry growth might not be a problem in double-shell slurry if the complexants had been kept out. But the offending organics were already present in SY-101.

Despite the lowered expectations of space savings from double-shell slurry and the discovery of the potential flammability hazard, Rockwell believed that slurry growth could be worked around by simply allowing a little extra tank space for the growth. All that was needed was a way to predict how much the slurry would eventually grow so they could accurately compute how much space to leave.

The evaporators kept making double-shell slurry feed into the early 1980s. It filled AN-105 in 1982 and AN-104 in 1984. Apparently everyone still assumed that most of the waste in the new double-shell tanks would eventually be evaporated down to double-shell slurry, opening up a lot of tank space. This space would be used to store waste eventually retrieved from the now-dormant single-shell tanks. The only tank other than SY-101 and SY-103 that got the final double-shell slurry product was AN-103 in 1984. But this time it wasn't mixed with complexed concentrate, so AN-103 remained relatively tame.

THE TANK BEGINS TO BURP

Worse problems than slurry growth finally put a stop to double-shell slurry production altogether. After the last fill, the waste level in SY-101 continued to grow. Then, in January 1981, it suddenly dropped a few inches. The waste kept doing this about every month or two afterward. Because the cause of level rise or slurry growth was now known to be a buildup of gas, engineers deduced that the level drops signified a release or "burp" of the accumulated gas. And, because the gas was part hydrogen, a burn or explosion was possible if the burp was big enough and a spark was there to light it off.

Storm Warnings: 1977–1989 | 43

In the late 1980s, Rick Raymond, manager of Westinghouse's Tank Surveillance Program, was trying to set criteria for tank leaks. He was mainly worried about the old single-shell tanks, many of which had leaked by the 1980s. He decided that a rapid, half-inch waste level drop was sufficient to indicate a leak. But he noticed that SY-101, a new double-shell tank, had rapid level drops of much more than a half inch every 100 days or so but was not leaking. In fact, the level grew right back and dropped again and again.

Rick asked Gary Dunford of the Process Engineering Group about these level drops. He was told it was a result of slurry growth, by this time a well-known phenomenon. Then he asked what caused the level drop if it wasn't a leak. The answer was "gas release." At that point Raymond decided not to use rapid level drops for tank leak detection in double-shell tanks.

This "level cycling" phenomenon was first noticed in late 1977. A sudden level "subsidence" in SY-103 was formally reported as an unusual occurrence in February 1979. Level drops began in a few other tanks and were also listed as unusual occurrences. But by the late 1980s, Rockwell decided, with good reason, that the level drops were so common it didn't make sense to keep reporting them as "unusual." Later, when the storm broke, halting unusual occurrence reporting for burps was roundly criticized as downplaying a serious danger. These burps were suspected to be hazardous because they involved a release of flammable hydrogen. But the tank domespace

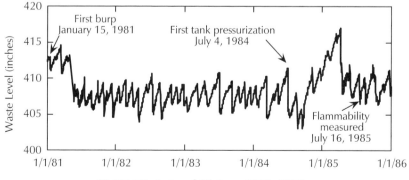

SY-101 Waste Level History: *1981–1985*

was so vast it would take a huge volume of hydrogen to make it flammable. The relatively small level drops did not signify a large enough release.

This is why the Independence Day burp, on July 4, 1984, grabbed everyone's attention. For the first time, a burp in SY-101 overcame the slight vacuum created by the ventilation system and pressurized the tank dome. This first high-pressure alarm revealed that a relatively small level drop could mean a relatively large gas release. The burp dropped the waste level only 2.9 inches, but, assuming the gas was 50% hydrogen (actually it was about 30%), and that it all came from the bottom of the tank where it would have been under 2.6 atmospheres of pressure, Rockwell engineer Leela Sasaki calculated that the domespace would have reached 3.5 percent hydrogen. A hydrogen burn just begins to be possible at a concentration of 4 percent.

Now the situation was getting serious! Rockwell immediately set up a task force and ordered them to meet daily until they found a short-term solution to prevent further tank pressurizations and identified the tasks necessary for a long-term solution. The task force worked hard and fast and came up with a short-term solution on July 23. They recommended that the tank be routinely water lanced to try releasing some of the gas before a burp. The task force also recommended that the next lancing should be done in stages to find from what depth most gas was released. In-tank photos should be taken as soon as possible because the tank level was at a 4-year low. All in-tank thermocouples should be read at least weekly because there appeared to be a correlation between waste temperature and level.

Water lancing was first tried as a means to control SY-101's gas releases on October 20 1983 when the crust was water lanced as an initial test. Water lancing is done with a 2-inch pipe inserted into the waste. The pipe has holes drilled along part of its length for jets of water to squirt through and break up the waste. It caused an immediate gas release and waste level drop. Water lancing of both crust and sediment began in earnest in September 1984. Hopefully, the lancing would release the gas in stages and prevent the next burp—at least make it smaller. Eventually four water lances were permanently installed in SY-101 to make the operation more convenient.

Each lance run used 200 to 500 gallons of water. From 1984 to 1988, 27 water lancing runs added a total of 8,000 gallons of water to the tank, raising the waste depth about 3 inches. Water lancing stopped in May 1988 because of concerns about further adding to the waste volume. Air lancing replaced water lancing until March 1989, but it soon became obvious that it wasn't doing any good. The waste level continued to rise, and the burps got bigger. There were also fears that injecting air might create a flammable mixture if it mixed with hydrogen in a large bubble supposed to exist under the crust. Tom Lane's 1979 warning that hydrogen and nitrous oxide could be flammable by themselves had apparently slipped into the background.

Water lancing has a long history at Hanford, though not usually for releasing gas. From the early 1980s, when single-shell tanks were declared inactive and pumping began to remove drainable liquid, 8-inch fiberglass tubes were installed for measuring the amount of liquid remaining in the waste. These "liquid observation wells" had a hollow "burrowing" ring attached to the bottom with vertical holes squirting water to dissolve and erode a passage through the waste. Central wells containing the jet pumps used to remove the liquid were also lanced into the waste with similar burrowing rings.

Water lancing was also used regularly after 1990 to sluice holes through the tough crust layer in SY-101 and other tanks to ease installation of various items of hardware and sampling devices. An X-shaped lance formed of two perpendicular horizontal pipes about 40 inches long and fitted with many small, high pressure nozzles was used to install the SY-101 mixer pump in 1993. This lance bored a large hole into the waste using 500 gallons of water in a 7-minute pass to the bottom. This same lance was used in August 1999 to bore through the much-thickened crust to make way for the new transfer pump and as an attempt at mitigating runaway level rise. This test used 1,200 gallons of water and released about 80 cubic feet of gas.

FLAMMABLE GAS!

The task force kept on working and reported their long-term recommendations 3 weeks later on August 13, 1984. Their wisest recommendation was that "no long-term solution should be implemented until more is known about gas buildup in the tank." To that end, they repeated the short-term recommendations for lancing, photos, and monitoring waste temperatures, but they also advised installing a hydrogen monitor with a strip-chart recorder to see just how serious the flammability problem might be during a burp.

Leela Sasaki, who was a member of the task force, got a gas monitoring system up and running January 11, 1985, and worked at monitoring the hydrogen concentration until September 11. There were five gas releases during these 9 months, two of which were big enough to pressurize the dome, but only three releases were monitored because of problems with the instrument. No hydrogen was detected during two of the monitored events (possibly because of the low sensitivity of the instrument), but the one on July 16, 1985, registered a spike in hydrogen concentration above 5 percent, well into the flammable range!

Even though the sharp rise and abrupt fall of the measured hydrogen concentration clearly showed this to be a small plume of gas that happened to enter the ventilation duct where the monitor was, it still proved that SY-101's gas releases could create flammable conditions at least temporarily. Also, by comparing the 3,800 cubic foot total gas release computed from the 3.9-inch level drop with the 1,000 cubic feet of hydrogen estimated to have exited the ventilation duct; Sasaki concluded that the gas must be about 26 percent hydrogen. This calculation was remarkably close to the 30 percent hydrogen fraction found later by more precise measurements. If all of the gas were instantaneously released and mixed with the air in the dome, the average hydrogen concentration would have been 2.5 percent.

The difficulties and frustrations encountered in this effort at gas monitoring reveals how very hard it was, and still is, to get data in the tank farms. The following is quoted from Leela Sasaki's report to Kelly Carothers on September 16, 1985.

"Several problems were encountered during the hydrogen monitoring period. After a few days of operation, the cold weather caused the reading to drift downward. Low concentrations of hydrogen would not have been detected during this period. . . . However, no gas releases occurred during this time.

> "On February 11, the chart recorder began malfunctioning; about one month passed before a replacement recorder could be found, tested, and installed.
>
> "In April, the sample tube was found disconnected from the tank vent line; it is estimated that the line had been disconnected for as long as a month.
>
> "Finally, due to other manpower requirements, maintenance of the recorder was suspended until the work was formally scheduled; about two months passed before maintenance of the recorder was resumed. As a result, hydrogen concentration data was obtained for only sixty-nine days . . . [of the 243 days] between January 11 and September 11."

Double-shell slurry did not look like such a good idea any more. A flammable atmosphere had now actually been measured in SY-101, double-shell slurry feed in several single-shell tanks was growing, and several double-shell tanks with double-shell slurry feed and the two tanks (AN-103 and SY-103) with double-shell slurry had began having small burps. With these factors and a shortage of funds, Rockwell and DOE decided in 1986 to halt laboratory and pilot-scale technology studies on double-shell slurry. This also included studies that might have developed a better understanding of burp behavior.

The size of the burps in SY-101 and the period between them kept increasing, so by the late-1980s, level drops were commonly 6 to 10 inches and roughly a hundred days apart. Most of these were now big enough to pressurize the tank and some of the larger ones probably made the domespace flammable. But the few hundred cubic feet per minute of ventilation quickly diluted the tank dome, so it was only flammable for an hour or so at most, and nobody could conceive a likely spark source to ignite the gas.

Nevertheless, to see how bad it might be, the initial tank farm safety analysis report of May 16, 1986, analyzed a burn in the tank dome after a burp. The analysis predicted that the pressure pulse could blow the inlet and exhaust filters, releasing a little radioactivity, but the tank would not be damaged. A burn would not be a catastrophe, just a bad day at Hanford with only local contamination. The minimal consequences predicted for this accident implied that SY-101's now demonstrated tendency to become flammable really was just another operational hazard to be wary of, like dropped crane loads, electric shocks, or chemical spills—not an emergency.

Several later assessments of hydrogen burn consequences arrived at much the same answer. Only the most wildly conservative, worst-case calculations predicted a possible tank failure with offsite radioactive contamination. Besides, the party line was that a waste vitrification plant would be operating at Hanford by 1989, turning radioactive waste into glass logs. SY-101 was to be one of the first tanks processed. Because we could live with the problem and it would apparently be with us only a little while longer, the SY-101 flammable gas issue went to the back burner. In hindsight, this was a perfectly reasonable attitude from a technical viewpoint. But the lack of concern would be seen as incompetence, concealment, or even conspiracy in the public wildfire soon to ignite.

EARLY BURP THEORY

So what did Hanford process engineers, scientists, and managers really know about SY-101 at this stage? They actually had a pretty accurate perception of the waste properties and burp behavior, even if they didn't understand the mechanisms very well. In-tank photos in the early 1980s clearly showed a broken, crusty waste surface. Below the crust was a layer of liquid with solid sediment on the bottom.

In September 1985, digging some old synthetic double-shell slurry out of a 4-foot-diameter test tank in the REDOX Plant helped confirm this concept. The tank held 700 gallons of nonradioactive double-shell slurry simulant that was prepared for a pilot-scale evaporator run in 1983. A cylindrical screen was driven into the tank to get ready for removing the waste. During this process the crust was found to be a spongy layer 6 to 8 inches thick and a stiff sediment layer 18 to 24 inches deep. Between the crust and sediment was 4 to 5 feet of slurry thick with sandy crystalline solids. The gas over the synthetic waste was found to contain both hydrogen and nitrous oxide, just as Cal Delegard had found 5 years earlier. Engineers assumed that this waste roughly represented the consistency, configuration, and composition of the waste and gas in SY-101.

The surface level rise or slurry growth preceding a burp was believed to be a symptom of accumulating gas bubbles generated by chemical decomposition of organics and radioactive breakdown of

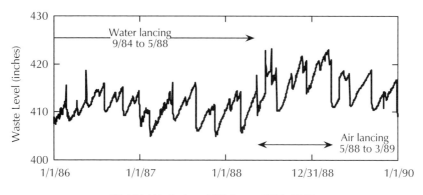

SY-101 Waste Level History: *1986–1989*

water. At some point, by some unknown mechanism or combination of mechanisms, the gas suddenly released and caused the surface level to drop.

So far, so good. But where did the gas accumulate? Was it in the crust or in the sediment? And what caused the release? Sasaki noted in her gas monitoring report of 1985 that "the fact that temperatures at the bottom of the tank rise and fall with liquid level suggest that gases accumulate at the bottom of the tank." But an internal report from a team of Pacific Northwest National Laboratory (PNNL for short), and Rockwell scientists issued the same year stated, "Slurry growth involves entrapment of gases produced by chemical reaction resulting in the formation of a surface crust and thus increased waste volume.... [These] gases are suddenly released through collapse of surface crust." This conclusion was based on small scale, nonradioactive experiments where a foamy crust was seen to build and collapse. The flammable gas task force report from the prior year also focused on the crust where a "confined layer of organic-containing waste produces gas which, unable to escape, lifts the surface crust until a fissure develops which allows gas release."

So direct evidence from the tank pointed to gas retention and release from the sediment layer at the bottom, while another theory based on small-scale experiments made the crust the culprit. A gas release model described in a draft position paper from December 1989 ingeniously combined the two theories: Gases build up underneath a floating crust, causing slurry growth. Eventually, the increasing volume and pressure of the gas under the crust breaks through the crust between the crust and tank wall and around equipment protruding through the crust. Because temperature data shows a large and sudden disturbance in the sediment layer during a burp, gas is apparently trapped there as bubbles that eventually grow large enough that their buoyancy exceeds the strength of the sediment and they escape towards the crust. The release of bubbles from the sediment may be what triggers the release of gas trapped below the crust.

This model still assumed that the crust was the ultimate barrier to a gas release. This implied that there must be large "pockets" or a large "bubble" of gas close to the waste surface, potentially accessible to a spark or heat source that could ignite it. Worse yet, poking a small hole into the crust could suddenly release all the gas! This possibility, even though it was proven incorrect only 2 years later, was the fatal flaw that brought on the full fury of the SY-101 storm!

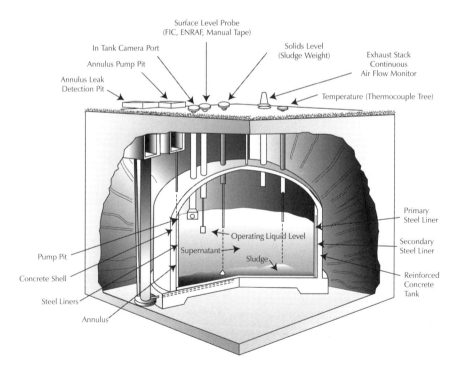

The early theories about SY-101's behavior had to be constructed from very limited data. Weekly temperature readings were available from 7 of 18 thermocouples covering the full depth of the waste (tank farm staff began daily recording of all 18 beginning in 1990). Twenty-two more thermocouples were welded to the bottom of the primary tank when it was built, but only a few were recorded.

The waste level was measured once a day both by lowering a weighted measuring tape (manual tape) and by using an electrical contact probe built by the Food Instrument Corporation (FIC) (an unlikely name to be associated with toxic radioactive waste!). While large level drops during burps were clearly visible on both devices, their measurements might be quite different because of the irregular and changing waste surface produced by shifting chunks of crust. Also, both of these probes tended to collect "stalactites" or "wastecicles" of dried saltcake that might be several inches long. Sometimes wastecicles broke off during burps, giving the appearance of a much larger level drop. At other times they were flushed off with water.

A "sludge weight" was part of the original tank instrumentation that was supposed to be lowered into the waste to find the top of the settled sediment layer. But there is no record of any measurements being recorded with the device in SY-101.

The only other measurement was the domespace pressure, actually a slight vacuum, that was recorded continuously on a paper strip chart. A sudden rise in pressure would indicate a burp. The ventilation rate was estimated infrequently by manually inserting a pitot tube just like an aviation airspeed indicator to measure the velocity at the center of the exhaust duct. But the actual ventilation rate would decrease considerably when high humidity restricted flow through the inlet filters.

There were no core samples to describe waste properties until 1991. There was no domespace gas monitoring, except the 1985 attempt already described, until 1991. Most importantly, there was no in-tank video until December 1991 to show what really happened during a burp. Researchers were literally "in the dark."

DARKENING SKIES OVER HANFORD

The 1980s were a trying period for Hanford for reasons other than SY-101. The old Cold War plutonium production reactors had all shut down by 1971, and hopes for diversification to replace the extinct production mission were fading. In 1982, cost overruns and bad projections of electric power needs caused cancellation of two of the three commercial power reactors being built at Hanford. Hanford would not become a "nuclear power park." In 1983, Congress cancelled the breeder reactor program, removing the mission of the recently started Fast Flux Test Facility and putting its future at risk. The 23-year-old, dual-purpose N Reactor went cold in 1987. By the mid-1980s the most powerful Hanford advocates, Senators Warren Magnuson and Henry Jackson, had left office. The Plutonium Finishing Plant ceased operations in 1989. The huge old PUREX plutonium separation plant, which had been limping along since 1983 after an 11-year shutdown, could not comply with new environmental rules and made its final run in March 1990. Hanford's mission officially changed to waste cleanup in May 1989, when DOE, Washington State Department of Ecology, and the U.S. Environmental Protection Agency signed the *Hanford Federal Facility Agreement and Consent Order*, better known as the Tri-Party Agreement. Ironically, the Hanford workforce during the late 1980s was one of the largest since the Manhattan days.

The political winds were shifting. The short-lived Energy Research and Development Administration issued the first environmental impact statement on Hanford waste management in 1975. This document described the huge volume of radioactive waste stored on the Site, the large volumes that had already leaked into the soil or were dumped there on purpose leading to the first flurries of negative publicity. In 1982, Hanford's Basalt Waste Isolation Project was named as a candidate for a national radioactive waste repository. The local Indian tribes and citizens groups from Spokane fought fiercely against it. Their questioning led DOE Richland Manager Mike Lawrence, with his philosophy of openness, to release piles of recently declassified documents detailing radioactive releases and secret monitoring studies. More came out under the new Freedom of Information Act. Hanford opened its secret files in 1986. Declassified documents further fueled public outcry by revealing what were seen as cover-ups, incompetence, or deliberate damage to public health and safety. In 1987 Hanford was dropped as a geologic repository candidate in favor of the Yucca Mountain site in Nevada. Articles in Spokane's daily *Spokesman Review* described the concerns of the "downwinders," people living just east of Hanford who believed their health was compromised by radioactive releases in the 1950s and 1960s. The treatment of several whistleblowers who were censured for raising safety issues further bruised Hanford's reputation.

These were all signs and sources of a great cultural change. Similar shutdowns, disclosures, and public outrage were happening at most other DOE weapons sites across the country. DOE in general and Hanford specifically began to be viewed as bumbling anachronisms at best, dangerous and evil conspiracies at worst. The political atmosphere was similar to the occasionally flammable atmosphere in the domespace of SY-101. All it needed was a spark in the right place at the right time to create an explosion.

October 1988–March 1990

Landfall

When the flammable gas hazard in SY-101 showed itself, the public was already primed to react. Not only had recent revelations of radioactive pollution caused a national outcry, but the news also followed hard after loud publicity of Hanford's mishandling of the ferrocyanide and organic-nitrate wastes safety issues.

The solid waste in many tanks was mostly sodium nitrate along with some dissolved ammonia. Nickel ferrocyanide was added to some of the tank waste in the 1950s to remove radioactive cesium. Then, in the 1970s, to recover cesium and strontium, significant volumes of water-soluble organic complexants were added. Finally, in the 1980s, DOE began pumping liquid out of the single-shell tanks to prevent leaks, allowing the upper part of the waste to dry out.

The added organic complexants in contact with dry sodium nitrate could burn if heated. Ferrocyanide could actually explode in the presence of dry sodium nitrate. Under certain conditions, sodium nitrate and ammonia could combine to form highly explosive ammonium nitrate. Starting such fires and explosions required a temperature of only 400 to 500 degrees Fahrenheit. It didn't need a flame or a spark, just a hot surface. Critics asserted that, because the tanks were known to contain organics and ferrocyanides along

Part One: Flammable Gas

with sodium nitrate and ammonia, it was therefore possible for the waste in the tanks to burn or explode.

Unfortunately, it is very easy to raise a theoretical possibility but nearly impossible to prove it won't happen. Worse, even the most bizarre conjectures breed accusations of incompetence for not discovering them earlier and of conspiracy for not agreeing they are dangerous. Worst of all, the public reaction to these accusations often makes systematic study of the hazard difficult, and sometimes impossible. That would soon happen to the SY-101 flammable gas problem.

> One of the most common industrial explosives is ANFO, granular ammonium nitrate fertilizer mixed with diesel oil. I worked in a silica quarry during several summers in the late 1960s where this explosive was used. The blasting crew shoved a couple of sticks of dynamite, trailing wires from the blasting cap, into the bottom of a 30-foot drill hole with a long wooden stick. They filled the remaining space with ammonium nitrate fertilizer from the farm store poured from sacks through a funnel. A few gallons of diesel fuel supplied the organic and a couple of shovels full of mud sealed the hole. The power of shooting off 20 to 30 such holes simultaneously was *very* impressive.
>
> An ammonium nitrate explosion at Texas City, Texas, April 16, 1947, was often mentioned to sensationalize Hanford hazards. A French freighter delivering ammonium nitrate fertilizer was attempting to dock when the deck of the ship caught fire. Oblivious to the danger, the crew completed docking. Later that morning, the freighter exploded with a force of several kilotons of TNT, creating a tidal wave that engulfed the shore. Many oil refineries on the waterfront also caught on fire and burned for days. Over 560 people were killed.

SAFETY ISSUES

The public perception of the ferrocyanide and organic-nitrate safety issues arose almost by accident. In the early 1980s Rockwell floated an idea to permanently isolate waste in single-shell tanks by first pumping out all possible liquid, which was already being done to prevent leaks, then filling the tank with crushed gravel to avert dome collapse and sealing them up. Some thought that this probably was not a very good idea because, if the waste was dried, buried and sealed, it might get hot enough to set off a runaway chemical reaction. Examples of such reactions included the already-mentioned organic-nitrate burn and the ferrocyanide-nitrate explosion. In 1983, Harold Van Tuyl, then manager of the PNNL analytical chemistry laboratory, wrote a report raising these concerns. This prompted Rockwell to investigate further.

Rockwell hired PNNL to search the scientific literature and old reports to see how serious the problem might be. PNNL scientist Lee Burger studied the ferrocyanide explosion issue, did some preliminary calculations, and submitted a draft report in November 1984. He found that the probability of a ferrocyanide explosion in a Hanford waste tank was very low. Nevertheless, he calculated that, in the absolutely worst case, ferrocyanides might theoretically produce a blast equivalent to 36 tons of TNT. DOE was taken aback, stopped publication of the report, and requested PNNL and Rockwell to do some more research to better define the risk. Another PNNL scientist, Earl Martin, did a similar study on the organic-nitrate reaction and concluded that the possibility of this reaction was "nonexistent to negligible." DOE cleared this report, and it was published in June 1985.

The proposed task to evaluate the ferrocyanide issue and flesh out the Burger report was never funded. Nevertheless, it was listed as "complete" in the Hanford Waste Management Technology Plan (HWMTP, pronounced "Humpty") for 1986 and 1987. At the same time, an environmental impact statement for Hanford defense waste disposal, issued in December 1987, discussed *both* the organic-nitrate reaction and potential ferrocyanide explosions and stated that *both* were of negligible risk. The ferrocyanide risk was erroneously referenced to the Martin report, which did not discuss ferrocyanides at all. This oversight would soon fuel the smoldering Hanford tank safety controversy and turn it into a conflagration.

Westinghouse Hanford Company (Westinghouse) replaced Rockwell as the prime contractor in June 1987, consolidating more of the operation back under a single company. Though it inherited the closed, secrecy cloaked Cold War culture of General Electric, Atlantic Richfield, and Rockwell; Westinghouse was different. As a prominent U.S. nuclear power plant supplier, Westinghouse had considerable experience in nuclear technology and, more important, in dealing with public reviews and government regulators. In April

58 | Part One: Flammable Gas

1989, Westinghouse also replaced DuPont in managing the Savannah River Site in South Carolina, the other major Cold War plutonium production site in need of cleanup. Savannah River, like Hanford, had a lot of radioactive waste in large tanks and had been wrestling with a hydrogen problem for a number of years. Looking back, Westinghouse actually did a pretty good job of handling the festering safety issues it took over.

In October 1988, Bob Cook, a former employee of the Nuclear Regulatory Commission who had become a Hanford critic, was asked to speak at "The Great Nuclear Debate" in Spokane. His speech pointed out the erroneous reference to the Martin report on ferrocyanide in the Hanford environmental impact statement and alluded to another unpublished PNNL research study he had found out about. This was the Burger report, with its suggestion of possible large explosions in the tanks. Intense public pressure finally forced DOE to release it. The Washington Department of Ecology eventually got a copy still labeled "draft" in July 1989. The consequent charges of cover up and mismanagement sorely damaged Hanford credibility and generated a public outcry that Congress had to attend to.

In October 1989, during Senate Armed Services Committee confirmation hearings on the chairman of the newly formed Defense Nuclear Facilities Safety Board (established to improve safety at DOE nuclear sites), the PNNL report on ferrocyanide reactions was brought up as revealing a "huge" safety issue. Senator John Glenn brought up the 1957 chemical explosion of a Soviet radioactive waste tank at Kyshtym as an example to show that "we're not dealing with an abstract problem here." It made the front page of the *New York Times* on October 17. Alluding to the unreleased Burger report, *Times* reporter Matthew Wald wrote, "In an effort to squeeze more waste into giant tanks . . . engineers in the 1950s added chemicals that created the risk of an explosion, according to a Government study completed five years ago but not made available until now [i.e., the Burger report]. Experts say that if such a blast occurred it would shatter the tank and disperse its intensely radioactive contents."

On December 29, 1989, at 5:00 AM, Tank SY-101 had a fairly large gas release. The waste level dropped over 6 inches immediately and continued to almost 8 inches after 2 days. The gas release pressurized the tank dome, but the hydrogen concentration was not measured. Changes in the waste temperature profile showed that waste was disturbed all the way to the tank bottom. The last such event had happened 109 days before, on September 11, 1989.

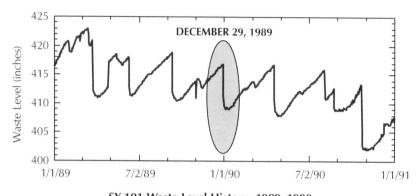

SY-101 Waste Level History: *1989–1990*

FLAMMABLE GAS GOES PUBLIC

In a effort to answer the flurry of questions coming down from DOE Headquarters, Don Wodrich, a Westinghouse senior manager, directed Nick Kirch and his staff to prepare carefully researched "issue papers" on each potential hazard, stating logically and sensibly what was known and why the tanks were safe. Work started in October 1989 and the group eventually turned out 15 "white papers" by January 1990. To bolster their credibility, Wodrich sent them out for review by experts at the Savannah River Site.

Nick Kirch has been involved longer and more intensely than just about anyone with the Hanford waste tank safety and operational issues, SY-101 flammable gas especially. A tall, soft-spoken gentleman of friendly demeanor, Nick managed Process Engineering, the group responsible for collecting tank data and understanding the waste. He meticulously recorded the developing body of knowledge about the tank in his planner and was always a great resource at meetings. Nick remained a guiding influence in tank safety even after he left to manage the group responsible for divvying up double-shell tank space.

Nick was totally unflappable. He could present the latest data from SY-101, which was often incomplete, flawed, or poorly understood, to a hostile audience and keep smiling while being loudly disbelieved and ridiculed. But reviewers and oversight groups respected Nick because he did no political filtering on the data; he just told it like it was. He often encouraged his staff with his own quiet self-assurance that "in spite of what anybody says, we know more about this tank than anyone else." Nick was right.

One of the white papers described the flammable gas hazard and the burping behavior of SY-101. It concluded that the probability of ignition was low because the tank domespace was so seldom flammable. Savannah River engineer Paul d'Entremont reviewed this paper. The gas composition concerned Paul, so he called Roy Jacobs, who had been studying the flammability of benzene, air, and nitrous oxide. Jacobs' data showed that the nitrous oxide made benzene more flammable than air alone. Not only that, but nitrous oxide was an oxidant, so it didn't even need air to burn. They decided that if hydrogen acted in any way like benzene, the composition of SY-101's gas was explosive! It wouldn't just burn—it would detonate!

On February 6, 1990, Paul returned his conclusion that the mixture of hydrogen and nitrous oxide trapped in the waste was dangerous because nitrous oxide supplied the oxidant for a burn. This factor, coupled with the theory that the gas was trapped in a large "gas pocket" under the hard crust layer, prompted Paul to comment, "I'd watch out for SY-101!" In fact, he considered this the most serious threat identified of the 15 issues reviewed. Though Rockwell and Westinghouse had been aware of nitrous oxide flammability for years, this conclusion by an outside reviewer had the weight of a new discovery of high importance.

Now, in addition to the recent embarrassment of the organic nitrate and ferrocyanide safety issues, Westinghouse had a new, possibly even more urgent, one to somehow explain to DOE and a suspicious public. Things began to happen very quickly. Don

Wodrich became a very busy man. He briefed the Westinghouse manager of Waste Management Engineering on the Savannah River concerns on February 27, 1990, and gave the same briefing to DOE Richland management on March 6. A day later Westinghouse halted waste or water additions and waste-penetrating operations in SY-101 and other suspect tanks to reduce the possibility of flammable gas ignition. These restrictions were formalized as official safety controls on March 20.

On March 21, 1990, Wodrich briefed the Defense Nuclear Facilities Safety Board consultants on the history of slurry growth and the associated flammable gas hazard. The board initially affirmed that gas accumulation in SY-101 was a serious concern. However, two weeks later, after further review of available data, they concluded that there appeared to be no real threat to the public from hydrogen buildup in the Hanford tanks. This comforting conclusion by such an august body might have helped tone down the crisis. But it was too late!

The situation exploded on Friday, March 23, 1990, with DOE Richland Manager Mike Lawrence's press conference announcing "new information" about flammable gas hazard in SY-101. The next day, the Hanford-friendly *Tri-City Herald* reported the news under the headline "HANFORD WASTE TANKS POSE EXPLOSION RISK." *Herald* staff writer Wanda Briggs wrote, "Hydrogen buildup in Hanford tanks containing highly radioactive nuclear waste could cause them to explode, a study by Westinghouse Hanford Co. scientists has concluded. However, Mike Lawrence, the DOE's Richland manager, emphasized the likelihood of one of the underground tanks rupturing and spewing contamination into the atmosphere is slim." But, further down, Lawrence was quoted saying, "The worst case is any explosion that could cause the dome to collapse and send the contents up to the air. I can't sit here and say it's not going to happen." Stated this way, the danger of a catastrophic release from an exploding tank sounded pretty real.

On March 27, 1990, Don Wodrich gave a very open, honest, and detailed presentation to Washington State Ecology (and presumably some of the Governor's staffers). It was a reasoned discussion of the safety issue, what the latest data showed, and what was being done about it.

But history had already marched past the point of reasoned discussion. Lawrence's news conference had already triggered an angry letter from Washington Governor Booth Gardner to U.S. Secretary of Energy James Watkins, dated Tuesday, March 27, 1990. Gardner wrote, "Last Friday, Department of Energy officials at Richland announced that hydrogen buildup in the Hanford waste tanks was potentially explosive. This is the second time in six months Washington citizens have had to consider the potential for explosion in waste storage tanks. . . . I am troubled that safety concerns frequently appear to be revealed only after they are raised by independent reviewers." The Governor's letter went on to formally ask Watkins to act immediately to mitigate any danger of explosions and provide all necessary funds to do it. He also asked for an independent public review of procedures and management for handling health and safety issues at Hanford. Further, he asked for the creation of a management structure including non-Hanford people to evaluate policy along with health and safety risks and to coordinate public information.

These requests, direct from the Governor, plus mounting pressure from Congress, commanded a strong response by the Secretary of Energy. He initiated a series of actions that would occupy most of Hanford's attention, efforts, and resources for the next several years. At the same time, Westinghouse mobilized their organization to try to reach higher ground ahead of the storm surge of public clamor that was about to wash over them.

March–October 1990

Blowing in the Wind

So much happened from March through October 1990 that it is clearer to describe the reactions of DOE, Westinghouse, and the public separately, outside of strict chronological order. Forces, personalities, and concepts that would shape the next 3 years were being formed and sorted out. It was a remarkable time. Hanford was blowing in the winds of change. Unremitting, strenuous effort was the rule; 12–16 hour days, shortened weekends, ignored holidays, and deferred vacations were common. Facts were scarce. Every piece of new data, every calculation, every tentative conclusion was exposed to intense, often hostile, scrutiny. Presentations prepared to enlighten and reassure brought criticism and ridicule. It was a time that brought the Hanford community to full realization that the SY-101 flammable gas safety issue was *very* serious business indeed!

DOE JUMPS IN

Secretary Watkins' first action was to rebuke Mike Lawrence for his frank and open statements at the news conference. During his testimony at Senate Armed Services Committee hearings on March 29, Watkins commented, "I'm sorry it was said that way. . . . We do not have an immediate safety problem for residents of Richland." Lawrence held to his openness policy, holding frequent news conferences announcing results of new analyses, relaying recommendations of various reviews, and gallantly admitting problems when they showed up. At the same time Lawrence tried to calm the hysteria by pointing out that *two* groups of independent reviewers had concluded that, even if the gas in SY-101 were to explode, the tank would not fail and radioactive waste releases would be negligible. Later analysis would show that this assessment was about right, but few believed Lawrence at the time.

Watkins delegated the DOE response to Gardner's request to mitigate any danger to Leo Duffy, his director of the Office of Environmental Restoration and Waste Management. Duffy acted immediately by forming an ad hoc review team to go to Hanford and get some answers. The team included Paul d'Entremont, whose review comments triggered the whole chain of events, combustion expert Scott Slezak from Sandia National Laboratories, and others who would serve on future committees and review teams.

In 1988, George H. W. Bush appointed Leo Duffy, a nuclear navy colleague of Admiral Watkins, Director of the DOE Office of Environmental Restoration and Waste Management. The label for this office, rather like a military call sign, is "EM-1." Like Watkins, Duffy also had a rather negative opinion of Hanford. In a letter of advice to a later successor, he wrote that Hanford in 1992 was extremely slow to recognize safety problems and had a safety mentality that was, at best, complacent. He found out more in local newspapers than in official site reports. Duffy left office in January 1993 along with Watkins as the Clinton administration took over and was not on hand to see the burp problem mitigated in July 1993. That honor would go to his successor, Thomas Grumbly.

The Headquarters review team arrived at Hanford on April 1, 1990. Paul d'Entremont remembers reading all about himself and the other team members in the *Tri-City Herald* that morning at breakfast. Used to working behind the scenes, he was quite surprised by the sudden publicity! The team saw photos of the inside of SY-101, revealing bent pipes and splash marks on the walls.

Nobody could figure out what would do this. Admitting they did not understand the process, the team adopted the Hanford view that a stiff, peanut butter-like crust sealed off a large bubble, or "pocket," of gas under it until enough built up to break through. Using this model, waste level drops recorded during past burps implied that the biggest *credible* gas pocket might be a foot thick across the entire tank. If such a bubble were to burn, the team's preliminary calculations showed it could crack the primary shell but would not break the second shell or the surrounding concrete. In fact, they thought the probability of igniting the gas was low. It turns out that the 4,400 cubic feet of gas involved in this hypothetical accident is actually less than half of the volume of gas released in some of the larger burps, and there were potential ignition sources nobody could then imagine.

The team reported their findings to Duffy at DOE Headquarters on April 9. The next day they flew back to Richland to brief Mike Lawrence, Westinghouse management, and the newly formed

A 1989 in-tank photo of the waste surface inside SY-101. Note the bent pipe on the right and white splash marks on the wall.
Courtesy PictureThis.

Defense Nuclear Facilities Safety Board, who happened to be meeting in Richland. They suggested no long-range solution to the problem but recommended, instead, that both the waste and the gas in the tank be sampled to find out what was really happening. They also stated the opinion that tank farm operations had a culture of tolerating inoperative equipment, discrepancies in data, and generally "accepting the unacceptable." This phrase was picked up and used to criticize Hanford operations for years afterward.

> On April 19, 1990, 1:00 AM, one of the larger gas release events occurred in SY-101. The waste level dropped 4 inches immediately and 7 inches after 6 hours. The hydrogen concentration reached 3.5 percent, and the gas release pressurized the tank dome to 0.1 inch of water gauge pressure. Changes in the waste temperature profile showed that waste was disturbed all the way to the tank bottom. The release occurred 111 days after the prior one on December 29, 1989.

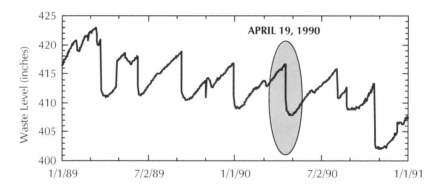

SY-101 Waste Level History: *1989–1990*

On May 14, Mike Lawrence declared the hypothetical ignition of the mixture of hydrogen and nitrous oxide in the waste tanks an unreviewed safety question (USQ). That officially put the issue on the books. A USQ is an administrative admission that "we've got a problem." Something potentially dangerous has been discovered that either is not covered in the current safety documents or is worse than what was analyzed. Usually the contractor who does the safety analysis and documentation declares the USQ. Having one declared by DOE Richland was unusual. Theoretically, it is pretty simple to close a USQ. The contractor reviews and analyzes the new condition and defines any necessary additional controls or limitations on operations. Of course, DOE and their reviewers need to be convinced that the analyses and controls are complete and adequate to ensure an acceptable risk. When political sensitivity is high, this may become impossible, as it was in the case of SY-101. In fact, the SY-101 flammable gas USQ was not closed until June 1996.

The pressure on Mike Lawrence increased. A report highly critical of Hanford management of safety issues was soon to be released, as well as a watershed report on historic radiation doses to the downwinders. A management shakeup was in the wind, and Mike had lost the confidence of the Secretary of Energy. He resigned on July 7, 1990, for a job with a private waste cleanup company in Denver. Watkins appointed John Wagoner to replace Lawrence on July 12. Wagoner served until he retired in January 1999. Keith Klein took over management of DOE Richland in May of that year.

> John Wagoner didn't know what he was getting into, but he got into it in a hurry. Secretary Watkins was about to hold a news conference announcing the release of the two Hanford-critical reports. He knew this would ignite an explosion in the media, and he wanted to announce the appointment of a new manager for DOE Richland at the same time to mitigate the damage. Two of Watkins' aides were sent to ask John Wagoner if he would accept the position. John told them he'd never thought about going to Richland and would like to think about it. The aides replied, "You have ten minutes." Faced with this ultimatum, John agreed but had to hurry and call his wife, telling her of their new plans before she heard it on the news!

DOE's Tank Advisory Panel

In late May, Leo Duffy made a decision that had a powerful impact on Hanford for the next 10 years. Feeling that he needed a more

Part One: Flammable Gas

permanent source of technical advice and counsel on the waste tanks, he authorized his subordinate, John Tseng, to form the Tank Advisory Panel or TAP. This panel, chaired by Massachusetts Institute of Technology Nuclear Engineering professor Mujid Kazimi, was to evaluate Hanford proposals and analyses and report their findings and recommendations directly to DOE Headquarters. If Westinghouse was the executive branch and DOE was the Congress, the TAP might have been something like the Government Accountability Office.

> John Tseng was born in Taiwan in 1948 and received degrees from Massachusetts Institute of Technology and Northwestern in the early 1970s. He gravitated to environmental work and rose to manage the Savannah River Site's effort to come into compliance with the National Environmental Policy Act. Tseng came to DOE in 1984 and began working for Leo Duffy in 1989. John Tseng was a goal-oriented, "hands on" manager who took a very active role in the SY-101 mitigation project.
>
> He had a rather low opinion of Hanford talent and made his feelings known. One Westinghouse engineer observed, after being assigned to Tseng's office at DOE Headquarters for 15 months, "I have grown a lot of scar tissue and taken numerous shots about Hanford productivity, technical competence, and fiscal responsibility." On the other hand, Tseng had a high regard for the capabilities of Harold Sullivan's group at Los Alamos National Laboratory and pulled them into the flammable gas mitigation team. Though Los Alamos performed admirably, the forced collaboration caused a lot of hurt feelings.

The newly selected TAP members got a briefing package on June 20, 1990, that included the report of the earlier Headquarters review team, the Ahearne committee report (discussed a little later in this chapter), and a number of Hanford reports from the 1980s. The TAP charter was to "review and evaluate issues related to the safe, efficient operation of high-level waste tanks at DOE facilities. Activities will include but not be limited to reviewing operations to independently determine hazard potential. The adequacy of mitigation actions is to be evaluated and recommendations will be provided where appropriate." Don Oakley from Los Alamos chaired the initial TAP. Members included Charles Abrams (consultant), Fred Carlson (consultant), Mel First (Harvard), George Schmauch (Air Products & Chemicals), William Schutte (Idaho National Engineering Laboratory), Scott Slezak (Sandia National Laboratories), and Richard Wallace (Savannah River). Dr. Kazimi came on as chairman in the fall of 1990.

They were a diverse group. The TAP members were nuclear engineers, mechanical engineers, and chemists. There were consultants, a professor or two, and several from the DOE national laboratories. Some were very reasonable, kind, and helpful. Others were belligerent and critical. Most had strong opinions about many things. One was very concerned with data validity and uncertainty. Another criticized any kind of local measurement as worthless because the waste might be very different elsewhere. Some typically ridiculed analyses as too conservative while others questioned whether assumptions were conservative enough! They challenged the quality and interpretation of the data, and they often were not very polite about how they said it! It took the firm, even hand of Dr. Kazimi to wring a useable consensus out of this group.

The TAP waxed and waned as a prominent institution on the Hanford Site over a period of 11 years. The TAP's first meeting, July 26, 1990, in the 2750E Building in the 200 East Area, was to review Westinghouse's safety analysis for core sampling. They apparently approved of core sampling because Leo Duffy told Watkins on August 3 that he was ready to do it. They did not meet again until January 9, 1991, where they again endorsed core sampling and agreed that increasing tank ventilation could not prevent flammability but could only reduce the time the domespace was flammable. They also began talking about options for mitigating the big gas releases. The TAP met several more times in 1991 and began meeting every month in 1992 and 1993 as mitigating the tank became more urgent.

The full TAP was split up into SubTAPs to address specific safety issues in 1993, and the full TAP ceased to meet. Flammable gas mitigation came under the Chemical Reactions SubTAP. Without the positive leadership of Dr. Kazimi, the effectiveness of the SubTAP gradually declined. Work progressed in spite of, instead of in consultation with, the SubTAP. But the full TAP was revived one last time in 1999 with Kazimi again in charge. Under his leadership, the TAP supplied vital high-level credibility for the SY-101 level rise remediation project. This momentum held through closure of the flammable gas safety issue in 2001. At that point, DOE decided the TAP was no longer needed and finally disbanded it.

The TAP's most powerful influence and their greatest contributions were indirect. They were the grindstone that helped sharpen our thinking. They forced us to defend our work and to make it defensible. The very thought of presenting our results to the TAP motivated us to strive for deeper understanding. They forced each task to justify its existence, which kept the team focused on what was important. They made us sensitive to the difference between what was measured (which they believed to be knowledge) versus deduction from models or theory (which they derided as speculation). They forced us to improve our tank monitoring systems. They also taught us to state every assumption and to rigorously quantify uncertainties. They forced us to look at data in different ways if only to confront their own creative interpretations. These good habits are the TAP's greatest and most enduring legacy.

The Headquarters Task Force

In August 1990, John Tseng also set up a Headquarters task force to complement the TAP. The task force not only reviewed and evaluated the tank safety problems but actually recommended solutions. There were about 20 high-level working scientists and engineers on the task force—half from Hanford and the rest from places like the Sandia and Los Alamos National Laboratories, universities, and Savannah River. One of the Savannah River members was Paul d'Entremont. Usually the task force met concurrently with the TAP at central places like Denver, Salt Lake, and Albuquerque, but also made many trips to Hanford.

The task force was instrumental in the genesis of the TWINS (Tank Waste Information Network System), an on-line database that gives us direct access to all the tank monitoring and waste sampling data available at Hanford. They also get credit for developing formal lists of the important safety issues at several DOE sites, including Hanford and Savannah River. They focused attention on hazards that might otherwise have been ignored in the uproar over SY-101.

The task force is last mentioned in December 1992. By then, the mixer pump was ready to go into the tank, and an effective technical team from Westinghouse, PNNL, and Los Alamos was hard at work to make it happen. The task force was no longer needed.

The Ahearne Committee

Watkins also formed an Advisory Committee on Nuclear Facility Safety, chaired by former Nuclear Regulatory Chairman John Ahearne, to do a higher-level investigation. Ron Gerton, a manager at DOE Richland who was to become most closely associated with SY-101, briefed the committee at DOE Headquarters on May 22. Gerton stated that the primary safety issue was that hydrogen and nitrous oxide in the waste could ignite and the resulting burn would release radioactivity. Secondary reactions of ammonium nitrate on crust and sodium nitrate-organic reactions on crust would exacerbate the release. The consequences of either of these events were outside the bounds of the existing safety analysis that had been written in 1986.

But the briefing also showed that Westinghouse was beginning to understand things a little better. Gerton showed "evidence that gas is not in one large bubble under crust." Lancing holes in the crust seldom vented gas, as it would have if the lance penetrated a large gas pocket. The waste temperature profile changes showed that gas was stored near the tank bottom. He used temperature profiles from the April 19 event and the previous one December 28,

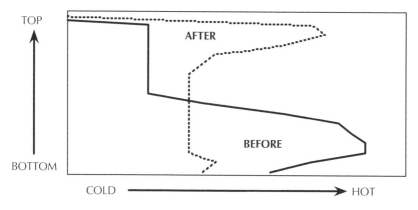

Temperature Profiles Before and After a Burp

1989, to show that hot waste on the bottom went to the top and was replaced with cool waste from above. Gerton's bottom line was "if gas is in small dispersed bubbles or foam [instead of a single big pocket] it will not combust!"

Hanford was also making plans to solve the problem. Gerton described four near-term actions that DOE Richland and Westinghouse were already planning. By September 1990, at the earliest, the waste would be diluted, reversing the concentration that started the problem in the late 1970s. This was the permanent solution, but waste would have to be removed to make room for enough water to accomplish the dilution. There was no room to spare in the 200 West Area and no usable pipeline to transfer waste to 200 East. The next best action was to stop gas buildup by stirring the tank with internal circulation from bottom to top. A decision was to be made on this by July 1990. In lieu of this, the accumulated gas could be released periodically by mechanically breaking the crust, assuming the gas was trapped in a pocket beneath it, which it wasn't. A decision on this idea was also to be made by July 1990. Finally, if nothing could be done about the gas releases, they were to be rendered non-flammable by injecting nitrogen.

> This "inerting" idea resurfaced often over the years. It seems simple but requires such a huge flow of nitrogen as to be impractical. Savannah River tried inerting a non-burping tank with nitrogen. It was generally successful but used up huge volumes of gas. There was a standing joke claiming that enough nitrogen leaked from the tank to suffocate seagulls flying through the plume!

Ahearne's committee visited Hanford on June 21, 1990, and reported back in a letter to Watkins on July 23, 1990. The main conclusion was, "The waste tanks are a serious problem." The letter cited an unconcerned operating staff, weak safety basis, lack of information about the waste, and operations having higher priority than safety. Top DOE Richland and Westinghouse management were accused of paying more attention to making tank space and to potential tank leaks than to tank safety. The letter raised an urgent need for waste sampling to determine what is in the tanks and for a better understanding of the hazards. They recommended more fundamental research and urged management to base safety decisions on a probabilistic risk assessment.

The Blush Report

To Governor Gardner's second request, Watkins directed the Offices of Nuclear Safety and of Environment, Safety and Health to jointly review the Hanford management structure and information handling. Nuclear Safety Director Steve Blush assigned a research team to go to Hanford and talk to people. They were on the site from April 10 through May 18, 1990, interviewing over a hundred old Hanford hands, their managers, and their managers' managers. They also dug through dusty old documents to see what these people knew, when they knew it, and what they did or didn't do about it.

During this onsite review, Blush's team made a fresh connection between a potential hydrogen burn and the organic nitrate and ferrocyanide safety issues that blocked progress on mitigating the flammable gas hazard for a while. They thought that a hydrogen burn in the tank domespace could raise the crust surface temperature above the 500 degrees needed to ignite the organics and nitrates that were supposed to be sitting there just waiting to cause trouble. Such a burn, while probably not explosive, could generate a prolonged, very serious release of radioactive material, especially if the preceding hydrogen burn had already breached the tank.

On May 24, this revelation prompted Secretary Watkins to direct DOE Richland, who then ordered Westinghouse, to develop a plan and method to spray water on the waste so the organics and nitrates in the crust couldn't burn. But just how flammable was the crust anyhow? How much water was enough? And how could one ensure the whole surface was wetted? These anxieties came up in review after review—only a waste sample could resolve them. Ironically, the worry about the perceived dangers of waste sampling halted all progress on waste sampling by September 1990.

Blush's report was issued towards the end of July 1990. It was not a good time for Hanford. Making full use of hindsight and the wealth of historic data his team found, he concluded that the safety issues should have been discovered and corrected years earlier. Not only that, but information on the safety issues may even have been deliberately suppressed! Though not received well at Hanford (it caused managers to "blush"), the report tells a pretty complete story

74 | Part One: Flammable Gas

of how knowledge developed step by step and how the knowledge was used or not used. A reader who can filter out the finger pointing will find it an excellent Hanford history resource. The Blush report won the Elmer Staats Award for Accountability in Government for 1991.

Maybe even more than Lawrence's March news conference announcing SY-101's flammable gas hazard, the Blush report was a lightning rod that raised Hanford vilification to an even more sensational level. Blush and others were called to testify at congressional hearings. Editorials loudly condemned Hanford. One in the August 5, 1990, *Seattle Post-Intelligencer* was headlined "HANFORD BRASS DUCKED REALITY." The article complained, "Those who run this abysmal facility give the impression of sleepwalking through everyone else's nightmare. They pride themselves on matter-of-fact tolerance of intolerable conditions. If the reservation is ever to be stabilized, Hanford's stewards must get a wake-up call from Congress—a bracing application of a 2 by 4 between the ears." It went on to raise the specter of "a disaster that might kill thousands of workers and affect people as far away as Spokane." The Portland *Oregonian* ran an editorial the same day accusing Hanford of knowing about potential hydrogen explosions in the tanks for 13 years but doing little to lessen the risk, calling it "arrogance run amok."

Actually, it was public hysteria run amok. It was probably the lowest point in Hanford's reputation. The organization was helpless to defend itself because each attempt at defense was viewed as continuing to ignore a grave public threat. The hysteria would have to burn itself out, and the congressional "2 by 4" applied. Then, suitably bruised and chastened, Hanford might be able, by careful work and politically astute communication, to regain a little credibility.

Paul d'Entremont told me that Steve Blush had the ability to read through an entire file cabinet of documents in a weekend (and he apparently worked a lot of weekends), extracting every hint of a safety issue. He was a pleasant fellow who smiled a lot but asked tough questions. If he didn't like the answer, he wasn't afraid to say so.

Blush flew from Washington, D.C., to Washington State with Paul and the rest of Duffy's Headquarters review team on April 9, 1990, when they came out to brief DOE Richland on their initial findings. Steve got hold of a copy of the team's report and read it on the plane, covering it with notes. When they changed planes in Salt Lake, he asked the team a lot of good questions. By the time they got to Pasco, he had thoroughly devoured the report and sucked all the useful information from the team.

Steve Blush stayed on as Director of the Office of Nuclear Safety until April 1993, when he resigned, protesting Energy Secretary Hazel O'Leary's reorganization that combined his office with four others into a larger department. He continued to be active in nuclear waste issues and authored *Train Wreck Along the River of Money: An Evaluation of Hanford Cleanup* in March 1995.

WESTINGHOUSE RAMPS UP

The year 1990 was a watershed year for Westinghouse as well as for DOE. They formed the team that would take the heat and do the work for the next 3 years and created the core of experts who eventually saw the whole safety issue closed. The team quickly figured out what was really going on in the tank and spent the next year or two confirming their theories and convincing DOE that they fit the facts. They developed the basic concepts and constructs such as "windows" for doing work in a burping tank and the "watch list" of tanks that were potentially dangerous for one reason or another. It was the year when all the fun began.

At the end of February, on receiving the Savannah River comments raising the hydrogen-nitrous oxide ignition hazard, Don Wodrich and the rest of Westinghouse jumped into action. Wodrich became "Mr. Slurry Growth," giving a series of briefings throughout March 1990 to Westinghouse management, DOE Richland management, the newly formed Defense Nuclear Facilities Safety Board, and the State of Washington. Westinghouse also immediately placed restrictions on SY-101 and other suspect tanks to prevent ignition and waste disturbances that might release gas. Wodrich also set up a consulting contract with industrial explosion expert Chet Grelecki, who would hang around Hanford giving advice for the next couple of years. Westinghouse also mobilized the forces of science, engineering, and management by establishing a whole string of their own task forces, panels, and committees to attack the

Part One: Flammable Gas

problem and defend against outside groups. It was almost as if there was a competition between DOE and Westinghouse to see who could set up the most oversight groups in the shortest time.

Waste Tank Safety Task Team

In late March, Westinghouse set up the SY-101 Waste Tank Safety Task Team. This core team drove the entire Westinghouse effort for the next 10 years. The team staffed the Hydrogen Mitigation Project that designed, installed, and tested the mixer pump. It became the Flammable Gas Program that finally resolved the overall safety issue. Fortunately, several excellent people were available from the recent shutdown of the Fast Flux Test Facility. Among these were Bill Leggett, who initially headed the task team, and Jerry Johnson, who became the leader of the Flammable Gas Program and who was primarily responsible for its success. Don Wodrich, Rick Raymond, and Nick Kirch were also on the task team, as well as Harold Van Tuyl and Larry Morgan from PNNL. The task team went right to work, meeting *every day* from 7:00 to 7:30 AM for 5 months straight!

One of the first things the task team did was to look at the other 176 waste tanks to see if any might be as dangerous as SY-101. Bill Leggett conscripted Rick Raymond, Rick's surveillance group, and a couple of other people to work over the weekend of March 31 and complete the survey. They were mainly looking for evidence of slurry growth and "level cycling," but they also used their combined engineering judgment to decide which tanks might be accumulating flammable gas. On Monday, April 2, they sent a list of 5 double-shell tanks, including SY-101, and 15 single-shell tanks, to the Westinghouse president. Don Wodrich named this the "watch list," and it has been called that ever since.

The "watch list criteria" were soon refined and formalized. Double-shell tanks were put on the list if they showed unexplained waste level increases, dome pressurizations, or waste temperature changes. Single-shell tanks were added if they had unexplained level growth, if photographs showed a surface crust, if waste samples had greater than 3 grams per liter of organic compounds, or if transfer records showed the tank got organic complexants from B Plant. These more formal criteria added only two single-shell tanks that were not on the initial list. Another tank was added because the ventilation system connected its domespace to several other watch list tanks. These additions brought the total number of watch list tanks to 23. In 1993 two more tanks were added for a final total of 25. SY-101, the first tank on the watch list, was the also the first tank to be crossed off, in January 2001. The other 24 tanks were removed in September 2001, when the Flammable Gas Safety Issue was closed.

Lofty, whip-thin Jerry Johnson is a master at wood turning and faceting gem stones. He was also made for the flammable gas issue. A technically astute project manager and phenomenal organizer, he strived to keep everyone happy and working enthusiastically towards the same goal. With that ambition and his easy-going manner, Jerry was always very pleasant to work for. At meetings, Jerry often warned us to shade our eyes as he bent to reflect some light off his polished pate, inviting others of similar condition to join his "shining" example.

He had an amazing memory along with a clear view of the big picture by which he guided the Flammable Gas Safety Program over, around, or through its many hurdles with a minimum of wasted energy. When discussions grew contentious or confused, Jerry firmly advised everyone to stop and "look at it from 50,000 feet." From that height we could see which trees defined the best path through the forest. Jerry was a master of the long-range plan. The strategy he laid out in 1996 to resolve the Flammable Gas Safety Issue served to accomplished its goal in late 2001 with little change. He will likely be remembered as the one most responsible for getting it done. I was honored to have Jerry's encouragement, aid, and advice in writing this book.

On June 6, Bill Leggett issued the safety improvement plan for hydrogen in waste tanks as the first comprehensive product of the task team. The plan claimed credit for the task team itself and the 23 tank watch list, and alluded to independently reviewed analysis that "has been found an adequate basis for concluding the waste tanks do not constitute an imminent danger." Nevertheless, he admitted that the available data was really not adequate for such an analysis and recommended a waste sampling and characterization program as the highest priority. The most important aspect of this program was to get core samples of the waste in SY-101 as soon as possible. If all went well, Leggett thought samples could be taken as early as June 29, after the next burp expected June 25, but they were weighing the benefits of sampling even sooner.

The plan also recommended a reasonable path to stop the burps. The waste would first be diluted with water to the extent possible to dissolve gas-retaining solids and to make it more pumpable. Presumably, added volume from dilution would be accommodated by transferring some waste into SY-102 and SY-103 because there were no other double-shell tanks in 200 West Area, waste could not be put back into leaky single-shell tanks, and there was no operational pipeline to other double-shell tanks in 200 East Area. Once diluted, mixer pumps were to be installed to circulate the waste from top to bottom to prevent gas accumulation, releasing it at about the same rate it was generated. Assuming that core sampling stayed on schedule and that sample analyses were completed 2 months later, the pumping could begin in early December 1990. And it would only cost $6 million!

While this schedule was wildly optimistic, the basic plan was entirely sound in hindsight. In fact, mixing stopped SY-101's burps in July 1993 without the benefit of dilution. Gas retention was halted altogether in 2000 by dilution at about 1:1 with water. The problem was that the available tank data and knowledge of chemistry and physics were not adequate to give confidence that this plan would work and would not cause some greater disaster. It eventually took 10 more years to build this confidence at a cost of more like $100 million. Maybe it was better this way. If the task team's plan had been accepted, I might not have written this book!

John Deichman replaced Bill Leggett as Tank Safety Task Team chairman in August. Leggett pleaded other commitments and probably had just plain had enough. Hanford critic Bob Cook sent Deichman a letter congratulating him on his new assignment. In Cook's words, "I am not sure whether it reflects Westinghouse confidence in you or a recognition that you are expendable. I hope its [sic] the former. The job you got is the toughest one Westinghouse or Rockwell has had at Hanford throughout their tenure here, in my estimation." Deichman led the task team for almost 2 years during some of its most productive times. Then, discouraged by severe budget cuts, he turned the responsibility over to John Fulton in June 1992.

Core Sampling—The First Challenge

Besides Leggett's recommendations, the earlier DOE Headquarters ad hoc review team and the Ahearne committee also pushed waste sampling as the most urgent step in understanding and mitigating gas releases from SY-101. Accordingly, preparations for core sampling proceeded at full speed. It was to be the first test of the task team's capabilities to take action against the forces that tried to smother it.

To establish immediate credibility, Leggett set up an internal Senior Chemists Panel on April 27, 1990. Dan Reynolds was on the committee. Their job was to define the chemical reaction hazards that should be considered before core drilling in SY-101. The panel's first meeting concluded that the greatest risk was that drilling could make a spark igniting the gas mixture in the large gas pocket assumed to exist under the crust. On May 22 they concluded that it was safe to spray water on the crust surface but did not pass any judgment on whether the wetting would actually be effective in preventing a spark or crust burn.

On May 31, 1990, an official safety analysis was issued for adding water to SY-101 in response to Secretary Watkins' order for a feasibility study of a crust wetting system. The official purpose of water addition was to "mitigate potential organic-nitrate reactions in the crust if a flammable gas burn occurs in the dome." The analysis claimed that adding water would not cause any new hazards though it unhappily increased waste volume. However, based on the large-gas-pocket-under-a-hard-crust model, if the water made the crust a better seal, thicker, or stronger, it might cause larger, less frequent gas releases. Beyond that, its only claim was that wetting the crust would only help mitigate a crust burn. Without a better knowledge of what the crust was like, analysis of how much water was actually needed to prevent a crust burn, or of whether the crust was even flammable, was not possible. They had to get a sample to decide whether it was safe to sample!

Another problem was how to get equipment into the tank to get the sample while the tank was burping. Nobody wanted to have workers operating the drill truck on top of the tank in the middle of a big gas release. To make tank work possible, Don Wodrich, Dan Reynolds, and Nick Kirch developed the concept of a "window" for safe core sampling. These evolving criteria would be used for any kind of work in the tank while it was still burping. The idea was that it was safer to muck around in the tank right after a burp released a large amount of gas than before a burp when the waste held a lot of gas. According to the big-gas-pocket-under-the-crust model, penetrating the crust with the 2-inch core drill would "pop" the bubble and let the gas out. If the bubble had already "popped," it would be safe to poke around for a while until a new bubble accumulated.

Wodrich sent a memo describing the criteria to Leggett on July 16. He assumed it would be safe to work in the tank while the waste level was close to the minimum it reached after a burp. SY-101's waste-level history showed that the minimum occurred 3 to 7 days after the burp, and the level rose at 0.1 inches per day afterwards. Wodrich proposed opening the "window" 7 days after a burp and closing it after 26 days, or when the level rose 2 inches from the minimum. This is the first informal definition of "window criteria" that would later become a formal part of the tank's safety documentation. Dan Reynolds issued the first formal Westinghouse report describing the window criteria on October 8, 1990.

On July 26, 1990, members of the TAP, John Tseng, and his DOE Headquarters task force met out in the Hanford 200 East Area offices for a presentation on the core sample safety analysis by Westinghouse staff. The sampling was to be done with the standard core drilling truck that had been used in other tanks for years. Sampling was planned for the window right after the next burp, expected in mid-August.

There was no argument that having a core sample would be of great benefit. The controversy was that, by triggering a gas release and igniting the potentially flammable crust or hypothetical subcrust gas pocket with a spark, core sampling might cause the very accident that the information to be gained was supposed to help prevent! At the meeting, Lewis Muhlstein explained how very conservative the core sampling safety analysis really was. Though the tank dome was well mixed by natural convection and very seldom flammable, the analysis assumed it was *always* flammable. Though the gas was clearly *not* trapped as a large bubble under the crust, the analysis assumed it was. Though the crust was believed to be moist and not very reactive, the analysis assumed it was a dry reactive organic-nitrate mixture. No wonder the calculated consequences were catastrophic!

To prevent this extremely unlikely catastrophe, the sampling plan wisely required the domespace to be non-flammable during drilling. This implied frequent, if not continuous, gas monitoring. The drill string also had to be electrically bonded to the tank to prevent static sparks, the drill bit was to be cooled with water, and the crust was to be sprayed with water to prevent any possibility of a reaction. Workers doing the drilling had to be dressed for toxic gas releases, and the tank exhaust was to be monitored for ammonia. Except for wetting the crust, these controls actually made good sense and generally became the standard for all waste-penetrating operations in flammable gas tanks.

The very act of publication inadvertently preserved the big-gas-bubble-under-the-crust model just as evidence was beginning to show there wasn't one. Hindsight can tempt us to call this scenario ridiculous. But this would be unjust. Only opinions were available in those days, and this one was about as good as another. Solid data to show what was true or false just wasn't there, and what was available was not fully comprehended. We need to cut these people some slack.

Based on the outcome of the June 26 meeting, Leo Duffy reported in a memo to Secretary Watkins on August 3 that they were ready for sampling after the next burp, estimated to be August 8. But there were still some high hurdles to jump. Duffy needed the concurrence of Steve Blush's Office of Nuclear Safety and the Office of Environment, Safety, and Health! He also stated the new mantra of the emerging DOE safety culture, "We may have to wait until future [gas release] events to properly complete all documentation and reviews. Safety will not be compromised by the need for action."

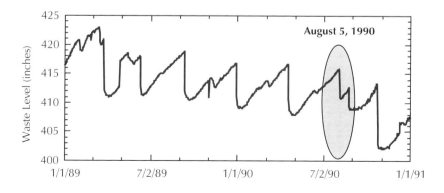

SY-101 Waste Level History: *1989–1990*

82 | Part One: Flammable Gas

> On August 5, 1990, at 5:15 AM, a relatively small gas release event occurred in SY-101 for 2 to 3 hours, with a waste level drop of 5 inches. The hydrogen concentration reached 1.3 percent, but the gas release did not pressurize the tank dome. Changes in the temperature profile showed that waste was not disturbed below 50 inches. The release occurred 108 days after the previous one on April 19, 1990.

After the August 5, 1990, burp Wodrich evaluated his criteria against the tank data. In an August 16 memo, he declared that the window had opened August 8 when the waste level reached its minimum of 410.8 inches. The window would close when the level rose to 413 inches, which he estimated would take about 30 days, until September 7, 1990. Wodrich added a temperature criterion that would close the window if the waste exceeded 136 degrees Fahrenheit, three degrees lower than the peak temperature just before a burp. Similarly, to avoid gas-filled waste regions, core drilling would not be allowed to penetrate a region whose temperature exceeded 132 degrees Fahrenheit. This was the first time that burp data were actually evaluated against window criteria.

Though the window was open, the work the task force had hoped to do was blocked. Various reviewers began to vent their anxieties and criticisms in letters to Steve Blush's Office of Nuclear Safety. One letter declared that core sampling at one location would not tell anything about the content of the whole tank. At best, it would be an experiment that would only show core sampling could be done safely. The letter went on to warn that adding water to the crust, as proposed, would not ensure safety because there was no way to know how deep the water penetrated. Another reviewer wrote to recommend that the crust not be wetted until details were provided on how the data would be used to reduce the hazards. He also advised that sampling should wait until after a bigger burp to ensure that a minimum of gas remained in the waste.

Steve Blush himself weighed in with an official letter to Duffy on August 20. He recommended against sampling SY-101 because he didn't believe that the safety analysis documents provided an adequate basis for sampling. In Blush's opinion, "They conflict, conclusions from experiments and hypotheses have not been verified, and they are drafts." Before sampling could be done, Blush recommended that tank ventilation system be made spark-proof and have a much higher flow rate. He also wanted an applied research program "to identify the precise data needed from the tank, safe ways to get that data and to specify how the research using that data will support resolution of the slurry growth and flammable gas problem."

Of course, the task force thought the need for the data was already perfectly clear. Besides, because nobody understood exactly what the waste was doing, it was impossible to write out precisely what calculations would be done with the data. They wanted to use the sampling data to confirm or adjust estimates of waste composition, which had been made based on the tank fill history. The sample would also tell them what the waste was physically like and how deep the layers were, or if the layers they expected were there at all! Only with all this information available could better explanations be posed that would, in turn, suggest analyses and calculations of a more precise nature.

Apparently DOE Headquarters agreed with the task force and still wanted to proceed with sampling despite Blush's objections. To keep things moving, DOE and Westinghouse organized a big public meeting for August 29, 1990, at Richland's Hanford House, a local hotel and conference center. Everyone could present their criticisms and proposals, and they could develop a consensus on the best course of action. According to those who were there, this ill-fated meeting was probably the nadir of the whole SY-101 mitigation effort. It was standing room only in the big conference room with 80 or 90 people attending. Every faction within and outside DOE brought in experts to challenge Westinghouse's presenters. The local media set up shop in the Hanford House lobby and tried to interview people. Steve Blush presented his fault-finding report. There was a lot of shouting. It got so bad at times that Deichman had to pull people off the stage in the midst of a presentation. One Westinghouse observer told me, "It was sooo bad! Everyone went nuts. People were panicked. They thought the tank was going to blow."

Nick Kirch related this example of what the meeting was like. Some of the reviewers believed that a very high domespace ventilation rate would suck gas out fast enough to prevent flammability during a burp. Calculations showed that this was not possible, but a ventilation rate of at least 5,000 cubic feet per minute could reduce the power of an explosion enough to keep the tank intact. Nick presented these results along with an estimate that such a high-flow ventilation system could be designed, built, and installed for $10 to $15 million. An engineering professor at the back of the hall stood up and yelled that if one of his students came up with an estimate like that for a simple ventilation system, he'd flunk him! Nick's response that the students were probably not designing systems for a flammable, radioactive environment had no noticeable effect on the professor's opinion.

This ugly meeting, along with lack of confidence at DOE Headquarters, killed core sampling for the time being. The window for sampling after the August 5 release was closed anyway. Leo Duffy and Tseng retreated to Headquarters to ponder the unresolved issues. On September 26, 1990, Steve Blush sent an e-mail to Duffy adding to what he thought must be done before core sampling could proceed. He repeated his demand that tank ventilation be increased as much as possible as soon as possible. He wanted more accurate measurements of surface level, flammable gas concentrations, pressure, and temperature, and new instruments to measure crust thickness and strength. He demanded improved data on potential ignition sources in the core drilling operation, especially heating the bit by drilling into dry saltcake. Blush again cited safety documentation that he believed was inaccurate, incomplete, and had too many unsubstantiated assertions. Finally, he insisted that sampling be integrated into a comprehensive remediation plan for SY-101 and other Hanford tanks with a flammable gas hazard.

The overall concern of Blush and the other reviewers was the question of whether the risk of sampling was balanced by a future benefit. The possibility of drilling into that fearsome big-gas-pocket-under-the-crust chimera and igniting some kind of explosion was the biggest issue. Bob Cook, acting as a consultant to the Washington State Department of Ecology, put it this way to a *Seattle Times* reporter: "It doesn't make sense to go in and, in haste, precipitate an accident that could cause a catastrophe just to get a sample." So, before full-scale sampling could be sold, a small sample of the crust material had to be dug out of the tank to show that the catastrophe could not happen. Accomplishing this baby step was to be the first order of business for the rest of 1990 and beginning of 1991. And the big gas bubble model needed to go away.

The Science Panel

In June 1990, to bring the combined expertise of a team of working scientists, as opposed to reviewers and advisors, to bear on the tank safety issues, Westinghouse formed the Science Panel. Its charter was to do research to find the physical and chemical processes that were causing episodic gas releases in SY-101. The Science Panel was a high-horsepower group of smart people from all over the country. Members were drafted from PNNL, Westinghouse, Argonne National Laboratory, Georgia Institute of Technology, Savannah River, and Idaho National Engineering Laboratory.

The Science Panel held several meetings in the summer of 1990 to get themselves up to speed. At their first meeting, on June 27, Westinghouse briefed them on the tank data and the history of the flammable gas issue. They agreed with the growing consensus that the gas was trapped in the sediment layer, and not as a big bubble under a crust. They suggested that buoyant forces probably triggered the burp and the crust had only minimal resistance to gas release. On the other hand, they could propose no mechanisms to explain hydrogen generation from the data available. In July 1990, the Science Panel reviewed a plan for chemical analyses on the hoped-for core samples of SY-101 waste. They asserted that a fundamental understanding of the physical and chemical processes causing episodic gas release was essential and strongly recommended that DOE take no action until knowledge improved.

The Science Panel got down to serious work in September 1990. Various groups began a concerted experimental effort to develop, from scratch, the fundamental understanding they sought. They initially looked at both physical and chemical phenomena, but the members' interests soon made the radiochemistry of gas generation their center of attention. Argonne and Georgia Tech scientists did experiments defining the complex chains of chemical reactions that led to gas generation and what chemical species were created and consumed in the process. PNNL digested and interpreted the results. This was ground-breaking research that uncovered the primary mechanisms for the generation of hydrogen and the other gases.

But their research never seemed to end. In February 1991, the Science Panel again emphasized the importance of fundamental understanding and cautioned against premature remediation. In July and November 1991, preliminary experimental results began to

86 | Part One: Flammable Gas

come in that suggested further studies. In March 1992, more results stimulated some inferences but also more questions. In January 1993, the panel complained about unreliable tank data and flawed experiments being planned by Westinghouse.

Though it was nice to know where the gas came from, it was gas retention and release that made flammable gas a problem in SY-101. In spite of this, the Science Panel's radiochemistry research just went on and on with less and less relevance to stopping burps in SY-101. The work lost focus and momentum. The last formal meeting of the Science Panel was June 1993. Visiting Chemical Reactions SubTAP member Scott Slezak's impression was that the meeting was unstructured and that discussions wandered. The Science Panel did not appear to have a clear direction. Another SubTAP member didn't think their work had much practical application.

In February 1994, Jerry Johnson discussed the future of the panel with members Denis Strachan, a PNNL radiochemist, and Wally Schultz, a former Hanford scientist with long experience in waste chemistry. They told Jerry that if there were science issues there should be a science panel, but they did not want the panel to act as consultants. Though the important science issues centered at Hanford, they thought a future Science Panel should have *no* Hanford members and that their meetings should have a small controlled attendance. They would focus on a few specific issues and be subject to a "sunset clause" to disband by a given date. But the sun had already set on the Science Panel, and they were not heard from again.

Understanding the Burps

With all the fresh effort focused on SY-101 after Lawrence's March 23, 1990, news conference, progress accelerated on figuring out what really happened during a burp and why. A series of insights combined to describe the phenomenon about as well as we know how today. Right after the April 19, 1990, burp, Nick Kirch was calculating how much gas came out. He estimated the gas release volume from the waste level drop and also by calculating how much was needed to match the measured pressure in the tank dome. He found that the gas volume needed to pressurize the dome was over twice that indicated by the level drop. This meant that the gas had to be coming from the bottom of the tank where the pressure on the gas was over 2 atmospheres. The gas could not have come from a big bubble under the crust, where the pressure would have been only a little above *one* atmosphere. This was pretty close to proof, but the gas could still have been temporarily (very temporarily) trapped under the crust before it broke through.

The second insight came from Dan Reynolds. Someone told him about Leela Sasaki's discovery from 1985 that the waste temperature profiles before and after a burp were radically different, especially near the bottom of the tank. Dan plotted the temperature profiles for the April 19 burp and saw that they did, indeed, change dramatically. The hot region in the bottom third of the tank suddenly cooled, and the cool layer on top suddenly heated up to about the same temperature as the hot stuff. After brooding about this a few days, he plotted profiles for several previous burps. They all showed a similar behavior. He concluded that the whole tank, not just the crust, had to be mixed up to cause the temperatures to change like this.

Dan also made the vital connection between the temperature profiles in the waste and the ones every undergraduate engineer has to plot in required transfer courses. Before a burp, the temperature profile near the tank bottom has a rounded shape that plots as a parabola. The temperature profile in a slab of heat-generating material cooled on both sides plots the same way. Dan realized that the sediment layer in the tank is a heat-generating material cooled on top and bottom, and the depth of the parabolic part of the temperature profile matched the estimated depth of the sediment layer. The temperature profile in the middle part of the waste almost all the way to the surface was uniform. In the heat transfer texts, this indicates a fluid mixed by convection currents that keep its temperature constant, just like the supernatant liquid in the tank. This led to the supernatant liquid being called the "convective layer" because convection created its uniform temperature, while the sediment was the "nonconvective layer" because it was not a fluid subject to convection. Finally, the temperature of the 3-foot-thick waste layer corresponding to the surface crust decreased in a straight line from the bottom next to the supernatant liquid to the top surface exposed to the domespace. The linear temperature decrease indicates heat conduction through a solid material.

This insight not only connected what had been assumed about the waste layers with the accepted laws of heat transfer, but it also provided a method to measure the layer thicknesses in real time! The top of the sediment layer matched the point where the parabolic temperature profile intersected the uniform temperature of the supernatant liquid above it. Likewise, the base of the crust was where the uniform temperature in the supernatant liquid intersected the linearly decreasing temperature in the crust. We still use this method today.

Dan's insight on the relationship of temperatures to waste configuration all of a sudden made the progress of a burp clear. After the prior burp, the sediment layer on the bottom accumulated gas and heated up into a pronounced parabolic temperature profile. After the burp, the temperature in the crust region became almost as hot as the sediment, and the temperatures near the bottom were about the same as the supernatant liquid. The only feasible mechanism that could cause this was hot sediment material physically rising to the top and being replaced by cooler supernatant liquid in what was termed a "rollover." The rapid drop in the temperature at the top of the waste a few days later occurred when the hot material, which had been floating at the surface, fell back to the bottom.

Dan's new theory was included in the May 31 safety evaluation for adding water to SY-101. He described the physical interpretation of the temperature profiles and concluded that gas builds up in the sediment and the waste mixes during a rollover. But Dan assumed that the gas escaped from the sediment through "channels" rather

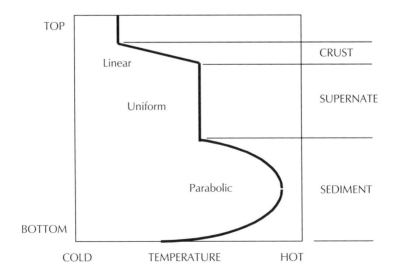

by riding a body of buoyant sediment to the surface. Even if no longer assumed to exist as a big gas pocket under the crust, the gas was still believed to escape in such a way that the crust was pretty much undisturbed. Unfortunately, the safety evaluation shelved the new findings as "one of many alternatives being proposed." It just happened to match the tank data better than the others.

So what could cause the sediment to rise, and how was this related to the gas release? The answer came from an unrelated source. In the summer of 1990, Jerry Johnson employed an engineer by the name of Howard Brager to investigate things for him. Howard was "a bull in a china shop" kind of researcher, fearlessly calling anyone who would answer the phone to get answers. He happened to describe SY-101's burps to some now forgotten high-powered professor who immediately replied that it sounded like Rayleigh-Taylor instability! In the early twentieth century, English physicist Lord Rayleigh and mathematician Sir Geoffrey Taylor independently studied thin layers of liquid heated from the bottom. When the heated liquid on the bottom became lighter (buoyant) with respect to the cooler liquid on top, the whole layer suddenly "rolled over." If this was happening in the tank, it was also buoyancy that made the sediment layer to rise to the top during a burp.

This was very close to what was later discovered to be the truth. In fact both Steve Eisenhawer at Los Alamos and Perry Meyer at PNNL later derived Rayleigh-Taylor models that gave reasonable estimates of burp sizes and the time between them. A month later the Science Panel also concluded that buoyancy was the trigger for gas release if gas was accumulating in a nonconvective sediment layer below a lighter layer of liquid. They also agreed with Dan's interpretation of temperature profiles (Dan was a member of the panel). Finally, PNNL engineer Rudy Allemann, working with the insights developed by Nick and Dan, along with inferences from the Rayleigh-Taylor theory and a comprehensive collection of tank data, put it all together into the explanation that is still accepted today.

For a buoyancy theory to work, the sediment had to be strong enough to lock the gas bubbles in place. But how? Rudy investigated this question by watching a drop of simulated waste under a microscope as he changed its temperature. He saw long, needle-like crystals form when the temperature fell below 127 degrees Fahrenheit. Rudy believed that these interlocking crystals gave the sediment its strength. Coincidentally, burps usually did not occur until the peak sediment temperature exceeded 127 degrees. Maybe dissolution of those crystals is what made burps happen? Years later, Phil Gauglitz showed that even tiny round glass beads could hold a lot of gas, but Rudy's conclusion seemed like a good explanation at the time.

By then analysis and observations had virtually eliminated the idea that gas was trapped under a hard, sealing crust. The crust theory could not explain the changing temperature profiles. Besides, gas monitoring showed the crust was not a seal because gas was seeping continually into the tank dome. The crust would also have to be impossibly strong to cap the huge gas pocket conceptualized in the old safety documents. Photos in the tank dome showed 2- and 3-inch pipes bent in different directions. This meant the crust did not move as a unit, and that a very large amount of hydraulic energy, more than just popping a gas pocket near the surface, was involved in a burp.

Besides putting the big-gas-pocket-under-the-crust theory away, Rudy found many more details in the data to better explain the burp process. He called bubble disengagement through channels into question. The complete reversal of the temperature profiles was consistent only with bulk motion of large volumes of sediment, not just bubbles. The brief surface level rise immediately before a burp showed a volume of gas expanding as it rose to the surface. Most burps showed a series of several distinct dome pressurizations. From this, Rudy inferred that a burp was actually several separate buoyancy events involving different volumes or "gobs" of buoyant sediment. This observation gave the new model the inelegant name "gob theory."

Courtesy Rudy Allemann

When Rudy first presented the theory to the TAP, he used the term "blob." Unfortunately, there was a 1958 horror movie titled "The Blob" starring Steve McQueen. In the movie, an alien meteorite discharged the Blob, which was described as indestructible and unstoppable. There were references to radioactivity. Some on the TAP had seen or heard of the movie and agreed that a new name was needed to allay potential negative public reaction. So blob was changed to "gob" on the spot, and the tank burps were described by the "gob theory" forever after!

Competition glider pilot Rudy Allemann was a senior PNNL fluid dynamics engineer when he was tapped to join Westinghouse's Tank Safety Task Team in 1990. He had developed an affinity and skill in managing large, high-visibility experimental projects.

His last one before tackling SY-101 was a politically sensitive project on aerosol dispersion and settling in a big steel tank simulating nuclear reactor containment. This project was planned and started by Rockwell. But when Westinghouse took over in 1987, DOE decided to turn management of the aerosol project to PNNL while Westinghouse staff would stay on to run the system.

Rudy was assigned as project manager. His technical competence and management skill soon gained the respect of the aerosol project staff, who had every right to hurt feelings. He also came in contact with some of the Westinghouse staff who would soon be sucked into the SY-101 vortex. Nick Kirch was one of them. That's why Nick called Rudy in the spring of 1990 to say, "We need you on this project." Rudy went, stayed, and finally retired in December 1993 after the mixer pump proved itself effective.

Rudy mentored me as his replacement during the fall of 1993. No matter how extreme the pressure of the work, Rudy always had a wide smile, as if he didn't quite believe all this was really happening, and invited others to share the joke. Even before the tank work, I'd see Rudy running, not just trotting, across the grass, taking the shortest path to the next meeting, briefcase swinging and necktie flapping over one shoulder.

Everyone liked and respected Rudy. His project management method was "fishing." We'd walk around talking to everyone working on a task to see if they had found out anything or "had a fish on the line," giving kindly suggestions if progress looked slow. Rudy always acted excited, even fascinated, about a person's work, whether he was squinting at a computer screen, watching some complicated mechanical device flail about, or hearing a presentation on some arcane aspect of waste chemistry.

CONGRESS ACTS

The revelations appearing in old Hanford documents declassified in 1986, the growing national criticism of the DOE weapons complex, and the breaking Hanford safety issues became opportunities for senators and congressmen to make points with concerned constituents. Two major pieces of legislation came out of this period that became the top-level driving force for much of Hanford's activity even to this day.

Defense Nuclear Facilities Safety Board

The Defense Nuclear Facilities Safety Board, or Defense Board, was formed because of the fear that safety hazards at the big DOE weapons labs, particularly at Hanford, were getting out of hand. At Hanford the 1986 safety analysis report and the 1987 defense waste environmental impact statement raised the possibility of several kinds of accidents that could blow a tankful of radioactive waste onto the downwind population. Congress established the Defense Board as an independent agency of the executive branch by an amendment to the Atomic Energy Act in September 1988. The five board members were finally sworn in October 18, 1989, after heated congressional hearings where the specter of tank explosions horrified the public. Senator John Glenn was one of the board's main proponents and one of the main Hanford critics.

Congress was serious when it set up the Defense Board. Their recommendations have almost the force of law, and DOE must take them very seriously. Congress required everything to be done in public, with each communication and response published in the *Federal Register* and inviting public comment. When the Defense Board makes a recommendation, the Secretary of Energy must reply, stating whether he accepts or rejects it in whole or in part, and describe actions to be taken within 45 days. If DOE rejects their recommendation, the Defense Board must either reaffirm or revise it and return it to DOE. DOE then must announce their final decision within 30 days, publish it in the *Federal Register*, and report it to appropriate members of Congress. If DOE accepts, which they are encouraged to do, the Secretary must prepare an implementation plan and submit it back to the Defense Board within 90 days. The implementation plan must be completed within 1 year. If it takes longer, the Secretary of Energy must explain the delay to appropriate members of Congress.

Two Defense Board recommendations have had a particularly big impact on the Hanford budget as well as our understanding of the waste in the tanks. Defense Board staff had already been poking around Hanford for some time, investigating the ferrocyanide and organic-nitrate safety issues, when the flammable gas hazard in SY-101 hit the papers on March 23, 1990.

On March 27, the board issued Recommendation 90-3 on ferrocyanide waste in the Hanford tanks. It meant a lot of work for the 22 tanks identified as having ferrocyanides in their waste. Temperatures had to be monitored to make sure waste was not getting hot, and domespace gas was to be analyzed to see if any suspicious reactions were going on. Waste samples were needed to see how much ferrocyanide had actually survived, and chemical reaction studies were mandated to see how bad the problem might be. Finally, emergency response plans were required in case an explosion actually happened.

DOE accepted the recommendation and submitted an implementation plan in August 1990. But the Defense Board didn't like it and issued a stronger recommendation, 90-7, in October, reemphasizing the urgency of the earlier recommendation. The board accepted DOE's revised implementation plan in March 1991. They finally closed Recommendation 90-7 on December 17, 1996, as a prelude to closing the ferrocyanide safety issue on December 24. The organic-nitrate safety issue was not closed until December 9, 1998.

The Defense Board works from the bottom up through its staff. The staff visit the sites, investigate, evaluate, and come up with recommendations. Then the actual Defense Board meets periodically, usually at the site where the problem is, to take public testimony and consider the information developed by the staff. If they agree that there is a serious problem, they may make a formal recommendation to DOE that can be a big embarrassment and cost a lot of money. For this reason, those who are asked to talk to Defense Board staff are encouraged to be on their best behavior.

From my own experience with various high-level review oversight groups over almost 30 years, the Defense Board staff are some of the most effective. These people are good! They

94 | Part One: Flammable Gas

> generally listen well and have technical knowledge broad enough to understand what they hear in the right context. Best of all, presentations to Defense Board staff are typically informal discussions among peers around a table rather than formal presentations to a critical audience. But you have to keep in mind that the opinion of the staffer who disagrees with you may find its way into a formal recommendation that DOE will have to deal with. So you have to have a *very* firm technical basis for whatever you say. Often the best thing to say is "I don't know."

A few years later, in July 1993, the Defense Board issued Recommendation 93-5 for waste tank characterization studies. The board saw that there was not enough technical information available on the waste to make good decisions about safe storage and associated operations. Tanks on the watch list were not being sampled soon enough, and the 222S Lab was way behind in analyzing the samples. The recommendation was simply for DOE to recalibrate their priorities and get all watch list tanks sampled and analyzed in 2 years with the rest of the tanks in another year.

DOE's implementation plan committed to analyzing at least two core samples from each of the 177 tanks and to conduct research to understand waste behavior. This was one big, costly recommendation. From 1993 to 1999, the total cost of all the required sampling and analysis was $420 million. But the results of this work contributed mightily to understanding the flammable gas retention and release behavior and to final closure of the safety issue. Without the information developed in response to 93-5, it would have been very difficult to deal with the SY-101 crust growth problem. The board finally closed 93-5 on November 15, 1999.

The Wyden Amendment

Oregon Senator (then Congressman) Ron Wyden was concerned that Hanford wastes could potentially get into the Columbia River and endanger his Oregon constituents. He was concerned when the ferrocyanide explosion and flammable gas burn hazards surfaced in late 1989 and early 1990. When public outcry rose to its crescendo at the congressional hearings on Blush's report, Wyden acted. He offered an amendment to the National Defense Authorization Act for Fiscal Year 1991 on September 18, 1990. The amended bill passed November 5, 1990. Senator Wyden testified on introducing the amendment that "scientists and engineers have publicly acknowledged the possibility that the chemicals in some of these tanks could actually explode. . . . We know too little today about the degree of hazard or how to reduce it. But we do know that the potential consequences of an explosion in the tanks are unthinkably great. We must begin to address this issue."

The so-called Wyden amendment or Wyden bill, officially titled *Safety Measures for Waste Tanks at Hanford Nuclear Reservation*, was the 2 by 4 the *Seattle Post-Intelligencer* had hoped to apply between Hanford's ears. It required four actions. First, within 90 days (by February 3, 1991), DOE had to list which single-shell and double-shell waste tanks "have a serious potential for release of high-level waste due to uncontrolled increases in temperature or pressure."

Dignitaries visit the SY Tank Farm in October 1990. Left to right: Phil Hamrick, Deputy Manager of DOE Richland; Leo Duffy, Assistant Secretary for Environmental Management; John Wagoner, DOE Richland manager; Admiral James Watkins, Secretary of Energy; and Washington Governor Booth Gardner *(with back turned). SY Farm is in the background.*
Courtesy John Wagoner.

Each tank so listed required continuous monitoring for waste release, excessive temperature, and pressure. The monitors were to be installed as soon as possible. Second, within 120 days (by March 5, 1991), DOE had to develop action plans to respond to excessive temperature or pressure or a release from the tank. Third, after the 120 days no waste could be added to a listed tank. Fourth and finally, DOE had to submit a report within 6 months (by May 1991) on the status of the first three requirements, other efforts to promote tank safety, and the timetable for resolving open issues on waste handling. It is interesting to realize that the text of the amendment did not define what "serious potential for a release of waste" really meant, nor did it prescribe how DOE could get a tank off the list. Maybe it was obvious to the good Senator, but it later created a lot of agony for DOE and Westinghouse.

The Wyden amendment institutionalized the famous watch lists that fortunately had already been constructed. Rick Raymond and his crew had compiled the flammable gas list in March 1990, and the list of ferrocyanide tanks had been written in response to the Defense Board's Recommendation 90-3 about the same time. SY-101 was the archetype of a watch list tank. It was removed from the watch list after its remediation in September of 2000. The provisions of the Wyden bill were finally considered satisfied for all tanks; the watch lists erased a year later, September 2001.

During this chaotic summer of 1990, the Hanford team of DOE Headquarters, DOE Richland, and Westinghouse realized it was too late to downplay the hazard, no matter how correct and reasonable it might be to do so. They finally recognized that the public and Congress were already on a runaway bandwagon careening towards a cliff. They had to claw their way into the driver's seat and try to steer it to safer ground. When 1991 arrived, the driver was named and began to take control.

1990–1992

Open Windows

This is the period where all thoughts, plans, and actions had to fit within "windows" after burps when it was believed safe to work in and around the tank. The burp had to let off enough gas to keep another burp from happening while people were working on the tank. Accordingly, the window opened after a burp, also called an event or a gas release event, only if the waste level drop exceeded a minimum value and stayed open until the level rose above another specified limit.

There always seemed to be far more work than the window duration allowed. Delaying important things for 3 months until the next burp was terribly frustrating. Even more frustrating, only about every other burp was big enough to open a window. The temptation to stretch the tank data to open a window or to keep one open was very powerful. Sometimes a window was declared opened only to be closed again when reviewers discovered what they believed to be an overly optimistic data interpretation behind it.

Because all work was planned, accomplished, and critiqued in the framework of these relatively short windows of opportunity, they began to be formally identified with letters of the alphabet, i.e., "Window A." The burp that opened Window A was called "Event A." There were nine named burps. Event A happened October 24,

SY-101 Gas Release Events and Windows

Event	Date	Max [H$_2$] (%)	Max. level drop (in)	Days since last	Work accomplished
	12/29/89	unknown	7.8	109	—
	4/19/90	3.5	8.8	111	—
	8/5/90	1.2	5.1	108	—
A	10/24/90	**4.7**	11.3	80	Remove sludge weight, waste samples from FIC and manual tape
B	2/16/91	~0.04	4.6	115	Run gamma coupon, install ventilation humidity and flow sensors and hydrogen probes
C	5/16/91	2.8	7.0	89	Sludge weight samples, obtain core sample, install radar gauge, color camera, and the black and white camera
D	8/27/91	0.4	5.8	103	Not opened
E	12/4/91	**5.3**	12.8	99	Replace color camera and lights, core sample, calibrate radar gauge with zip cord
F	4/20/92	1.5	7.2	138	Not opened
G	9/3/92	**5.1**	9.8	136	Repair camera, remove two air lances and bent TC tree, install MIT and two VDTTs, install vent duct gas monitoring probe
H	2/1/93	2.7	8.2	152	Size riser for pump, repair camera, install pressure relief riser, install pump load frame and cover block
I	6/26/93	3.4	9.8	144	Install pump and additional MIT

FIC   Food Instrument Corporation
MIT   multifunction instrument tree
TC   thermocouple
VDTT velocity-density-temperature tree

1990, and the last one, Event I, that opened the window in which the mixer pump was installed, was on June 26, 1993. The five burps up to Event E, December 4, 1991, cover what we might call the knowledge-gathering period. The last four events were totally consumed with preparations for installing the mixer pump.

WINDOW A—CRUST SAMPLES

Crust sampling was the primary objective for Window A. Showing that the crust was moist and that flammable organics were sparse would allay the worry about a crust burn, so the more vital core sampling could proceed.

On October 4, Jerry Johnson distributed a memo summarizing a brainstorming session on how to get information about the waste in SY-101. The summary identified some very simple and obvious possibilities. The manual tape and the Food Instrument Corporation (FIC) contact probe that measured the waste level both lowered cylindrical metal "bobs" or "probes" onto the waste surface every day. The FIC was an automatic device that lowered its bob until it made electrical contact with the waste, closing a circuit indicating it had touched the surface. The manual tape was lowered manually until an ohmmeter indicated waste contact to the operator. It could also be run in the "slack tape" mode, where the bob was simply lowered until the tape went slack. Both the FIC and manual tape bobs were known to collect waste residue, sometimes to the point of forming a "wastecicle" several inches long that had to be flushed off to get a good reading. Why not just raise the bob and scrape a crust sample off the probe?

The tank was also built with two "sludge weights." They were conical metal bobs on cables intended to measure the depth of sediment or "sludge" layer on the tank bottom in the same way as the manual tape was used to measure the surface level. They were seldom, if ever, used, but hung embedded in the crust. The sludge weights were expected to have a good volume of crust material adhering to them. Besides, they had to be removed anyway to make room for other instruments. So, why not remove one a little earlier than planned and get a crust sample in the process?

Another potential source of a much larger waste sample were the four dormant air lances that had been permanently installed in the tank in the mid-1980s. These 2-inch pipes would probably be full of waste up to the depth of the sediment layer. But extracting dormant equipment like this out of a tank had proven difficult. Long pipes that had hung in the waste for years were often bent or encrusted with hard saltcake that would bind in the riser. In that case, they might have to be cut off and pushed back down into the tank, becoming a piece of debris to complicate future waste

100 | Part One: Flammable Gas

retrieval. Also, a pipe full of radioactive waste would probably be far too hot for workers to get near, necessitating a lot of frightfully expensive remote handling.

> Just as windows constrained the schedule for work on the tank, risers constrained the work location. A riser is a pipe that extends 10 to 20 feet from ground level down through the gravel covering the tank and into the tank dome. The bottom end of the pipe is welded to the tank liner and may be flush with it or extend a foot or so into the dome. At the surface, the riser may connect to a concrete box or "pit" or protrude from the soil by itself.
>
>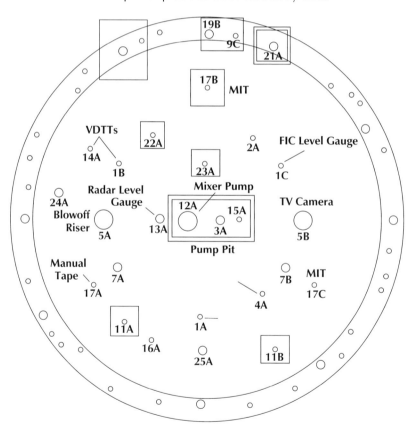
>
> SY-101 has 24 risers with diameters ranging from 4 to 42 inches. There are fifteen 4-inch, four 12-inch, two 20-inch, and three huge 42-inch risers. Each riser was identified by a number and letter assigned without location, size, or purpose.

For example, 4-inch Riser 1B was in the northwest quadrant of the tank just inboard of 4-inch Riser 14A. These matched two 4-inch risers, 7B and 17C, placed symmetrically in the southeast quadrant. After a while you just got to know them by heart. In the late 1990s, all risers got new identifiers of three digits, equally devoid of relationship to position, size, or function. Most found the old number-letter names easier to memorize. From here on, I'll use the old riser names. The diagram shows where they are.

Every instrument, pump, pipe, or ventilation port must use a riser, and allotting them to the various needs was a continual struggle. Installing any new device usually meant removing or moving something else already occupying the riser. Taking a core sample or making some other kind of one-time measurement often required temporarily removing the device already there and replacing it when the job was done. A riser may be crooked or out of round, so wise designers allow plenty of clearance around anything they intend to poke into one.

SY-101's burps moved chunks of crust around and tended to bend, even break, 2- or 3-inch-diameter pipes like lances or thermocouple trees. It is very difficult to pull a bent 2-inch pipe out of a 4-inch riser. The field crews have to use hydraulic jacks to force them through. This will be the main event in Window G.

Though core sampling would have to wait a while, Dan Reynolds issued the first report formally documenting the window criteria on October 8, 1990. To ensure that a burp released enough gas to gain the assumed safety benefits of a window, he required a minimum level drop of 5 inches and a minimum temperature drop of 3 degrees Fahrenheit on thermocouple #4 (76 inches above the tank bottom) to open the window. He kept Wodrich's original 2-inch level rise from the minimum or 20 days maximum duration as the window closure criterion. A revision in November 1990 added provisions for a longer window for work that did not penetrate the waste, like installing a camera in the domespace. This window would close after a level rise of 3 inches from the minimum or 30 days.

If the working windows were going to be opened based on tank data evaluated against strict criteria, good tank data had to be collected during the burp. Accordingly, operators were stationed in the SY farm on "burp watch." When the waste temperature indicated the tank was about ready to roll over, two or three people took up residence in the 242-S evaporator building, with one person stationed in the tank farm in the 271-SY instrument building. Besides one or two tank farm operators, the crew had one health physics technician to monitor radiation and a technician to gather gas samples. Inside 271-SY were electrical switches for the older equipment, alarm panels for ventilation flow, tank vacuum, and annulus leak detectors. In those days the SY farm was about the same as any other with only the basic instrumentation to meet the basic technical safety requirements for monitoring leak detection, ventilation, waste level, and temperature.

When a burp happened, everyone left 242-S and went out in the farm to go to work. Sample technicians went out and pulled gas "grab samples" at specified intervals to be analyzed later back at the lab. The tank farm operator dialed up the data monitoring frequencies to capture the maximum amount of information on what the tank was doing. He also had to be ready to restart the ventilation if it shut down. A "DP switch" measuring pressure difference across the exhaust high-efficiency particulate air filters turned off the fans if the pressure changed too much. Dan Niebuhr remembers doing this twice in the 2 years he was on burp watch. On one burp they

SY farm in November 1990. SY-101 is under the crane on the left, 271-SY instrument building in the foreground. Large pipes on the right are ventilation ducts. SY-102 is behind the 271-SY building and SY-103 is right of the tall light pole.
Project photo from *Window C Activities*.

hung a microphone in the annulus to see if the tank was making any noise prior to the burp. All they heard was the ventilation fan.

At first the operator waiting in 271-SY needed to be suited up in "whites" at all times. Besides being uncomfortable, there was no smoking, eating, or drinking allowed there or in 242-S. Needless to say, the crew became quite unhappy after a few days of waiting. Later the rules were relaxed so operators could be in street clothes except when they went into the farm. Beginning in late 1992 the burp watch crew could wait comfortably in the newly installed data acquisition and control system (DACS) trailer and monitor data. But the DACS data were not yet "official," and someone had to go into the farm every hour anyway to take readings.

> Tank farm workers sometimes got rather interesting jobs. Secretary of Energy Watkins once came out to check progress on SY-101 in October 1990. A visit to SY farm was on his agenda and, for several days prior to his arrival, much of the crews' time was spent sprucing things up. This included painting everything in the farm. But they only had time to paint the sides facing the access road and viewing stand! It looked really nice to the Admiral and the TV cameras, but as ugly as ever from the backside!

Meanwhile, on October 24 and 25, John Deichman brought in some high-powered senior scientists from the Westinghouse Corporate Science and Technology Center to Hanford. According to an e-mail from Deichman to Westinghouse staff, the experts were coming to "assess their ability to support the waste tank safety program." Their fame was in chemical process engineering, waste management, organic and polymer materials, sensors and ultasonics, systems instrumentation, and controls. The corporate people went back home without suggesting anything of use. Nick Kirch's assertion that "we know more about this tank than anybody else" again proved accurate.

> On October 24, 1990, at 6:05 AM, Event A, a large gas release, happened in SY-101. The waste level dropped over 8 inches immediately and the drop reached 10.2 inches the next day. The hydrogen concentration was 4.7 percent, making the domespace just flammable. The gas release pressurized the domespace to 2.3 inches of water gauge pressure for about 3 minutes and

drove the indicated ventilation rate above 1,000 cubic feet per minute. The total gas release was estimated to be 7,600 standard cubic feet. The waste temperature profile showed the waste turned over all the way to the tank bottom. Event A occurred only 80 days after the previous release on August 5, 1990.

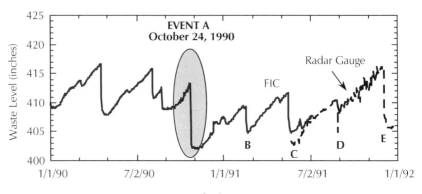

SY-101 Waste Level History: *1990–1991*

Window A opened October 29 when the waste level hit its minimum of 402.2 inches. The "20-day window" for waste-penetrating work required to remove the sludge weight would close November 18 or when the waste level rose 2 inches. The "30-day window" for non-waste-intrusive work where the FIC and manual tape probes could be lifted out would close November 28 or when the level rise reached 3 inches.

Exchanging the bobs on the FIC and manual tape and removing one of the sludge weights to get samples of the crust was on the schedule for Window A. But the paper work was becoming a serious roadblock. There was the safety analysis describing possible accidents and controls to prevent them, an environmental assessment of what kinds of and how much toxic gas and vapor would be released, and a collection of detailed work procedures. They all had yet to be written, reviewed, and approved by a string of managers. The approval process also required buy-in by Westinghouse's internal Safety and Environment Advisory Council (SEAC), passing a formal readiness review by DOE Richland, and getting formal approval from Leo Duffy at DOE Headquarters (and he asked concurrence from Steve Blush out of courtesy) as well as from the Washington State Department of Ecology!

Operator working on a riser on SY-101 in "whites" on mask November 1990 installing the radar level gauge.
Project photo *Window C Activities*.

DOE Headquarters waited until Friday, November 16, two days before the 20-day window closed, to approve the crust sampling! Even so, with heroic effort, the work began Sunday morning, November 18, at 7:30 AM. The sludge weight was easy to break loose from the crust, but it stuck in the retrieval container so the whole works had to be trucked to the hot cell. They got from 1/2 to 3/4 of a cup of crust material that looked like gray mud, dry on top and liquid on the bottom. By 4:45 PM operators had changed out the FIC and manual tape probes and everything was buttoned up. About a half a teaspoon of crust came off the FIC probe, but none was visible on the manual tape probe. This was the first real accomplishment in SY-101 since Leela Sasaki did the gas monitoring in 1985.

The crust samples turned out to be about 25 percent water, much wetter than originally assumed. One observer likened the crust to "pudding" and another called it "peanut butter"—a much tastier picture than the explosive nitrate salts that horrified reviewers back in August. This was major progress! But the SY-101 team

106 | Part One: Flammable Gas

needed more samples to counter accusations that a single sample would not tell anything about the crust as a whole. At their January 10, 1991, meeting at Hanford, the TAP also agreed that chemical analysis showed the crust would not burn, but they recommended additional samples to prove it generally.

A relatively easy way to get more samples soon revealed itself. Because the sludge weight extracted in Window A collected so much waste, somebody hatched a plan to create a new sampler designed to look like a sludge weight. But, instead of finding the sediment level, it would be used specifically to get crust samples. An urgent push began to get the new "sludge weight samplers" and to get all their documentation approved in time for Window B, expected in February.

NEW MANAGEMENT

Responding to DOE Headquarters, Blush report criticisms, Washington and Oregon state officials, and the public, Westinghouse launched a wholesale reorganization to concentrate on waste cleanup and waste tank safety. From November 1990 through January 1991, management positions from the cognizant engineer of the SY farm all the way to the Westinghouse president changed occupants. To make things interesting, none of these new managers had direct experience with SY-101 problems! However, the Waste Tank Safety Task Team stayed pretty much intact and was able to keep the work going while management played musical chairs around them.

The music stopped on November 19, 1990, when Westinghouse President Roger Nichols named Harry Harmon as Vice President of Waste Tank Safety, Operations and Remediation and appointed Steve Marchetti as Director of the new Tank Farm Project reporting to Harmon. John Deichman's Waste Tank Safety Task Team would report to Harmon with Marchetti's organization in charge of the field work. But there was another big change. The group would from now on be identified as the "Waste Tank Safety *Program*," implying a more formal, long-term, higher-priority effort. Whether the members of the team felt elevated by this change in status was not reported.

Both Marchetti and Harmon were recruited from Westinghouse's Savannah River Site, where Harmon managed the Chemical

Processing and Environmental Technology Department and Marchetti the Project Management Department. Soft talking, southern gentleman Harmon and abrupt, hard-driving Marchetti were a good match. It was said their "white hat-black hat" routine was very effective.

Harry Harmon's job was to make progress and communicate the progress. He said it was "both the best job and the worst job I have ever had." The differences between Hanford and Savannah River struck him forcefully. Hanford waste tanks were viewed as a national issue, the biggest safety issue for DOE. The Savannah River Site had nothing like Hanford's stakeholders, and it was new to devote so much time to contentious public meetings at Hood River, Portland, Seattle, or Spokane.

Criticism from the world and from DOE Headquarters was severe and demoralizing. He had to treat it "like water off a duck's back" to survive. DOE seemed to have little faith in technical positions developed at Hanford. It wasn't surprising that Harmon found the Tank Safety Program pretty demoralized. They had been beat up a lot and some were afraid to take a technical position out of fear of intense public criticism. Harmon and Marchetti held a lot of all-employee meetings and roundtable discussions trying to convince staff that waste tanks, with its promise of long-term employment, was a "good" place to be—at least it was secure.

On January 15, 1991, Tom Anderson replaced Roger Nichols as Westinghouse Hanford Company President, completing the management shakeup. A shakeup was probably in order. In early December, an employee complained of being harassed by a manager for raising a safety concern. Around the same time there were three incidents of employees disabling radiation warning alarms. On December 28, Westinghouse announced it got the poorest performance rating in its Hanford history from DOE, probably as a result of these safety issues plus the flood of management criticism over the last summer. In announcing its new president, Westinghouse also pledged it would not pursue the Hanford contract if

DOE was so dissatisfied as to open it for competition at renewal in 1992. As it turned out, the team of Anderson, Harmon, and those around them improved performance enough for Westinghouse to keep the contract until 1996.

Before Anderson took over, the Westinghouse president had stayed aloof from the agony being suffered by the SY-101 team, except what he got with the general condemnation of Hanford from outside critics. Now it would be different. Whether he was intentionally attentive to a serious problem or forced by DOE Headquarters micromanagement, Anderson's name started appearing on memos, letters, and meeting minutes dealing with this work. He was definitely working hands-on with SY-101!

The new management team and the Waste Tank Safety Program started making progress towards regaining credibility. The crust samples from Window A began to confirm what most believed the crust was really like: not the dry, dangerously explosive salts but a moist mush that was generally benign. They had found a model to explain the burps that, although incomplete and unverified, was based on simple physical principles and matched detailed tank data quite well. On February 8, 1991, a few days after the 90-day deadline of the Wyden amendment, Harry Harmon submitted the list of tanks with a "serious potential" for radioactivity release to DOE, thereby showing "serious progress" in responding to congressional direction. On February 13, John Deichman, recognizing that pressure might force rough drafts to go public before proper review and polishing, ordered all new documents to be "prepared from the beginning as external publications." Finally, by early 1991, the worst of the public name-calling died down.

But, despite the progress, the burden of the work remained intense, especially on the operators who had to bear the incessant training and uncertainty of waiting for a burp, endure the grinding rush of urgent work when the window opened, and then shoulder the unspoken blame when the window closed with jobs undone. There was no relief from window work until the mixer pump finally went in over 2 years later. Window B would be another stressful time.

WINDOW B—OPEN AND SHUT

Event B, expected in early February 1991, was to open the window for work to sample the crust with the new samplers, install probes in the dome for a permanent gas monitor, place instruments in the ventilation ducting to measure humidity and flow rate, and briefly dangle a lithium fluoride chip in the dome to measure the gamma radiation. A new-fangled radar unit advertised to measure the waste surface level more accurately and reliably was also to be installed. Things did not happen quite that way.

The design and safety analyses for the new sludge weight samplers progressed at a frantic pace. There were two designs. One used six hollow tubes a few inches long fastened below a weight. When the operators let the sampler down onto the crust, the weight was supposed to drive the tubes into the waste and fill them with a good sample. The designers hoped that the crust sample would stay packed inside the tubes until technicians dumped it out in the laboratory hot cell. But tests weren't promising, and this design was not used.

The other sampler was a hollow cone like a foreshortened tip for a jousting lance fastened below a weight. Like the tube sampler, the weight was supposed to push the cone point-first into the crust. The designers hoped a sample of crust would crumble back into the hollow cone so it could be lifted out and sent to the lab. Trials showed this concept worked pretty well and plans were drawn to drop it down five different risers. They expected that analyzing this many samples would show conclusively that the crust was not going to burn, or do anything else unpleasant, during the core sampling that they hoped to do next window.

Design of the sludge weight sampler was ever-youthful Carl Hanson's first job on the tank. He attended his first meeting on the sampler December 10, 1990, and never really left SY-101 afterward. In fact, he served as chief design engineer for both the mixer pump in 1992–94 and the transfer and dilution system in 1998–2000.

Carl is always totally pleasant, helpful, and friendly. Though unimaginably busy as chief of the mixer pump design effort and other SY-101 responsibilities, he always found time to give good advice or grab a drawing to answer a question. Carl never raised his voice or used harsh

110 | Part One: Flammable Gas

words, even when driving his team to meet an inhumane schedule. Instead, he lead by example and gentle, tenacious persuasion. He wore a permanent, wry grin that seemed to say, "putting up with all this is almost funny, isn't it?"

Carl put family first, taking three weeks off during one of the busiest times running up to Window B to be with his son, born New Year's Day 1991. He was also an enthusiastic company participant, riding 34 miles to work and back from his home in Prosser on a Westinghouse "Bike to Work Day." He was recognized for the farthest ride.

His notes reveal the gentle, positive sense of humor that his co-workers got to know. During a dreary discussion at a meeting in late 1992, Carl recorded a survey of the attendees' facial hair. There were 31 people in the room; seven had beards and four had moustaches. More interestingly, six of the seven with beards and half of the four with moustaches were from Los Alamos. In fact, Los Alamos accounted for 75 percent of those having facial hair while composing only 26 percent of the attendees. His survey has never been published until now.

Fears that even a small crust intrusion might trigger a big gas release had not yet been put to rest. Worse yet, the sludge weight sampler swinging on its cable was believed to be a potential ignition source. The worst-case accident was that the sampler would come loose and fall 15 feet onto the crust, point down, penetrating the crust, releasing gas and igniting it at the same time. There wasn't much time to analyze these hazards and write a safety analysis report, but it had to be done. Bob Marusich tells how it was. "The problem was trying to understand safety issues without information. You had to be conservative because you didn't KNOW anything, but you couldn't be too conservative because then you couldn't DO anything. It was a balancing act and every once in a while we fell off." The safety analysis finally balanced conservatism with progress by dictating that the samplers be prevented from free-fall onto the crust, a window allowing waste penetrating activities be open, and tank ventilation be operating.

Bob Marusich related one experience where he almost fell off the balance between action and conservatism. Though initial samples from Window A had been compared to pudding and peanut butter, the crust was still assumed to be very hard. Bob calculated how hot the tip of the cone sampler would get if it absorbed all the kinetic energy of a free-fall to the hard crust surface. He assumed the impact energy all went into a tiny volume the tip of the cone. This gave him a peak temperature approximating the surface of the sun, definitely an ignition source!

But Bob didn't believe it and decided to do a test. He found a stairwell where he could drop a prototype sampler from the right height and mixed up a box of a rock-hard simulant to drop it on. A colleague dropped the sampler at his signal. When it hit, Bob rushed out and felt the tip of the sampler with his finger. It wasn't even warm, but the impact bent the shaft between the weight and the cone. Bob still has the bent sampler.

The "radar gauge" filled a perceived need to improve upon the two point measurements of waste surface level then available in the FIC and manual tape. Besides a tendency to grow a "wastecicle," their readings could suddenly vary as much as an inch or two as pieces of irregular crust drifted under the bob. There was a nagging suspicion that the crust might rise up and down differently in areas where the level was not being measured. The radar gauge used a radar beam to continuously measure the distance down to the waste surface. Instead of a point, the radar beam covered a small area that tended to average out surface irregularities. Everyone hoped that this new device would give more reliable, higher-quality data that would increase their understanding of the burp process. At any rate, it would function as a spare in case the FIC failed.

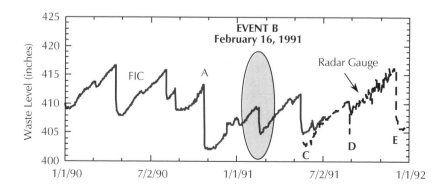

SY-101 Waste Level History: *1990-1991*

112 | **Part One: Flammable Gas**

On February 16, 1991, at 3:30 AM, a small gas release, Event B, occurred in SY-101. The waste level dropped 2.3 inches immediately and reached 4.4 inches on February 19, and finally hit a maximum of 4.8 inches. The hydrogen concentration was only 4,500–5,000 parts per million, and the domespace was not pressurized. The waste temperature did not change at the bottom three thermocouples. Event B occurred 115 days after Event A on October 24, 1990.

The official opening of Window B was somewhat of an embarrassment. The waste level had to drop at least 5 inches to meet the current criteria for opening a waste-penetrating window. Based on the experience with Window A, the entire community worked under the expectation that the window would open. After all, they had just received a letter from Watkins himself authorizing the planned work. To everyone's dismay, the level drop didn't quite make the 5-inch drop criterion. The level drop was 4.8 inches February 19 and then started moving up. But the level data were a little noisy, fluctuating up and down 1/8 to 1/4 inch as the floating crust moved around. On February 20, a single fluctuation in the measurement just happened to push the level drop to exactly 5 inches, even though it pulled back on the very next measurement.

Though the data were not very good, the window was declared open as of February 21, one day after the minimum level, as specified in the window criteria. The 20-day window for waste-penetrating work would close on March 14, or when the waste level rose 2 inches above its minimum. By March 6 the safety documentation was approved, and Westinghouse formally requested DOE Richland to authorize crust sampling because the 5-inch level drop allowed waste-penetrating work. DOE Richland passed the request to DOE Headquarters the next day. On Friday, March 8, Leo Duffy informed Steve Blush that he had authorized Window B sampling. With DOE moving that fast, the sludge weight sampler could have been run that weekend.

But the waste level was already nearing the 2-inch rise that would close the window! The Waste Tank Safety Program requested an extension from the Westinghouse SEAC, who had to approve keeping the window open. The SEAC heard the request Friday afternoon. The case for extension was that the minimum level, the starting point for the 2-inch rise criterion, was not really the true minimum because the data were fluctuating. The true value was at the approximate mid-point of the fluctuations, 1/8 to 1/4 inch higher. But using this higher level as the minimum meant that the initial level drop was less than the 5 inches required to open the waste-penetrating window in the first place! The SEAC saw through this right away and objected, saying, "Proceeding will violate the safety document. This sends a message to employees that procedure compliance is not mandatory." Window B should not have been opened at all!

On Monday, March 11, DOE Richland authorized *only* non-waste-intrusive activities. Steve Marchetti instructed all concerned with the work that "Sampling activities are not authorized and must not be performed." Only a week later, on Monday, March 18, 1991, the non-intrusive tail of Window B closed when the level rise exceeded 3 inches. The radar gauge was to have been installed March 20. Only the ventilation flow and humidity instruments and the gas monitoring probes were installed during Window B. It was disappointing to leave work undone and embarrassing to get caught stretching the data to meet the window criteria.

Reform and Regroup

A "lessons learned" document captured the difficulties the crews ran into during Windows A and B. Its introduction said, "After completing Window A and B activities, it became apparent that improvements were necessary in preparations and conduct of work as well as communications." One of the main problems with window work so far was that no single person was in charge, and "in the end it became the responsibility of everyone to do everything." To mend this deficiency, Marchetti formed the Joint Test Group that would meet frequently during pre-window planning and daily or more often during the window to keep work organized and flowing. The Joint Test Group was an improvement but not the final solution. A single test director was not assigned until after Window G.

The window criteria and use of the criteria were also a problem. No one organization had responsibility for collecting and interpreting tank data to evaluate against the criteria. This caused the "on/off" crises just described for Window B. At the same time the window criteria were criticized because they were "not sufficiently

114 | Part One: Flammable Gas

flexible to effectively use time available for safe operations." Dan Reynolds issued a revision to the window criteria on April 8, 1991, to try to fix these problems. The revised document made the newly formed Joint Test Group responsible for decisions on opening and closing the windows with veto power given to the Westinghouse Director of Environmental, Safety and Quality.

The revision also formalized the method of setting reference levels so it was harder to stretch the criteria beyond what the tank data allowed. The maximum (pre-event) surface level was defined as the mode (largest number) of the set of level readings from the FIC gauge taken every minute over the 2 hours immediately before any indication of a burp (pressurization, rising hydrogen concentration, waste temperature changes, etc.). The sudden level rise usually observed immediately before a burp was excluded from the mode calculation. The minimum (post-event) surface level was similarly defined as the lowest mode of the sets of FIC readings taken every minute for any 2 consecutive hours after the burp. This means that the minimum level could not be established until post-burp growth started.

In addition to level drop, a temperature drop of at least 5 degrees Fahrenheit was required on either thermocouple #4 (54 inches off the bottom) or #5 (76 inches off the bottom). Two independent indications a burp had happened also had to be documented. But the revised criteria left the Joint Test Group a powerful way out if the data were unclear, anomalous, or missing. The group was charged with reviewing all the existing data and evaluating the consequences. Then, "if the JTG [Joint Test Group] is satisfied that the *preponderance of data* (primarily from corroborating data) provides sufficient justification, then this group can declare the window open or closed as appropriate." This clause would theoretically have given ample opportunity for management to pressure the test group to open a window or keep one open in spite of the numerical criteria. But in their review for final approval the SEAC insisted that the Joint Test Group not act contrary to the level and temperature drop criteria if data were available. At any rate, it appears that the group stood firm against whatever pressure they felt and made decisions to open and close windows based on the criteria, not on the need to do work. Window H was an example of technical integrity prevailing against almost overwhelming pressure to open the window and get the mixer pump installed.

This is the first recorded mention of the concept of "preponderance of data" or "preponderance of evidence." It is a vital concept for tank work because it is usually impossible to prove a theory or validate a model with only the available data. Scientists don't like to say, "I know," until they have made an observation or series of measurements that directly confirms a theory. And they are more comfortable if, after the first data come in, they can revise the theory and take more data to validate the earlier conclusions.

But this kind of science is not possible in the tanks. Any measurements involving radioactive waste are incredibly expensive and time consuming. If done at all, they are usually done only once, and the data more often reveal more about the inadequacy of the instrument than the phenomena being measured. Direct observations of waste behavior are limited to videotapes or anecdotes from operators. Experiments must use either simulant material that might not represent the waste very well or very small samples of real waste that do not represent tank behavior. In either case, their short duration cannot represent the multi-decade processes in the tanks.

For these reasons, many Hanford scientists stand aloof from tank work, complaining about "barriers to science." Only a few dare to plunge in and make the best of the vague hints and hidden clues. But the lack of information does limit what science can claim. One may only "suggest" a model or theory and claim it to be "consistent with" the sparse and imperfect data and observations or, even better, "supported by" some measurement or experimental result. At the very least, it must conflict as little as possible.

So understanding the mechanisms underlying tank behavior truly develops as the "preponderance of evidence." This method can be hard for traditionalists to accept, especially when the result bears on safety issues. One particularly arrogant reviewer even denigrated it as "pseudo science." Nevertheless, the preponderance of evidence has been used successfully to stop the burps in SY-101, close the global flammable gas safety issue, and update the safety basis for Hanford tank farms. From this experience, the concept is a good model for attacking problems where little is known and data are scarce.

116 | **Part One: Flammable Gas**

The Big Progress Report

In late March 1991, DOE Richland, Westinghouse, and the Waste Tank Safety Program were called to present their progress to John Tseng and the TAP along with Tseng's Headquarters task team and representatives from the Defense Board in Washington, D.C. The TAP was now fully geared up. Their meetings had a full agenda of formal presentations, frequently interrupted for comment and criticism, followed by "the word" from DOE Headquarters and a TAP closeout summary. This meeting was the first major conference with new data and some real progress to report. The Waste Tank Safety Program staff gave presentations on the Event A and B data, crust sampling results, and the new crust burn calculations, tank ventilation, and revised window definition. Staff from Los Alamos National Laboratory gave their own presentation on crust burn and outlined a safety analysis report.

> Wait a minute! Why was Los Alamos at this TAP meeting? And why did they talk like they had the lead on safety analysis? John Tseng had a good feeling about the abilities of the bright young folks at Los Alamos. At the same time, both Tseng and Duffy thought that the Westinghouse safety analysis was too conservative, often untimely, their controls were too restrictive, and the window definitions too tight. Tseng had handed Los Alamos the Westinghouse safety documents to review and their comments were very critical. Tseng very badly wanted to get Los Alamos involved with SY-101. He could not command it just yet, but he began to apply increasing pressure. This is the advent of the "Los Alamos factor" that soon fully involved them in the Waste Tank Safety Program.

The chemical analysis of the crust samples from Window A showed that the crust would not burn unless it were absolutely bone dry. But the samples also showed that even the driest part of the crust was *not* dry and could not be ignited by a hypothetical flammable gas burn. So Tseng, Duffy, and the TAP wanted to proceed immediately with core sampling, but only in "push mode." They still thought rotary mode drilling was unsafe because of the experience with red-hot drill bits in a test with extremely hard simulant.

The TAP recommended that hydrogen monitors be installed immediately and that sampling stop at a hydrogen concentration of only 0.2 percent (2,000 parts per million), 1/20 of that needed for flammability! They also chided Westinghouse to better define their windows and suggested a hydrostatic pressure measurement in the liquid as more accurate than physically lowering a bob to the irregular waste surface. But they agreed with the Westinghouse conclusion that high ventilation flows could not prevent the domespace from becoming flammable and even agreed that the time at risk, when the domespace was flammable, was short. This was comforting because some at DOE Headquarters still believed that more powerful ventilation fans were required to make the tank safe.

Because the waste level change was the most direct indicator of a change in gas volume in the waste, an accurate and sensitive waste level measurement was desirable. The TAP was always dissatisfied with measuring the waste level by physically touching it at a single point and kept offering, even insisting, on other methods. Unfortunately, none of them were practical.

One recurrent suggestion was to calculate the waste level from the hydrostatic pressure at a point under the crust. Accurately measuring the hydrostatic pressure at a known elevation is the same principle SCUBA divers use to measure depth. The diver trusts his depth gauge because sea water or fresh water has a nearly constant and uniform density. In the tank, especially after a burp, the rising and resettling gobs of sediment added a huge load of solids to the liquid, greatly increasing its density. It would probably make the pressure go up, even as the surface level dropped by several inches. And how do you make a super-accurate pressure sensor that can withstand a hot, highly caustic radioactive environment for years?

The radar gauge already mentioned was the closest thing to a workable alternative. But, even after calibration and months of comparing with the traditional point gauges, we never really knew what it was measuring. After a burp it would tend to vary wildly as the crust got wet, then drift as it dried out again. After the mixer pump was operating regularly, the radar gauge sometimes appeared to react to pump runs in an inexplicable way. Mercifully, the radar gauge died of radiation damage in about 1996. It proved that the person with more than one gauge never knows what the waste level is.

Later, when crust growth became an issue, the TAP pushed for a "free liquid level" measurement. The idea was to bore a hole clear through the crust so the liquid would rise inside it like a sight glass on a nineteenth century boiler. A sensitive float gauge would measure the liquid height in the hole. The crust would have to be prevented from re-forming inside the hole, maybe by mechanical agitation, heating, water spray, or even ultrasonic vibration. Carl Hanson's crew drew up a conceptual design, but it was very expensive and nobody was persuaded it would work.

John Tseng described sweeping changes in the Waste Tank Safety Program at DOE Headquarters. Secretary Watkins had made tank safety, meaning SY-101, his *#1 priority*. He really meant it! John Deichman was to have a robust 5-year program funded at over $30 million per year. This size of program on tank safety was unheard of at Hanford, and many people's minds would have to change to make it successful. Tseng then proclaimed, "Hanford needs more tanks in the SY and BY farms starting in FY 1992!" At this, everyone probably stood up and shouted, "Amen!" But enthusiasm and need were not enough. No new tanks have been built yet.

Besides making the process of fixing SY-101 much easier, quicker, and cheaper, there was a real need for new tanks. Many of the old single-shell tanks still had to be drained of liquid, and progress needed to be made on removing their solid waste. But the Hanford Waste Vitrification Plant was being delayed, so no waste would leave the double-shell tanks very soon. At the same time, SY-101's behavior had proven that there were limits to how far the waste could be concentrated.

A new AQ farm with four tanks was proposed about 1990 to handle waste from PUREX, had it restarted. These tanks were designed to new seismic requirements and had some kind of mixing system. The cost came in at $64 million per tank. This was way too expensive, and the project never started. Harry Harmon pushed hard for new tanks when he arrived in late 1990. The Tank Waste Remediation System, when it was set up in 1992, was charged with building new tank farms. The first Multi-Function Waste Tank Facility was to be completed by 1996. The new tanks had to conform to applicable environmental standards and be capable of withstanding a large-scale flammable gas burn. These hyper-conservative criteria raised the

> cost to about $100 million per tank. But the Washington Department of Ecology didn't want any more tanks, and DOE desperately needed money to fix problems at the new waste vitrification plant at Savannah River. The new tank program was cancelled in mid-1993. Even so, new tanks were still being proposed as late as 1995. The shortage of double-shell tank space still plagues Hanford today.

The overall conclusion of the meeting was that core sampling could and should be done with all speed, hopefully in Window C. The core sample analysis results should be used to develop and begin a plan to remediate the flammable gas problem, which DOE admitted was their top priority. In spite of this encouraging progress, Defense Board staff present at the meeting were not impressed. In their view, there was too much emphasis on trivia like sampling while the key issues were, "Why is this tank releasing gas?" and "How is it going to be stopped?" Apparently they didn't believe that sampling was necessary to understand the waste and answer these key questions.

Having SY-101 at the top of DOE Headquarters' priority list generated a lot of activity on top of what the Waste Tank Safety Program was already planning. In addition to the large expense of core sampling, installation of the gas monitoring system had high priority, a special video camera was being ordered for in-tank use, and new instruments to improve waste temperature measurement were in the works. It looked like the SY farm would need a whole new ventilation system. A host of analytical efforts were also underway. As of April 1991, the Waste Tank Safety Program budget was estimated at $53 million for 1991, $62 million in 1992, and $65 million in 1993. Capital funding was to add another $20 million in 1992 and 1993 for a total of $220 million! Duffy's "robust five-year program" was going to cost some real money!

WINDOW C, CORE SAMPLES AT LAST!

Steve Marchetti set the tone for Window C work in a message to other Westinghouse managers on April 11. It said, "Our first work priority at this time is to support Tank SY-101 work and Window C activities. The second priority is to support Tri-Party Agreement milestones, which include interim stabilization of SSTs [single-shell tanks]. If your priorities don't match these, revise them accordingly." John Tseng transmitted the same kind message, asking, "Why can't we do more?" and "Why can't other people or more equipment be brought in?" This obviously referred to the Los Alamos crew. Ron Gerton, DOE Richland's "handler" for the Waste Tank Safety Program, got into the act too, urging that the radar level gauge installation be expedited for Window C and complaining that planning for the majority of Window C activities was running four to six weeks behind schedule!

But not so fast! On May 10, 1991, just before Event C, John Wagoner issued a terse "chewing out" to the presidents of Hanford contractor companies warning, "overspending will not be tolerated." Harry Harmon passed this message down to the Waste Tank Safety Program, admitting "we have pushed the system to the limit. . . . From now on we must demonstrate completion of milestones with disciplined financial control." So, get Window C work accomplished at all costs, but at no more than planned cost!

The SY farm ventilation system was becoming a headache. Headquarters, especially Steve Blush, had been very sensitive to ventilation from the start and demanded that a backup ventilation system be provided in case the main system happened to fail at an embarrassing moment. This was very difficult. A ventilation system for a radioactive waste tank has to suck air out of the tank to maintain a slight vacuum in the tank domespace. That way, all leaks are *in* and no tank air, potentially carrying radioactive particles, gets out unless sucked by the fan through high-efficiency particulate air filters. To do this, the ventilation fan has to be downstream from the tank, possibly exposed to a flammable concentration of hydrogen after a big burp! So the fan, shaft, bearings, and seals, plus whatever electrical components accompany them must be spark-proof, even during a catastrophic failure, to prevent lighting off the explosion that the ventilation system is supposed to mitigate!

The primary ventilation system on SY-101 at that time was not spark-proof, but the system had passed several "flammability tests" in the past when big burps created a flammable gas mixture and the fans had sucked it through without igniting it. This at least showed the system was non-sparking if not spark-proof. The portable exhauster planned as the backup system was not designed to be non-sparking either, but no other system was available in time for

Window C. So Marchetti, in concert with DOE, decided that the portable exhauster would be made ready but would *not* be turned on immediately if the primary system failed. Instead, tank intrusive work would stop and the Joint Test Group would decide whether to start the portable based on the hydrogen concentration, waste level, and temperature. Marchetti promised to install a backup power supply for the primary fan before Window C to lessen the probability of failure. He also promised to design and install a whole new spark-proof ventilation system for the SY farm. The new system was not ready until after the mixer pump went in and was never used.

The work planned for Window C was ambitious. It included installing a color video camera in the dome, dropping the new sludge weight samplers into the crust in several places, installing the new radar level gauge, and taking a push-mode core sample. The safety analysis documents were ready for each of the operations planned. In hindsight it is disappointing that the hazards analyzed still included igniting gas pockets under the crust and burning the crust. But the controls placed on the operations were not too onerous. All equipment entering the dome was to be electrically bonded to the tank to prevent static electricity buildup, and backup ventilation was required. The camera assemblies were to be pushed in gently, stopping if resistance got too high. Core sampling would stop if the hydrogen concentration exceeded 2,000 parts per million or the dome began to pressurize. On May 6, 1991, Duffy sent a letter to John Wagoner at DOE Richland notifying him that the *Secretary of Energy himself* had approved Window C activities. Headquarters oversight of SY-101 work had ascended to a whole new level! Would the President have to sign off on the next window?

On May 16, 1991, at 8:17 PM, a moderate gas release, Event C, happened in SY-101. The waste level dropped 4.8 inches immediately, and reached 5.9 inches May 18 and the maximum of 7.2 inches May 22. The hydrogen concentration peaked at 2.8 percent, and the total gas release was about 5,300 standard cubic feet. The domespace was pressurized to 0.24 inches of water gauge for about 40 seconds, and the ventilation flow increased from 580 to 650 cubic feet per minute. The waste temperature dropped down to 50 inches. Event C occurred 89 days after Event B on February 16, 1991.

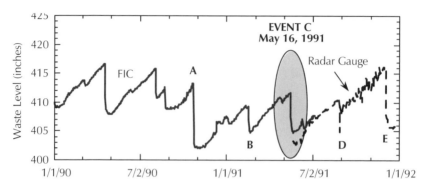

SY-101 Waste Level History: *1990–1991*

Dan Reynolds' new window criteria that attempted to alleviate some of the complaints from Window B were not actually released until the day after Event C, so the data were evaluated by the old criteria. Thankfully, there was no doubt about opening Window C. The tank data clearly met all the criteria. The waste level dropped past the required 5 inches in a day. The waste temperatures also changed sufficiently, though not all the way to the bottom. There were plenty of other signs of a good-sized gas release. The window opened at 9 PM on May 18. John Wagoner, DOE Richland manager, sent a letter to Westinghouse President Tom Anderson authorizing Window C activities the same day!

Operators began working hard to get everything done inside the 20-day window. Field crews were on the tank 24 hours a day and monitored the tank for hydrogen every hour. Steve Marchetti shut down all other work in the tank farms for 2 weeks. It was intense! At the daily meeting of the Joint Test Group May 23, it was reported that the field crew wanted a 3-day weekend. People were getting uptight, but they didn't get any rest until May 28, when the window closed with all the work done.

Harry Harmon sent a message to all Westinghouse staff congratulating everyone for their good work in completing all Window C activities ahead of schedule! Apparently this compliment wasn't entirely effective. In a memo to management June 7, a worker spoke of "worker frustration with high-stress conditions and management inattention" and warned of a "mass exodus on a biblical scale" if conditions didn't improve. His name was Jim, not Moses, and there

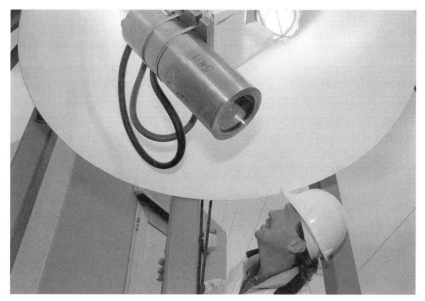

Color video camera to be installed in Window C.
Courtesy *Hanford Reach.*

is no evidence that the Columbia River parted so that crews could exit Hanford on dry land. But it was clear that window work was becoming a sore point. In fact, the work got even more intense, and many did leave. But those who stayed began to take pride in their ability to survive the stress.

Unlike the field crew, the engineers, scientists, and managers had stress all the time, not just during a window. The window work just added to their anxiety. There was frustration one afternoon when the relative humidity went below the National Explosion Prevention Association limit for static electricity on clothing, stopping work with the sludge weight sampler. There was disappointment when the new color TV camera died from radiation exposure on May 22 after only 72 hours. There was embarrassment, too. The gas chromatographs (GCs), installed in Window B to measure the concentration of hydrogen and some other gases very accurately, had been turned off during the Window C burp! An opportunity to get good data on a large gas release was missed!

But there was also a lot of satisfaction with Window C's tally of accomplishments. The crew managed to install a black and white camera after the color camera died. It worked like a charm, and gave us more information about the gas release mechanisms over the next several burps than any other instrument in the tank. The radar gauge went in without incident, and its first readings matched the other two gauges pretty well. This was very fortunate because the FIC went bad a few weeks later, making the radar gauge the primary measurement! The sludge weight sampler got crust material from four risers. Most rewarding, the drill crew pulled a full-length core sample from the tank. After a few months of analysis in the laboratory, we would finally begin to know what the stuff causing the burps was really like!

Steve Marchetti was a long-time Westinghouse manager with experience at the Waste Isolation Pilot Project in New Mexico and the West Valley Fuel Reprocessing Plant in New York. His first tour at Hanford was in the 1970s when the plants were still running. On returning to Hanford with Harry Harmon in 1990, his main job was as SY-101 mitigation project manager before Jack Lentsch arrived the following year. He knew all the right working people at Hanford and how to get commitments out of them, whether the work was doable or not. All you had to do was mention Steve's name and doors would open and work would get done. He was very unpleasant with anyone who put up obstacles. He was a strong advocate for using a mixer pump to mitigate SY-101.

Steve was a large, gregarious, and forceful man. He lived in a camper trailer in a tough part of Pasco. When he had people over for dinner, he always told them to bring their own pistol for protection! His pet project at home in South Carolina was restoring a classic XKE Jaguar. Steve was always dressed for success, in the Westinghouse mold, and was a polished presenter. He managed by loud but good-humored threats and intimidation. He was always very approachable and sociable in private, and spent a lot of time outside smoking and joking with the workers.

Even Leo Duffy was happy with the progress. On June 21, he sent a memo bragging to Admiral Watkins that "Steve Marchetti has made smooth progress and is well ahead of schedule." Harry Harmon passed a copy around with his own handwritten note in the margin, "Good Work! This is the first time I'm aware of Leo bragging on us to the Secretary!" Six months later, in a memo that chewed out Westinghouse in other areas, DOE Richland's Ron Gerton held Window C up as one of their few "moments of greatness."

In looking back over the whole Window C experience, several "lessons learned" were apparent to everyone. The Joint Test Group functioned well, but they had no fixed meeting place. Many of the Joint Test Group members had responsibilities for important Window C work, and it was a pain for them to have to drive five miles to the 2750E Building in 200 East or even farther for meetings. Somebody even suggested parking a bus at SY farm for the Joint Test Group. Reviews of the safety documents were a real burden. The main safety assessment was not ready until just before the window, delaying many activities. Because DOE Headquarters had estimated that their review and approval would require 67 days, the problem would only get worse.

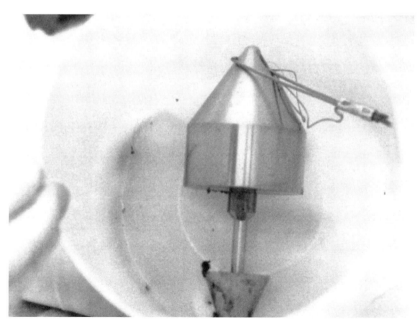

Sludge weight sampler with traces of crust attached being examined in a "glove box." Crust samples obtained during Window C.
Project photo *Window C Activities.*

126 | Part One: Flammable Gas

There was still confusion and frustration about tank data. Several organizations were doing data, and there was no clear plan for data management, coordination, or evaluation. Different values and different interpretations flew all over the country. This was a chronic ailment that never was completely cured. Responding to a DOE Richland complaint about conflicting data reports, Jerry Johnson explained, "The problem is that some people are running with the data to provide a quick response to Westinghouse and DOE management, but these people are not performing a sanity check on the data before it is released." Moreover, as valuable as the in-tank video camera was, there was no plan or schedule for its use, and no management plan for the videotapes. It was extremely fortunate that the video camera had a continuous time stamp labeled "SY-101." Data management came under better control in late 1992 with an official data management plan that named Nick Kirch's group, specifically Jeanne Lechelt, as the funnel for tank data and keeper of the archive.

Kind, grandmotherly Jeanne Lechelt is one of the unsung heroes of SY-101 mitigation. In a tight little corner of an office upstairs in 2750E, with computer screen blinking and trusty telephone propped on her shoulder, Jeanne calmly refined and sorted out the great ugly piles of raw data coming from SY-101, creating neat orderly files convenient for all to use. She was not only a virtuoso with a spreadsheet but knew the annoying habits and quirks of the devices that recorded the data. If something looked strange, Jeanne could usually tell if the strangeness was waste behavior or bad electronics and wasn't bashful about calling someone to fix it.

Jeanne had a rare work ethic. Nick Kirch tells that he was at a conference in Pittsburgh when Event H gushed up at 2 AM February 1, 1993. The tank farms shift office called Nick's wife who relayed the news to him. Nick told his wife to call Jeanne Lechelt and ask her to start processing the data in the morning. The shift office also called Nick's boss who went right back to sleep, confident that Nick would put together the data report. But he woke up in a sweat realizing, "Oh no! Nick is in Pittsburgh!" Fortunately, Jeanne couldn't sleep after Nick's wife called and made the 30-mile trip out to 200 East to compile the data package. Nick's boss used it to answer the stack of urgent queries that were waiting when he got to work. Jeanne's insomnia saved the organization's bacon and made Hanford look good!

On June 6, Secretary Watkins issued his 6-month report to Congress on waste tank safety issues, just as Senator Wyden's amendment required. Watkins wrote that he had made resolution of the safety issues one of the highest priority activities in DOE. He vowed that low funding and low priority for tank safety would no longer be tolerated. At the end of the report, Watkins warned that resolution of all issues would probably take more than 10 years, but the flammable gas and ferrocyanide issues would be resolved much sooner. As promised, the ferrocyanide safety issue was to be the first closed. But flammable gas was the last one closed, a little over 10 years after Congress got this report.

Apparently Watkins' report satisfied Congress and the news media that he now had the tank safety problem in hand and was making sufficient progress. The media did not indulge in the kind of sensational commentary on tank issues they did following Mike Lawrence's news conference and Steve Blush's report a year earlier. The *Tri-City Herald* described the crust sampling in Window A and core sampling of Window C in positive terms. The replacement of Westinghouse's president and the embarrassments that brought it about got only brief national notice. But the pressure was still on and everyone from the Secretary of Energy on down knew they had to do more than sampling and monitoring. They needed a plan to actually remove or reduce the hazards, and they needed it pretty soon.

After Window C, dissatisfaction began building with the rate that Westinghouse's 222-S analytical chemistry laboratory was analyzing the core and crust samples. Tseng kept asking why samples couldn't be sent to other labs with hot cells, like PNNL, just a few miles away. He also suggested Los Alamos, of course, and even Oak Ridge National Laboratory in Tennessee! However, for reasons unknown, samples never left Hanford and few left the 222-S lab. Though PNNL's radiochemistry labs in the 300 Area were not staffed to handle a large volume of chemical analysis, and their hot cell space was limited, PNNL was eventually tasked with measuring viscosity and strength of waste samples over a range of temperatures and dilutions to provide data for mixing predictions. Unfortunately, their report didn't come out until after the mixer pump went in.

Getting data out of radioactive waste samples is very difficult. Waste samples are collected by shoving a series of 19-inch-long, 1-inch-diameter tubes into the waste one by one. In SY-101 it took 22 of these 19-inch segments to sample the whole depth of the waste. Full samplers are hauled in shielded containers to a hot cell in the 222-S laboratory near the old REDOX Plant, not far from the SY farm. A hot cell is a sealed and shielded box with highly filtered ventilation. Using a hot cell, workers can safely study radioactive material with electro-mechanical remote manipulators that duplicate the motion of an operator's arms, hands, and fingers on robotic arms and fingers. Video and still cameras look through the thick lead-glass window recording every step. Because the yellow-tinted lead glass alters colors, a standard color card is placed in view so photographers can correct the hue later.

The first step in the analysis process is to extrude the waste from the sampler out onto a tray. Video tapes of the extrusion process tell a lot about the waste. The distance the cylinder of waste protrudes from the sampler before it bends or breaks off onto the tray shows its strength with fair accuracy, as does the way it slumps or holds its cylindrical shape on the tray. A trained eye can roughly estimate the particle size by whether it looks like a "snow cone," sand, or mud, and by the amount of liquid that puddles on the tray. The color may show whether the waste is sludge, saltcake, or some mixture of the two. If the waste is gassy, bubbles may be visible. One bottom sample of SY-103 waste held so much compressed gas that it shot a glob of waste against the hot cell wall when the sampler was opened.

After extrusion, the operator scoops up the solids from the extrusion tray and collects the liquid that drained from it into bottles in accordance with the analysis plan. A multitude of visual, chemical, radiochemical, and physical analyses can be done on waste samples. The number of samples analyzed and the number of analyses done on each sample depends on how much money the project sets aside for the laboratory. Even a "bare bones" analysis plan takes a very long time. The core extrusion video might be available within a week or two after samples arrive at the lab, and physical properties like density and liquid fraction come out in maybe a month. Raw chemical and radiochemical analysis results are not reported for at least 2 months and usually 3 or 4. The final result is a tank characterization report, tabulating all the analysis results along with the tank fill history and other pertinent facts. The report for Window C was first issued in February 1992, 9 months after sampling.

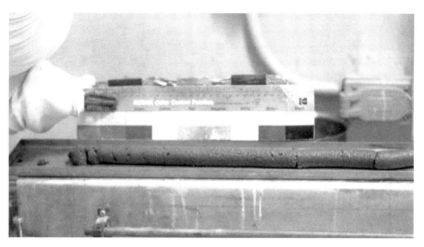

Core sample segment in the 222-S hot cell. Sample was taken in Window C from the upper part of the sediment layer. Manipulator hand (left) holds a color chart. Project photo Window C Activities.

WINDOW D, AN UN-WINDOW

Planning for Window D work began before Window C opened. Planning this far in advance was necessary to accommodate the lengthy DOE Headquarters reviews of all the plans and safety documents. As early as May 8, over a week before Window C, the tasks to be done in Window D were selected. The list included another core sample, calibrating the radar level gauge with "zip cord" measurements, and installing an inlet high-efficiency particulate air filter. Tseng wanted spark-proof ventilation fans installed, and Steve Blush demanded that the backup exhauster be in place by Window D. After the color camera failed, there were demands to replace it with a radiation-hardened color camera, even though the black and white camera was already showing sharp images of the inside of the tank.

By mid-August, Dan Reynolds predicted Event D for August 22. The list of tasks had solidified into a core sample, hooking up the portable exhauster, calibrating the radar gauge, and replacing the color video camera. The paperwork was still not approved. The safety analysis on backup ventilation was still being reviewed and redrafted by August 21. It was signed off August 23, but the readiness review by DOE Richland was not yet completed as of August 26. Tseng and Duffy had still not approved any Window D activities.

On August 27, 1991, at 9:35 AM, a small, slow gas release, Event D, took place in SY-101. The FIC level gauge was not working. The radar gauge showed an immediate drop of 7 inches, then rose back to only 3.3 inches by August 29. The manual tape showed a net drop of 5.5 inches but a wastecicle had fallen off, so the reading was faulty. The hydrogen concentration peaked at only 2,500 parts per million, and the domespace did not pressurize. The waste temperature on the bottom four thermocouples did not change. The in-tank video showed many scattered bubbling events. Event D followed Event C by 103 days.

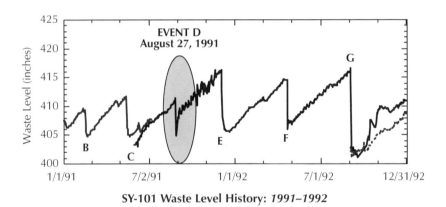

SY-101 Waste Level History: *1991–1992*

As of August 30, the last "punchlist items" from the readiness review were still being worked, and DOE had not yet approved the safety analysis and environmental assessment. But it didn't matter. On September 5, after thoroughly evaluating the tank data against the new window criteria, Nick Kirch sent a memo to Steve Marchetti concluding, "Window D cannot be opened." The work would have to wait until Window E, predicted for late November.

Event D was the first to be recorded on video. For the first time, everyone could see what a gas release, albeit a small, slow one, looked like. The video clearly showed small surges of dark liquid waste "blurping" thickly up, spreading out and quickly contracting again accompanied by a lot of bubbles. This happened all over the tank over more than an hour. The liquid was clearly coming up, and it sure looked like the crust was being penetrated with ease. One viewer described the surface as "wet, moving islands." The video could not prove, but it certainly did not disprove, the developing theory of buoyant "gobs" of waste rising from the tank bottom.

MODELS AND TESTS

The Science Panel's experimental work on gas generation was now well under way. In his notes of their meeting on July 10, 1991, Jerry Johnson recorded his opinion that the panel "appears to have an excellent team, really going now." On the physical side, Rudy Allemann had synthesized the "gob" model of buoyancy-induced gas release by the fall of 1991, based on the accumulated observations and calculations begun by Sasaki, Reynolds, Kirch, and others as early as 1985.

There was more research going on now that would strongly influence mixer pump work over the next couple of years. Science Panel member Don Trent from PNNL authored the TEMPEST computer program back in the early 1980s to model the detailed fluid dynamics inside liquid-metal-cooled breeder reactors. He had continued to refine the code and apply it to new situations such as the flow of molten glass inside electric melters and sewage outfalls in bays driven by tidal cycles. From what he heard at the early Science Panel meetings in mid-1990, Don believed TEMPEST could also be applied to reveal what was happening when SY-101 burped!

Courtesy Joe Brothers

Don Trent experienced a different kind of Hanford history than anyone else working on SY-101. The original Manhattan Project plant was protected by 16 gun batteries scattered around the periphery of Hanford. In the 1950s, the gun batteries were replaced by four Nike-Ajax and Nike-Hercules missile batteries designed to shoot down any stray Soviet "Bear" bombers that might growl over at 40,000 feet. Don was in the Army Air Defense Command assigned to Nike-Ajax C-Battery from 1958 through 1960, years of peak production at Hanford.

Sited on the rising bench above the Columbia River to the east and north of White Bluffs, C-Battery had a barracks, mess hall, and a few trees—all the comforts of home. Today, you can drive to the ruins from the road into the boat launch in the

Wahluke wildlife refuge. Back then, the only way in or out was on a government bus. They never fired any missiles from the battery but went to New Mexico once a year to fire one training missile. It must have been a very lonely place of epidemic boredom.

But there was other excitement. On Labor Day in 1958, a brush fire broke out north of C-Battery on the slope of Saddle Mountain. The wind was blowing the fire towards the Hanford 100 Areas, and the Atomic Energy Commission called the military out to fight the fire. Big green transport helicopters hauled Don and his company to the fire. They were given a wooden pole with a piece of rubber mat attached to beat the flames and a single canteen of water. The weather was hot and the flames made it hotter, and the crews were getting dehydrated. The canteens quickly emptied and urgent calls went out for water. A helicopter whopped in loaded with, not water, but cases of canned peaches! The firefighters sat opening cans of peaches as fast as they could, drinking the juice and throwing the peaches away! They could not put the fire out and it burned all the way down to the river. Don believes you could probably still find piles of rusting cans on the slopes of Saddle Mountain!

TEMPEST needed an overhaul first. It somehow needed to track both rising gas bubbles and settling sediment particles. Don decided to use a single, homogeneous "pseudo-fluid" that carried bubbles, particles, and several other components as "passive scalars" that simply ride the main flow without affecting its motion directly. He needed a completely new viscosity model too. The classic model calculated the viscous force as the product of the viscosity and the velocity gradient. But this model did not allow the sediment layer to remain stationary. Eventually all of it would begin to move and join a burp. The calculation broke down when the fluid viscosity varied orders of magnitude between that of water and peanut butter. To alleviate these problems, Don found a different way to compute the viscous force directly from the fraction of solid particles in the fluid and its shear rate. This way he could specify a threshold strain rate below which the fluid would stay put. It was a stroke of genius that in Don's words "worked like a charm."

The overhaul took the whole fall and winter of 1990. Don and his cohort, Tom Michener, had to work 60 and 80 hours a week every week. Finally in January 1991, he was able to report "preliminary TEMPEST simulations for a postulated turnover caused by gas volume growth yielded remarkable similarity of temperature profile behavior with the actual tank measurements." Don and Tom continued working on TEMPEST, improving and automating the graphics output, making use of each new bit of data to better define the waste, and just building experience with burp behavior. In April 1992, the Joint Test Group asked for TEMPEST runs to evaluate Event F to see if a window could be opened. When Los Alamos took over the safety analysis, they used TEMPEST as their workhorse for gas release predictions under various assumptions of pump behavior and waste conditions.

TEMPEST soon got competition. Even though PNNL had been given the mandate for research by DOE, Westinghouse still had engineers who wanted to study SY-101's burps and help figure out how to stop them. Westinghouse manager Don Ogden had a contract with a local firm, Numerical Applications, Inc. (NAI) to run computer programs to calculate waste temperature distributions and ventilation flows in the tanks. Like Don Trent, Ogden and his NAI cohort, Marv Thurgood, thought their computer program might reveal burp behavior. In mid-1991, Jerry Johnson supplemented Ogden's and NAI's funding to start modeling SY-101 with their FATHOMS computer program.

NAI had formed only a few years before when Marv Thurgood and several other PNNL staff decided to start their own company. While still at PNNL they had developed their mainline computer code for simulating the violent boiling water and steam flows during major accidents in nuclear reactor cores. Because the code was government-sponsored it was public property, and Marv could take it with him to NAI.

Needless to say, there were hard feelings when NAI plucked several very talented staff from PNNL. The hurt was aggravated when NAI began competing with PNNL on SY-101 work. Even worse, the line manager that Thurgood and his NAI team had worked for at PNNL was Don Trent! Rudy Allemann, Tom Michener, and I were also in Don's section.

Ogden and Thurgood have since formed their own smaller company and still use a derivative of FATHOMS for Hanford work. Don Trent retired in 1996, but Yasuo Onishi still applies improved versions of TEMPEST to investigate jet mixing and

Part One: Flammable Gas

solids settling in the tanks. But the two computer programs have found different niches and don't compete so much any more.

FATHOMS was fundamentally different from TEMPEST. It used a multi-field model where the quantity, motion, and temperature of gas bubbles, liquid droplets, continuous liquid pools, and gas spaces were calculated separately. But, like TEMPEST, FATHOMS had some serious difficulties. One was the lack of knowledge about how these separate fields interacted with each other. For example, how did the continuous liquid field drag on the bubbles? And how could the sediment layer be modeled? One of Ogden's first tasks was to "benchmark" FATHOMS against data from past burps to show that its simulations were reasonable.

The most debilitating problem for FATHOMS was the huge number of computations required to solve four sets of equations compared to only one set for TEMPEST. It took forever for FATHOMS to complete a simulation run, while TEMPEST's relative speediness made it the workhorse for safety analyses and for any urgent investigation DOE or Los Alamos demanded. Nevertheless, Ogden and Thurgood tenaciously kept running FATHOMS. Despite the differences in the methods, the two codes always gave pretty much the same results. This gave TAP and DOE some tentative confidence that TEMPEST results might have some shade of validity. But this all changed about the time of Window H.

Computer simulation was not the only area of competition between Westinghouse and PNNL. Both were also working on scaled experiments mixing up sediments in small tanks. Rudy Allemann wanted to do scaled tests with "chemically correct" waste simulants to validate the "gob" model. Westinghouse's Shih-Chih Chang wanted to do tests with yeast and starch to define bounds on conditions that would cause burps *before* going to synthetic waste tests. He didn't think TEMPEST (or FATHOMS) simulations were of much value. It was perfectly clear to Jerry Johnson that both groups were proposing the same thing: tests, analytical studies, models, and numerical analyses. He had to try to scope out programs so both groups would have a role they were happy with.

Jerry was never really successful in crafting the kind of program that kept both groups happy. There was a lot of overlap and considerable tension between them. Once Shih-Chih and Rudy

Open Windows: 1990–1992 | 135

privately agreed to divide the work so that they wouldn't compete. But at the next project meeting, Rudy's jaw dropped when Shih-Chih again claimed everything for Westinghouse! It got so bad by June 1993 that PNNL's Max Kreiter attempted to arbitrate the arguments by having eminent professor Graham Wallis from Dartmouth College review the work. Wallis strongly recommended experimental verification of TEMPEST and went on to suggest in-tank measurement of gas fraction and other properties. But because the pump was already built and would be installed a month later, it was a little late. The need for modeling work went away after the mixer pump proved itself in early 1994. Most would now probably admit that few if any of the small-scale burp or mixing experiments by either group gave grand insights or enhanced project progress very much.

The one experiment that grabbed headlines was Shih-Chih Chang's burping corn meal mush. In early 1991, Chang was searching for a method he could use at meetings to show visually how a burp really happened. It had to be non-toxic, non-violent, and easy to see. He chose to generate carbon dioxide with common yeast and sugar. For the sediment, Chang used corn meal. When mixed with the ingredients water and warmed it up to get the yeast going, the sweetened corn meal produced very realistic looking burps.

The *Tri-City Herald* featured the recipe May 30, 1991, under the headline "HANFORD SCIENTISTS SIMULATE BURPING A-WASTE." Here's a recipe you can try out at home:

Ingredients
clear glass or plastic jar or other container holding
    at least a quart
water to fill the container to 3/4 depth
corn meal to fill the container to 1/3 depth
2 teaspoons dry yeast
3 teaspoons sugar
2 teaspoons corn starch

Mix yeast, sugar, and cornstarch with the water and heat to about 100 degrees Fahrenheit. Pour heated mixture into the jar and add the corn meal. After the yeast gets started, a burp should occur about every 10 minutes.

Shih-Chih once wheeled a great vat of this stuff into a TAP meeting to demonstrate the current burp theory. They said you

136 | Part One: Flammable Gas

could smell the yeast all over the conference room. The TAP was initially impressed but later ridiculed the demonstration as not properly representing the waste. Nevertheless, it showed the physical mechanisms at work in the tank quite accurately.

WINDOW E—REVELATION!

One good thing about Window D not opening was that all the safety documentation, procedures, crew training, and much of the readiness review paperwork were all ready for the next window. Therefore, the work plan for Window E was simply to do the work originally planned for Window D. DOE agreed, and on October 22 John Tseng promised an official letter authorizing the approved work to carry over to Window E. Everyone hoped that the work could start in a little over a month.

> On December 4, 1991, 11:10 AM, SY-101 had its largest recorded gas release, Event E. The FIC level gauge was broken. The radar gauge dropped 13 inches immediately, and then rose back to 8 inches. The manual tape showed a net drop of 14 inches. The hydrogen concentration peaked at over 5 percent with a gas release that may have exceeded 10,000 cubic feet. The domespace pressurized to 6.8 inches of water gauge pressure for 3 minutes and the vent flow spiked above 1,200 cubic feet per minute for about a minute. The waste temperature decreased from over 130 to 118 degrees all the way to the bottom. The in-tank video showed violent surges of waste. Event E occurred 99 days after Event D.

There was no doubt that this huge gas release satisfied all the criteria for opening a window. Nick Kirch quickly fired a message to Marchetti declaring Window E open as of noon on Thursday, December 5. All the planned work was done without serious difficulty. The drilling crew got another full-length core sample, the color camera was replaced, the radar gauge was calibrated, and the portable exhauster was plumbed into the existing ventilation system.

Though the work went well, the post-window critique suggested some improvements. These included assigning a secretary to keep all the documents organized for the readiness review and that the Joint Test Group should use better discipline in their meetings.

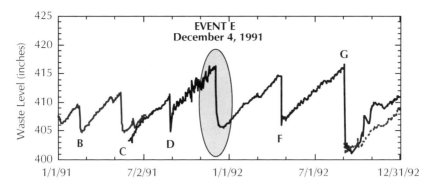

SY-101 Waste Level History: *1991–1992*

The Window E core sampling operation was criticized as "really shoddy." Some segments were either empty or partial and valves on some samplers were open when the lab received them. Chain of custody papers, listing who had handled the samples, were not filled out properly, and the segment labels did not agree with papers. The labeling error rate approached 85 percent. Though these defects compromised some samples, there was enough information, including the cores from Window C, to characterize the waste pretty well.

The crust samples from Window C were interesting because the driest, flakiest material attached to hardware above the waste level still had 10 percent water. This is apparently as dry as the salt waste gets. Even the harder dry-looking pieces of crust had 20 percent water, and the wet samples were up to 40 percent water. A truly dry crust that could burn did not exist! Another interesting, but expected, observation was that the composition of the waste was very similar from crust to tank bottom. Both Window C and Window E cores also showed essentially the same composition. Everyone expected this because tank burps had been mixing the waste up for many years and the core samples were taken shortly after a large burp. However, it was comforting to have an answer for those who warned of "hot spots" or other nonuniformities in the waste that might cause dire accidents.

Overall, the composition and properties of this waste held no surprises. It was perfectly capable of generating and holding a lot of gas. And the Window E video showed that the gas releases could be very fierce!

A GREAT SEA CHANGE

Event E was a real monster. The black and white video showed the awesome power of giant boils of black liquid surging across the tank like surf. Five or six of these great floods welled up in alternating quadrants of the tank with only brief pauses between them. The hydraulic forces were so strong that thermocouple trees and water lances bent back and forth in the current. The video dramatically displayed the violent nature of the burps in SY-101 that had been seen only as lines on charts and illustrated with cute cartoons. But now it could be seen that the tank was really doing things that made mitigation very urgent.

The video evidence now clearly supported the "gob" theory and discredited the old model of a big-bubble-under-the-crust. TAP member Dave Campbell wrote his colleagues "the concept of a hard crust and under-crust explosion can clearly be laid to rest." But invalidating one hazard revealed another, potentially more serious, one. Jerry Johnson and Westinghouse consultant Chet Grelecki, an explosion expert, were enjoying a beer in John Deichman's basement while they watched the Event E video. Seeing the thermocouple trees and lances swaying back and forth, banging against their risers, Grelecki exclaimed, "there's your spark source!" Now the safety analysis had to assume that the action of the gas release produced not only the fuel for a burn but its source of ignition as well! Because there had not yet been a burn even though the tank domespace was known to have been flammable several times, ignition was not a sure thing. But it was obviously more probable than anyone had thought before.

There was an ominous change in SY-101 after Event E. The average waste temperature went down 5 to 10 degrees Fahrenheit and the gas releases got bigger and farther apart, coming about every 140 days instead of about every 100 days. Some thought the cooling trend might be an artifact of the bent thermocouple tree raising the lower thermocouples to higher, therefore colder, elevations. But the actual bending angle would not have raised them enough to matter. Others suggested radioactive decay or slowing thermal reactions as they depleted their organic "fuel." But this kind of cooling would be very gradual, not abrupt like it was after Event E. After some months of study, Dan Reynolds provided this explanation of the cooling trend: "During this burp [Event E], cool material, perhaps old crust material, was thrust down to the bottom of the tank . . . and formed a cool zone around the thermocouple tree. . . . The cooler material had too much strength to be dislodged [in later burps]."

> Dan's explanation, though plausible, did not satisfy the critics, and the question was never resolved. However, new evidence of a cause for the cool-down was literally uncovered in February 2000 when the waste level was drawn down about 90 inches lower than it had been before Event E, 8 years earlier. With the side of the tank now exposed, a video scan showed a layer of waste about 3 feet thick adhered to the wall all the way around the tank. When water began to flow back in during the second dilution stage, there were several huge splashes as slabs of this waste fell off the walls.
>
> This may explain the Event E temperature drop. A thick layer of waste on the tank wall might have been knocked down by the extra violence of Event E. This large volume of material coming from a cool wall would have dropped the average tank temperature sharply when it mixed with rest of the waste. The solids from the wall added to the gas-trapping sediment layer would also explain the increase in burp size and decrease in burp frequency. Similar, abrupt, and more-or-less permanent cooling episodes occurred in 1985 and 1988, possibly due to a similar cause. It could be that Dan had the right idea all along. We'll never know for sure.

The change in burp behavior completely confused future burp predictions. On February 27, 1992, Dan Reynolds issued predictions for the next six burps using the historic 101-day period. The table below shows that the period was much longer after Event E and actual burp dates got later and later than the prediction.

The new insights into the tank's behavior were not the only changes. An even greater one was about to take place in the project. A new organization, almost a new company, would soon be formed to focus even more tightly on tank work. Mitigation of the burps would replace understanding the burps as the main thrust of the Waste Tank Safety Program. A new project manager would appear to drive the mitigation project that would come to personify Hanford's battle against SY-101. But the tanks had a few more burps and a few more surprises in store for us before the mixer pump temporarily tamed it.

SY-101 Predicted Burp Schedule

Event	Prediction	Actual Date	Days Late
F	March 14, 1992	April 20, 1992	37
G	June 23, 1992	September 3, 1992	72
H	October 2, 1992	February 2, 1993	123
I	January 11, 1993	June 26, 1993	166
J	April 22, 1993	did not occur	—
K	August 1, 1993	did not occur	—

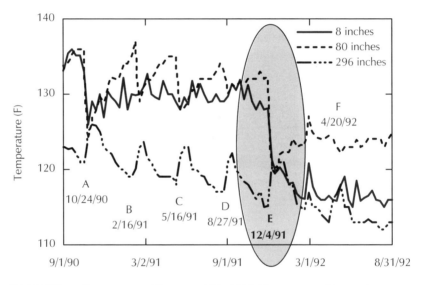

SY-101 Waste Temperature History: *1990–1992*: Temperature history shows the waste near the tank bottom (8 inches), the middle of the sediment layer (80 inches) and the middle of the convective layer (296 inches). Note the large drop in sediment temperatures at 8 and 80 inches in Event E.

PART TWO
The Mixer Pump Era

December 1991 – Rudy Allemann documents the "gob" theory explaining SY-101's burps

February 1992 – The "Fiasco in Pasco" workshop launching the SY-101 mitigation project

April 1992 – Mixer pump located in 200 Area warehouse and modifications for tank installation begin

September 3, 1992 – Event G, as big as Event E, bent more hardware and foiled mixer pump installation

June 26, 1993 – Event I, last big burp in SY-101, opened Window I for mixer pump installation

July 3, 1993 – Mixer pump installed

August 10, 1993 – "Rock-on-a-rope" contamination incident shuts down tank farm work

October 21, 1993 – High-speed mixing tests begin and mixed the tank by December

June 21, 1996 – DOE closes the flammable gas unreviewed safety question in SY-101 because the mixer pump prevented burps

1991–1992

Deciding to Mix

From the fall of 1991 through about April 1992, mixing was chosen as the primary mitigation method for SY-101's burps, and a specific mixer pump was selected for the job. This is when the SY-101 mitigation project really accelerated and put everything into getting the mixer pump built, installed, and turned on. While the pressure was severe in the bad days of 1990-91, the stress intensity rose several notches with a clear goal in view. This period begins the "glory days" of SY-101. Big burps were still disturbing the U.S. Department of Energy (DOE) and dictating the schedule while the world demanded progress. But a powerful team formed up, dug in, and prepared to do the work in spite of all the obstacles and frustrations.

THE MARCH TO MITIGATION

Fixing the flammable gas hazard in SY-101 had always been the overall objective. Way back in June 1990, Bill Legget's safety improvement plan for hydrogen in the tanks called for pumping waste out of SY-101 and mixing water back in. This would dilute the waste, reversing the concentration process, and return the tank to a non-burping state, permanently remediating the problem. Harry Harmon's initial position when he arrived in late 1990 was Vice

| 143

144 | **Part Two: The Mixer Pump Era**

President, Waste Tank Safety, Operations and *Remediation*. His *job* was to fix the flammable gas problem in SY-101.

But it couldn't happen immediately. For the first year or so, the tank safety team had to learn enough about the waste and its behavior to intelligently discuss how to stop the burping. Back in January 1991, the Tank Advisory Panel (TAP) mentioned mixing, transfer, destruction of organics (to reduce gas generation), and ultrasonic agitation as possible mitigation methods. After the core samples came out in May 1991 during Window C, and the "gob" model gained credibility as the mechanism for the gas releases, the team was ready to start thinking about mitigation and remediation. Also in May 1991, at a meeting considering additional waste samples, John Tseng predicted that "we should be able to stop after two cores and go directly to remediation!"

Tseng's expectation demanded a method or at least a plan for deciding on a method. At a joint meeting on June 12, 1991, the TAP and Science Panel recommended that working groups be formed to look at specific methods for both mitigation and remediation of the flammable gas hazard. In response the Tank Safety Program chartered working groups on July 31, 1991. There were initially three groups. One considered ultrasonic vibration, the cure-all pushed by Los Alamos scientist Steve Agnew. A group of engineers assessed mixing methods other than by ultrasonics; this kept the engineers busy. In the third group, chemists cogitated on organic destruction, chemical addition, etc. These and a succession of other committees and groups continued to hold meetings and workshops, brainstorming, investigating, categorizing, evaluating, and recommending for the next six months.

Every meeting around this time seemed to require some kind of argument over remediation versus mitigation. What did each imply? Which was the subject of the meeting? Which *should* be? Even John Ahearne's Advisory Committee for Nuclear Facility Safety had this discussion. Jerry Johnson observed that people "got really spun up over it." The dictionary describes mitigation as making something less harsh or severe. Remediation is completely correcting a problem. Mitigation had a temporary flavor while remediation was viewed as a, permanent solution.

There were also "active" and "passive" methods to argue about. An active method required periodic operator action and

> an annual budget to do something to the waste. One with
> permanent results that required no further action was passive.
> Some insisted on arguing about the semantics of what "passive
> mitigation" meant and if there could even be such a thing as
> "active remediation." The mixer pump was classed as active
> mitigation because, if the pump stopped, burps would return,
> probably worse than before. Dilution was remediation because,
> after taking out enough waste and adding back enough water,
> burps would stop permanently.

The TAP and Science Panel held a major in-depth discussion of the results of the working groups' studies of mitigation and remediation methods in Chicago on September 5 and 6. Chicago was chosen because nearby Argonne National Laboratory was heavily involved with gas generation chemistry work. Because of the chemists' interests, chemical mitigation concepts were discussed first. PNNL chemist Denis Strachan made the presentation.

There were three different chemical concepts: two reduced the gas generation rate and one changed the fluid properties. Topping the list was heating the waste to just below boiling. This was supposed to destroy the organic fuel responsible for about half of the total gas generation. Heating could theoretically be done by blowing hot air into the tank annulus without mixing or otherwise intruding into the waste. But early experiments showed that gas generation would rise to extremely high rates before the waste got hot enough to break down the organics. Also, if heating was discovered to be exacerbating instead of mitigating, it would take a long time to reverse.

The second method was to remove the cesium-137, the organics, or both by chemical methods. But this would require some kind of mixing that would in itself mitigate burps and adding chemicals would create more waste and maybe worse problems. After all, one of the causes for SY-101's bad behavior was the addition of organic complexants in prior attempts to remove cesium-137. The final method involved adding chemicals to reduce the strength and viscosity of the sediment, so it would not accumulate as much gas. This method likewise required mixing and created more waste. A novel proposal was to add microbes to eat the organics, after adding acid to neutralize the waste enough for the bugs to survive.

146 | Part Two: The Mixer Pump Era

Most of the TAP didn't like the chemical mitigation concepts very much. Scott Slezak cautioned that these were really full-scale experiments that were difficult or impossible to stop and reverse if they started to go bad. John Tseng agreed, concluding that "chemical treatment is so difficult that physical methods may be the only near-term solution." The consensus appeared to be that chemistry manipulation should be reserved for pretreating the waste as it entered the future vitrification plant.

George Mellinger from PNNL described the physical mitigation concepts, including mixing and ultrasonic agitation. The primary mixing concept employed a horizontal hydraulic jet from a pump to stir up the waste and release trapped gas. Use of ultra-high-frequency sound (ultrasonic) waves was supposed to liquefy the sediment to allow the bubbles to rise freely. Other, indirect methods to change the waste properties were also listed, including heating (again) and dilution. Both were intended to dissolve the solids making up the sediment layer, thereby reducing the total amount of gas that could be trapped. As with all such meetings, there were objections of uneven validity to every method, but ultrasonics collected the most criticism.

The TAP's recommendations simplified the task ahead quite a bit. They urged that adding chemical agents to the waste be discarded and that organic destruction and cesium-137 removal be sidelined for waste pretreatment under the purview of the vitrification plant. They believed that ultrasonic agitation was possible, but wanted their criticisms answered before endorsing it. This left mixing, heating, and dilution as the priority options for which the TAP wanted formal feasibility studies.

Also included in the high-priority list of mitigation/remediation methods was the "do nothing" option. Just let SY-101 keep on burping until its waste could be pumped out to the glass plant. Remember, this was before the in-tank video of Event E revealed the potential ignition source of pipes bending and banging on risers during burps. Before this revelation, the actual risk of a flammable gas burn was thought to be very small (though the publicly perceived risk was huge) while the risk of something bad happening during the mitigation process was believed to be high. After all, the tank had been burping uneventfully for 10 years.

However, the "no action" option was officially discarded after the primary success criterion for mitigation was agreed upon.

> Mitigation would be deemed successful if the hydrogen concentration stayed below one-quarter of the flammability limit. Because SY-101's domespace had already been measured flammable at least once, it was clear that the no-action option could not be a success by this definition.

With these marching orders, the three working groups spent the rest of September 1991, studying mitigation and remediation concepts, developing detailed technical descriptions for each, and setting up evaluation methods and criteria for selecting winners for further study. They completed their work and issued a draft report for review by the TAP and Science Panel on October 5, 1991. That the final report was not issued until April 1992 is a witness to the volume of comments or the difficulty of resolving them!

The four concepts that ended up scoring highest were heating (to dissolve solids), dilution (to dissolve solids), ultrasonic agitation, and horizontal jet mixing. Specific engineering studies were recommended to strengthen the technical basis for these chosen options so that the costs and schedule to actually do them could be worked out intelligently.

Each working group generally built a good technical basis for their recommendations. Unfortunately, they associated "remediation" with eliminating or greatly reducing gas generation while reducing or eliminating gas retention and release was classed as "mitigation." The report described only three remediation methods—cesium-137 removal, organic destruction, and "mineral forming" (precipitating aluminum out of solution). Dilution, which eventually *did* remediate gas retention, and removing the supernatant liquid, which also would have stopped large gas releases, were both placed in the mitigation bin. This stacked the deck and led the authors to conclude, "Remediation, if needed, is higher in complexity and cost, and takes longer to accomplish compared to implementing the mitigation concepts." With this, the working group consigned all the remediation concepts to waste pretreatment in the Hanford Waste Vitrification Plant. The SY-101 work was now and forever identified—for good or for ill—as "Hydrogen *Mitigation*," and the SY-101 Hydrogen Mitigation Project was born!

The introductory meeting of the Mitigation Project was November 6, 1991. Attending were project manager Jerry Johnson, Nick Kirch, Tom Burke, Tony Benegas, and Norton McDuffie from Westinghouse and Max Kreiter, Ben Johnson, Rudy Allemann, and Jim Colson from PNNL. Tseng kept the pressure on. At a Science Panel meeting in November, he demanded that the project must "demonstrate progress this FY [by the end of September 1992]" and wanted a definitive plan on mitigation that would "put a prototype in the tank!" This became the immediate and all-controlling object of the Hydrogen Mitigation Project for 1991, 1992, and half of 1993.

MORE MANAGEMENT CHANGES

The thorough Westinghouse restructuring of the fall of 1990 hastened the change in culture from plutonium production to waste cleanup and brought Harry Harmon and Steve Marchetti from Savannah River to oversee tank safety issues. New Westinghouse President Tom Anderson increased upper management interest in tank safety in general and SY-101 in particular. In a June 1991 message to Hanford employees, Anderson said, "I am particularly pleased with our tremendous accomplishments in the past month on Tank SY-101 [Window C in May], as well as continuing improvement across the site." He had also improved Westinghouse's performance rating with DOE Richland to "satisfactory" and got their contract extended to 1994.

As the Hydrogen Mitigation Project branched off the Waste Tank Safety Program, more management changes happened that raised the priority of SY-101 work right to the top. On December 3, DOE Headquarters announced that Leo Duffy's Office of Environmental Restoration and Waste Management was raised to an Assistant Secretary position. This position, one of the maximum of eight allowed by DOE's 1979 enabling statute, was unique in being subject to oversight by more than one congressional committee. In the new position, Duffy was now responsible for directing environmental restoration, environmental compliance, waste management, and related technology development for cleaning up the entire nuclear weapons complex.

A major change was happening at Hanford, too. On December 18, 1991, Duffy informed DOE Richland that he had decided to form the Tank Waste Remediation System (TWRS, pronounced 'tours' or 'twirrs') at Hanford. TWRS was responsible for safely managing the waste in the tanks and preparing it for disposal. Tank safety was TWRS' highest priority, and the top job was mitigating the primary safety concerns: flammable gas burns, ferrocyanide explosions, organic-nitrate reactions, and high heat generation. Supporting mitigation, TWRS was also responsible for sampling and characterizing all the watch list tanks. To prepare for final disposal, TWRS was to resolve the technical and regulatory concerns with grout (mixing the waste in a kind of concrete) disposal concept while keeping the Hanford Waste Vitrification Plant on schedule to start up in December 1999.

Westinghouse reorganized itself to mirror the DOE TWRS organization. As of January 1, 1992, Harry Harmon became Vice President, Tank Waste Remediation System Division of Westinghouse, responsible for both long-term remediation efforts as well as the day-to-day waste tank operations. In this job, Harmon managed 1,000 employees with a budget of $155 million for fiscal year 1992. Mike Payne was assigned to manage the tank farms under Steve Marchetti, who stayed in place.

As the mitigation effort ramped up, the flammable gas task became too big for Jerry Johnson to handle effectively. The Hydrogen Mitigation Project split off in November 1991, and Marchetti and Harmon recruited a new manager. Jack Lentsch, senior plant manager at the Trojan nuclear power plant near Portland, Oregon, was the man. Jack came to Hanford with 20 years in nuclear safety analysis and regulation. As Trojan plant manager, he was responsible for maintenance outages where, in his words, "you tear the whole plant down and rebuild it quickly and efficiently." Jack decided to make the move when Portland General Electric decided to shut down Trojan. The Trojan reactor core vessel followed Jack a few years later, shipped up the Columbia River on a special barge to its final resting place in Hanford sand in 2001.

There is a fictional power plant engineer of heroic skill featured monthly in *Power* magazine. Identified by steely eyes, wire brush moustache and occasional black cigar, he solves power plant problems all over the world with miraculous mental acuity. If anyone could personify this man, sans moustache and cigar, it would be ramrod straight, steely eyed Jack Lentsch. In casual conversation, Jack is one of the most kindly, soft-spoken gentlemen one could meet. But in a project status meeting where the schedule is all, the kindly gray eyes harden and penetrate one's very soul! Picture a marine drill instructor or a scary police interrogator and you get the idea.

Those daily, 7 AM meetings were to be approached with fear and trembling. Jack would go down the schedule and ask each task leader for a concise status. All strived to be on or ahead of schedule because falling behind required some very serious explaining. And woe to those who didn't *know* their status!

In project management parlance, "float" is the number of days between when a task can be done and when it has to be done. If you're behind schedule, which happens often, your float can go negative. At one meeting someone asked about the difference between positive and negative float. Jack answered, "Positive float is when you float face up!" Jack also had this warning: "Blow the budget: no raise. Blow the schedule: no job!" The only excuse for missing the schedule was a death in the family—your own!

Jack was the right man at the right time for SY-101 mitigation, but he was not very impressed when he first took the job. The Trojan nuclear power plant ran at a pressure of 2,000 pounds per square inch and a temperature of 500 degrees Fahrenheit. The reactor core contained 2 billion curies of radioactivity. Compared to that, SY-101's waste at only 28 pounds per square inch, 120 degrees F with a thousandth of the radioactivity was not exciting at all.

THE LOS ALAMOS LANDSLIDE

John Tseng had never made a secret of his rather low opinion of Westinghouse technical capability and pushed hard for more particiation by Los Alamos, his favored source of expertise. During 1991, Tseng had managed to involve Los Alamos in safety assessments, the ultrasonic mitigation method, and designing new waste temperature probes. Here's how it happened.

Criticism of Westinghouse safety documentation actually began with Blush's report in late 1990 and did not abate. In October 1991, Tseng demanded better quality safety documentation to stop the stream of negative comments Blush kept sending him. Shortly afterwards, while testifying before the full Defense Nuclear Facilities Safety Board (DNFSB), Tseng publicly blamed the poor quality of Westinghouse's safety assessments for schedule delays on SY-101 window work. In December 1991, Harry Harmon's own Waste Management and Environmental Advisory Committee (WMEAC) criticized Westinghouse's defensiveness and "bad attitude" about safety assessments as a problem to be fixed. In January 1992, DOE Richland chewed out Westinghouse for submitting safety and environmental assessments of poor quality. Los Alamos had also been reviewing Westinghouse safety analyses all through 1991 and were usually highly critical. The stage was set for a changing of the guard!

Since late 1990, Los Alamos scientist Steve Agnew had promoted using ultrasonic waves to stop SY-101's burps. It had a scientific flair that interested the TAP, and Tseng pushed it at every opportunity. In late May 1991, Tseng asked whether an ultrasonic system could be installed as early as Window E. In August 1991, he demanded to know when the project would make a decision on ultrasonics. At a Science Panel meeting in November, his patience apparently wearing thin, Tseng directed that a team of Hanford and Los Alamos staff be formed to develop an ultrasonic mitigation system. Bowing to the inevitable, Westinghouse and PNNL met privately November 27 to divide up the work. Los Alamos would do proof-of-principle research, PNNL would handle lab testing, including designing a prototype for a full-scale test, and Westinghouse had in-tank installation and testing. The Hanford team was skeptical that the concept would work while Los Alamos hoped it would become the primary mitigation method. In any event, Tseng would ensure that it would not be dismissed lightly.

In the fall of 1990, both the Wyden Amendment and Defense Nuclear Facilities Safety Board Recommendation 90-7 required Hanford to monitor and record waste temperatures continuously in the watch list tanks. In the required progress report providing the watch lists to Congress, Harry Harmon admitted in February 1991 that monitoring capabilities are not adequate by current standards and need to be upgraded. Two months later, Westinghouse developed a schedule to design and install "multifunction instrument trees," or MITs for short, to measure temperature, pressure, and hydrogen concentration. Los Alamos was interested and, by September 1991, had convinced Westinghouse that they should design the MITs. The plan was to have Los Alamos design and build one MIT for Tank BY-104 (on the ferrocyanide watch list) and then contract the rest out to a private company. Things began to change December 17 when DOE Richland decided that all 24 ferrocyanide

152 | Part Two: The Mixer Pump Era

tanks plus SY-101 needed the new MITs. Westinghouse decided that Los Alamos's MIT prototype should go into SY-101 instead of BY-104. Since it was now under the aegis of mitigation, John Tseng volunteered to fund it from DOE Headquarters. Through this open door Los Alamos's role in MIT work would soon expand!

By January 1992, with these initial entries, Los Alamos was poised to plunge much more deeply into the SY-101 Hydrogen Mitigation Project. Like a landslide or avalanche, the great weight of Los Alamos's enormous capabilities and John Tseng's enormous influence had begun to move and the acceleration could no longer be stopped. The Los Alamos landslide hit Hanford February 4, 1992, at the seminal mitigation project workshop at the Red Lion Hotel in Pasco, Washington.

THE FIASCO IN PASCO

Because of growing diversity of opinion and because no clear winner was named in the initial mitigation alternatives report, Tseng decided to get all parties together and bang heads until a consensus came out. He wanted a large, but tightly focused and well-orchestrated, workshop that would produce a definitive, detailed plan to demonstrate feasibility of one or more mitigation concepts.

Carl Hanson got the job of composing the music to orchestrate the meeting. On December 5, 1991, Carl's boss, Jim Thomson, instructed him to "prepare a working package to facilitate a . . . meeting to initiate planning for SY-101 mitigation." The package required a summary of mitigation options and a description of each, as well as a rough work breakdown structure and time line. On January 23, 1992, assignments were made to complete the logistics for the meeting. Jerry Johnson reserved the Red Lion Hotel in Pasco for February 3–7. Attendees were identified by the team leaders from each organization: Marchetti for Westinghouse, Ben Johnson for PNNL, and Harold Sullivan for Los Alamos. Jack Lentsch developed project logic, goals, and objectives and arranged for scheduling, tank drawings, photos, secretarial help, etc. Steve Marchetti did the agenda and program structure.

Carl Hanson and Tony Benegas exercised their sense of humor and called it the "Fiasco in Pasco." The dictionary defines a fiasco as a total failure, especially a humiliating or ludicrous one. The meeting was actually anything but a fiasco. This meeting kicked off the

SY-101 Hydrogen Mitigation Project, formed the project team, and defined their responsibilities. Several more-or-less personal agendas became part of the official project plan. Decisions were made that swept aside much, though not all, of the past bickering and uncertainty. After the meeting, John Tseng could stand tall and announce confidently to his boss and the world a concrete and well-reasoned plan to mitigate gas releases in SY-101.

> Imagine putting 10 argumentative scientists from Los Alamos, the opinionated PNNL troop, and the wily lead engineers from Westinghouse together in a big room for a week along with taskmasters Jack Lentsch and Steve Marchetti, autocratic John Tseng from DOE Headquarters, and a couple of skeptical TAP members. Jack remembers that the mutual distrust and contention was palpable. The groups sat on opposite sides of the room and would not even look at each other. Westinghouse hated PNNL and Los Alamos, PNNL felt threatened by Los Alamos, and Los Alamos didn't hold a very high opinion of either. On top of that, Tseng was getting desperate for a decision, and the project managers knew they had to make it happen.
>
> Attempting to sweeten the mood, Jack Lentsch spent over $200 to buy donuts every day for 50 people. He was never reimbursed. Most of the group had lunch together at the Red Lion every day. It was said that the Los Alamos faction met at the Heidi Haus (a little sausage place about 10 blocks away), returning late in the afternoon well lubricated, having developed their own mitigation strategy. All, including Los Alamos, were treated to a catered Mexican dinner one evening.

On the first day, Jack guided the group through the process of setting team objectives. It took the whole day to agree on a two-sentence statement! There were heated discussions and strong opinions loudly expressed. But a sense of purpose took hold and turned the contention into team building. This was a great milestone, and one of the most positive products of the meeting.

Then the newly built team had to agree on how to reach the objective, choosing from more than 20 mitigation concepts plus several new variations. To further complicate things, Mark Hall from the Hanford Engineering Development Laboratory, Westinghouse's technical development arm, gave a good presentation on a *sonic* agitation method. It was based on the same principle as Steve Agnew's ultrasonic concept, but used a lower frequency sound wave that could penetrate the waste much better. It was added to the list by general acclamation.

While the technical staff were conferring, Jack Lentsch and the Westinghouse upper management met in a back room to refine the mixer pump design. They decided that the mixer pump should suck from well up into the convective liquid layer and discharge it into the sediment at the bottom. They thought this would force the fluid to lift the sediment off the bottom, thereby doing a better job of mixing. Though this was a management decision with little or no technical input, it later proved advantageous.

The field finally narrowed to four or five approaches, and Jack split the group into teams to prepare data to rank them against each other. The organizations were not mixed very much among the teams to prevent excess acrimony and maintain momentum. It took a couple days to evaluate the approaches. Designers, a scheduler, and an estimator were right there to fill in the fiscal details and bring "pie in the sky" down to Earth, so to speak.

When the groups reported back, fierce disagreement began afresh. PNNL generally favored dilution or heating, while Westinghouse, especially Steve Marchetti and Harry Harmon, lobbied hard for the mixer pump. The TAP folks were upset. Only a month earlier they had recommended that criteria be developed so intelligent decisions could be made. In their opinion, the procedure for rating the concepts was poorly thought-out and favored preconceived Westinghouse ideas. Except for the faction favoring the ultrasonic agitation concept, Los Alamos did not want to mess around with the waste at all. They thought it would be most prudent to just let it burp until everyone really understood what was happening and found a really good solution. Doing something without adequate knowledge had a high risk of causing a worse problem.

The group could not agree on a winner! Still, it was clear to many that the mixer pump was the only workable, near-term option. Before Harry Harmon left Savannah River, he had seen a big slurry pump mixing demonstration and thought that was the best solution to mitigate SY-101. Westinghouse engineers were used to working with pumps in tanks, and PNNL experiments had shown that a little pumping could do a lot of mixing. To push the team toward the obvious choice, Marchetti defused Los Alamos objections to the mixer pump with a masterful political move. He asked Harold

Sullivan to do the safety assessment on mixing! Harold knew this would be several million dollars of work for Los Alamos and held his peace. As a result, consensus eventually converged around mixing, heating, dilution, and Mark Hall's "sonic probe." The ultrasonic method was not discarded but rated far behind.

With an equally clever move, Jack Lentsch pulled the meeting to the needed conclusion by proposing to try out all non-mixing methods in a "test chamber," a small isolated volume within the tank, while mixing would be tested in the tank at full scale as the mainstream method. The test chamber conceptually hung a relatively large cylinder right in the tank that would agitate, heat, and dilute a small volume of waste using lots of instruments to measure densities, viscosities, hydrogen concentration, etc. This concept allowed everyone to keep their pet mitigation method alive while the mixer pump could proceed at full speed. All came away from the Fiasco in Pasco relatively happy and motivated! A big, difficult project needs skillful politics just as much as profound technical insights to ultimately succeed.

The plan that came out of the meeting was to start building the mixer pump, designing the test chamber, and doing initial feasibility work on the sonic probe. Westinghouse's Mike Ostrom took charge of the test chamber, which was to be designed and built by EBASCO Services of New York. To populate the test chamber, Los Alamos would supply an ultrasonic probe, and PNNL would write its test plan and design systems for heating and dilution, and chamber instrumentation.

Westinghouse was in charge of the mixer pump with PNNL writing the test plan and Los Alamos doing the safety analysis (SA). The mixer pump was only supposed to be temporary. It would mitigate the tank until the waste could be pumped into new tanks that were assumed to become available in a few years. If the mixer pump failed to stop the burps, the sonic probe and the lead concept from the test chamber would be tried next in hopes that something would work.

A week after the meeting, Jack had written a formal project plan and reported the meeting's conclusions to the Westinghouse senior managers. Jack started to become a little more enthusiastic. Now he had mission, objective, and a team as well as a budget and schedule.

On February 11, 1992, at a TAP meeting at the Clover Island Inn at Kennewick, Jack briefed Tseng on the results of the Fiasco in Pasco. Harry Harmon, Steve Marchetti, and Westinghouse President Tom

Anderson attended. Tseng was not entirely comfortable with the plan and made his suspicions felt. He wanted the whole project plan peer reviewed by a couple of DOE Headquarters experts and the TAP with the Defense Board sitting in. He questioned the experimental plan for the test chamber. He didn't believe much could be learned from more data and wondered why any computational modeling was needed. To avoid the appearance of favoritism, Tseng demanded that Brookhaven National Laboratory review the mixer pump safety assessment that Los Alamos would be writing.

> Jack gave the cost and schedule summary himself. Marchetti had told him in no uncertain terms not to tell Tseng the cost, no matter what he asked. Jack was to "feign a heart attack" if he had to. But when Tseng specifically asked what the mixer pump would cost, Jack had to tell him, in spite of Marchetti making slicing motions across his throat from the back of the room! The mixer pump cost and schedule were estimated at $15 million and 12 months. Under pressure from Tseng, they cut it to $6 million and 6 months. It actually ended up at about $25 million and 12 months.

In spite of Tseng's suspicions, he allowed the plan to proceed. Though several methods were ostensibly being worked on, the primary push was on mixing. To mitigate SY-101 according to the plan, at least four things had to happen: 1) the pump had to be designed, built, tested, and ready to install; 2) the tank had to be prepared to accept the pump by removing old hardware and installing new; 3) the required safety analyses, environmental assessments, procedures, readiness reviews, and the rest of the paperwork had to be approved; and 4) there had to be a burp of sufficient size at the right time to open a window wide enough to fit the pump through! If we thought things had been tough up until now, we hadn't seen anything yet.

WHERE CAN WE GET A GOOD PUMP?

John Deichman led a meeting with TAP to confirm the mitigation options selected at the Fiasco in Pasco. Several participants recall that each TAP member and observer got three "red sticky dots" to vote for a concept. Accustomed to observing and criticizing, the TAP members were nervous about such active participation in an actual decision. John Tseng voted too; perhaps he was also uncomfortable about the implied accountability of placing his sticky dot on the line. In any event, the vote confirmed mixing was the winner.

> Harry Harmon grabbed the TAP support and ran with it. He called a meeting to find what pumps could be had at Hanford and whether they could be modified for the mixing job. The question fell to Tony Benegas, who had just been briefed on available pumps by pump expert Craig Shaw. Tony said, "We have pumps and we can do this here." But Steve Marchetti objected. He didn't believe the existing Hanford culture was capable of such a project. His strategy was to try to send jobs off of the site to get them out of Hanford hands. In fact, he thought EBASCO would be a good contractor for the pump as well as the test chamber.
>
> Undaunted, Tony and Craig went out to the warehouses and found what looked like a good candidate pump. Tony told Carl Hanson that there was a pump, and he could put together a team to do the modifications. Carl went to look at the chosen pump with George Vargo. They agreed with Tony and, after some gentle but persistent persuasion, finally convinced Marchetti that Westinghouse could do the job.

On February 17, 1992, just two weeks after the Pasco meeting, Carl Hanson, Tony Benegas, and George Vargo sat down with Steve Marchetti to get the pump building project going. Tony and Carl had to figure out what design modifications were needed to accommodate Jack's top-suction, bottom-discharge concept. The pump was to be sent to the manufacturer, Barrett, Haentjens and Company of Hazleton, Pennsylvania, to make the changes. They estimated that this work would take six months.

They also asked PNNL to do some calculations to confirm that "Tony's pump" would work. The horsepower needed to effectively mix up the sediment was in dispute. John Tseng's own pump expert recommended more power, while Shih-Chih Chang and some other Westinghouse engineers thought a much smaller pump would do the job. A hand-written note from Steve Marchetti to Ben Johnson at PNNL on February 18 illustrates the problem and a possible political solution. Marchetti wrote, "Ben—Need to fix on the H.P. scaling factor ASAP. I'd really prefer not proceeding with as much risk on H.P. as we have. When can we get an answer? SM" Ben replied, "We can numerically evaluate the system in four weeks. . . . If we stress that the 75 hp pump is a limited test, we may help John [Tseng] with his problem and still avoid creating expectations that generate risks we want to avoid."

Ben's recommendation was in line with the strategy stated in the project plan. The whole project was framed as a test where the most likely option would be tried first with alternates activated in sequence if the predecessors failed. Besides, the available pump had 75 horsepower, and it wasn't really very productive to argue for a larger or smaller one. If less power were required, the chosen pump would provide plenty. If more power were needed, a bigger pump could be put in later.

The project pressed ahead with this strategy. On February 26, 1992, John Deichman formally summarized the SY-101 mitigation plan to DOE Richland project monitor Ron Gerton. A separate mixing pump with horizontal jets would be installed by October 1992. They planned to install a "shrouded test assembly" with hardware for heating, diluting, circulating, and ultrasonically agitating an enclosed waste volume by January 1993. The estimated cost for both the pump and chamber was $16.4 million, with additional Los Alamos work funded separately by DOE Headquarters.

The project team pulled the mitigation schedule into the anticipated window schedule this way: pull air lances and install the mixer pump in Window G, operate the mixer pump and install data acquisition and control system (DACS) in Window H, and install the test chamber in Window I. This schedule assumed, of course, that each burp would be big enough to open a waste-intrusive window for the work.

But they chose the wrong pump! In March, Walter Haentjens, president of Barrett, Haentjens and Company, expressed worry about his liability if he modified the pump they sent him and it somehow caused the tank to explode. The pump was not explosion-proof but was to be installed in a hydrogen atmosphere. Duffy and Tseng began talking to Walter about these concerns. Haentjens wanted to sell the project a new pump built to their specifications. That didn't fit the project's schedule or budget, so Tony and Craig went out again to look for another pump. By early April, they had one that would solve their flammability problems and had twice the power to boot! It was a spare slurry pump that the Hanford Grout Program had purchased to mix up radioactive waste into concrete. It was a rough-and-tumble pump that could pass two-inch rocks through its impeller without wincing. The grout pump also came from Barrett, Haentjens and Co.

But the new pump was submersible! The big 150-horsepower, 480-volt electric motor was to run right down there in the waste! The original pump had the electric motor on top of the tank and drove the impeller in the waste with a long shaft. However, the long, spinning shaft might become unstable and shake itself to pieces in a shower of sparks. This potential hazard disqualified the original pump. The submersible motor, on the other hand, used a non-sparking brushless design and was encased in mineral oil inside a sealed steel can. On purely technical grounds, it could not be considered an ignition source. Regardless of technical correctness, sticking an electric motor down in the waste didn't sound very good. From day-one everyone had made a great issue about

Original mixer pump as found in storage. The pump inlet impeller is in the foreground and one of the two nozzles is just behind it pointing upward.
Courtesy Tony Benegas.

potential spark sources in the tank. John Deichman kept getting statements of concern about a submersible pump as late as the end of April. He instructed Jack to make sure the SA would thoroughly address pump failure modes clearly and explicitly to avoid future criticism. Little did he know what that would entail!

Walter Haentjens himself endorsed the submersible. It was the same kind of pump he recommended for use in explosive atmospheres. However, he characteristically described a host of expensive new castings that would be needed and which he, of course, could supply. But Tony believed Westinghouse could make the necessary modifications to the new pump themselves without Walter's help. Eventually the project accepted the change to the new pump and moved ahead.

Harry Harmon convinced the grout program to part with their pump. Steve Marchetti told Tony Benegas and George Vargo to get a flat-bed truck and go load it up. A photographer was sent along to take a picture of the grout staff waving goodbye to their pump. With the pump in hand, Tony kicked off the modification work with the designers. By mid-April, the project issued a "white paper" justifying the choice of pump style and work was accelerating. Confidence that the pump would stop SY-101's burps soon led to questions about what to do if the pump failed.

To answer the what-if-the-pump-failed question, the team reiterated the original plan that the pump Tony was modifying was only the "test pump" or "prototype." Running the test pump would generate sufficient information to write a procurement specification for a "permanent" pump. In the meantime, Barrett, Haentjens and Co. was under contract to provide an identical pump to replace the one borrowed from the grout program. This would be the spare pump. Hopefully the "test" pump would keep running until one or the other was available.

But the immediate task was to get the first pump ready to go into the tank. The earliest possible time was Window G, then expected in July 1992, though Event G actually waited until September. This ambitious schedule put essentially everything on the critical path and tested everyone to the limit.

Original pump in the 200 Area shop prior to modifications. The impeller and nozzles have been removed. Tony Benegas (left) stands beside the electric motor by George Vargo (right).
Courtesy Tony Benegas.

1992

Preparing the Pump

Hectic preparations for installing the mixer pump occupied the spring, summer, and fall of 1992. During this time, Westinghouse and Kaiser Engineers built the mixer pump, Westinghouse ran it through exhaustive tests at the huge Maintenance and Storage Facility at the Fast Flux Text Facility, and Los Alamos put together the initial safety assessment and supplied the data acquisition and control system (DACS) trailer. The schedule aimed at installing the pump in Window G, then expected in August 1992. Everything had to be finished at the same time, forcing the work into an urgently parallel mode.

This is when the SY-101 Hydrogen Mitigation Project imposed itself on the Hanford consciousness, transforming careers and destroying, diminishing, or delaying many other less urgent projects. Though mitigation did not even have its own budget through all of 1992, the work had to be done anyway. The money and manpower had to come from projects that did have budgets. The managers who got tapped to donate the resources were not happy, but most saw the need and supported, even participated, in the work.

Courtesy Rick Raymond

Harry Harmon knew that, because Jack Lentsch was new to Hanford, he needed help getting work done. So Harry arranged for a vice president from Kaiser Engineers to provide all the project construction support Jack might need. All he had to do was call. The best thing Kaiser did was to assign their most senior scheduler, Bruce Morrison, to the project.

Bruce was one of the most valuable members of the SY-101 team. Always friendly and good natured, he worked well with everyone, from high-level managers to junior technicians. Bruce was most comfortable in jeans and sweatshirt or flannel work shirt and often talked about his latest trip down to his cabin near Baker City, Oregon, in his big one-ton dually pickup.

His reputation for painlessly gathering and organizing schedule information so that the project's status could be checked accurately every day led Rick Raymond to use his talents for the later Surface Level Rise Remediation Project. SY-101 tended to attract good people.

DOE Headquarters watched the work intently. So did DOE Richland. The desires and directives of the two customers sometimes conflicted. It may also be true that the system was over-designed and over-analyzed, and that the safety documents were too thick. There was probably a lot of unnecessary activity in general. But nothing less would have been acceptable. After all, if SY-101 was DOE's top priority safety issue, Secretary Watkins had to show Congress and the public a lot of effort, necessary or not. And, though hindsight may show high cost and wasteful inefficiency, it also shows that the mixer pump stopped the burps. And that was the primary goal!

BUILDING THE PUMP

The mixer pump was the chosen mitigation method, the pump was physically in hand, and Jack Lentsch had all the priority and authority he needed. He started working on the mixer pump at full speed. Even though he freely delegated really big responsibilities to his team, few worked as hard as Jack did. He launched daily progress meetings held early (6:30 AM) in a cramped conference room in the Energy Technology Center (ETC1) in North Richland, so everyone could stop in and report on their way out to the 200 Areas. He also had to brief Harry Harmon, Westinghouse President Tom Anderson, and DOE Richland every day and send a weekly report to John Tseng. Jack also organized two-day extravaganzas every month to update the TAP on their progress. PNNL, Los Alamos, and Westinghouse all gave presentations. Each presentation required a stressful "dry run" before a group of Westinghouse and DOE managers.

> The morning meetings were an institution where Jack constantly reminded people of their responsibilities. If a person fumbled or failed his task, Jack transferred it to someone else. Thus the more capable got even more responsibility. But having too much work was never an excuse. They all had more than they could handle anyway.
>
> Because of all this, the morning meetings were necessarily terse, always tense, sometimes even brutal, but there was also occasional comic relief. Jack had copied the whole mixer pump schedule on a big, long white board in the ETC1 conference room. One day, Tony Benegas crept in and wrote a new line on the schedule to "continue beatings until morale improves." After several days, Jack finally noticed and immediately realized who did it. He told Tony to "erase that garbage!" I don't know how much longer the beatings continued. Another time, Tony made some irreverent remark that caused Jack to chase him around the room and catch him in a choke hold. This relieved the tension but probably didn't stop Tony's commentary.

Jack was harsh on the Westinghouse organization. He had to shake people up to keep the work on track. The old "Hanford way" of doing one's own job by one's own priorities at one's own speed, even inventing reasons for not doing one's job, was still the norm. From his experience managing Trojan nuclear plant maintenance outages, Jack expected everyone involved to be as energized and dedicated as he was. SY-101 mitigation could not tolerate otherwise.

Fortunately, Westinghouse President Tom Anderson, Vice President Harry Harmon, and Harry's deputy, Steve Marchetti, gave Jack the clout he needed and did not allow him to fail. Intolerance of unwillingness had not been enforced as a management policy at

Hanford since the Manhattan Project days, and it was not well received. But when the mitigation project began to show startling progress and the project team began getting accolades, others started emulating it. For a decade, the SY-101 mitigation project has epitomized the new Hanford way in which a dedicated, multi-organization team works together to accomplish a difficult, urgent mission. This change may have been of more value and lasting importance than taming the tank.

> Here is an example of the challenge Jack faced regularly. One day the concrete cover blocks over an access pit on the tank had to be removed with a crane. Jack was told that work had to stop because there was too much wind to do a crane lift. But Jack didn't notice any wind. A call to the Hanford Meteorological Station confirmed that the wind had been less than 10 miles per hour all day. People had lied! Jack named names to Harry Harmon, and several people had their sense of priority soundly recalibrated. Other cases of resistance or unenthusiastic support, even by high-level managers, brought similar results.

Organizational foot dragging was not the only difficulty to be overcome. The central one was to actually do the work they had committed to. They had to get a first-of-a-kind mixing system built and ready to install in just a few months. Tony Benegas and Craig Shaw, with Walter Haentjens' endorsement, had been able to get the chain of command to agree to use the 150-horsepower sub-mersible pump from the grout program. The pump with its motor was a relatively compact device, only about 4 feet long and a little over 3 feet in diameter. The two opposed jet nozzles, the inlet suc-tion port, impeller, and the electric motor were all close together. But the mitigation configuration Jack Lentsch and Harry Harmon fig-ured out during the Fiasco in Pasco lowered the nozzles about 20 feet below the suction, and the motor hung from the tank dome over 20 feet above.

Tony had a brilliant and simple plan to accomplish this major modification. The two discharge nozzles came bolted directly to flanges at the base of the pump. It would be easy to simply bolt pipes of sufficient length to these same flanges and bolt the nozzles to similar flanges on the other end of the pipes. These large

Preparing the Pump: 1992 | 167

"downcomer" pipes would probably be large enough to supply structural rigidity for the nozzles. The pump with its new downcomers could be easily hung from the dome by another large pipe through which all the power, control, and instrument wiring could be routed.

The apparent simplicity of these adjustments is what made Tony confident that it could be done at Hanford. He had the support of pump wizard Craig Shaw, instrument expert Tom Lopez, structural guru John Strehlow, electric power sage Ray Merriman, and several other of the most capable Westinghouse engineers. Tony spent most of his time being a barrier between his engineers and everyone else so they could work unmolested. They were running their own "skunk works" pushing the envelope to build the pump in time for Window G.

Tony's team completed the mixer pump design in early May and sent the pump to the Westinghouse shops in the 200 Areas to be torn down in preparation for modification, and to make sure every part was in working order. At the same time, they started the plant forces work review required by the Davis-Bacon Act. The work review, normally completed before work begins, is to make sure the union shops get all the work their contract allows. But, towards the end of May, before the shops made much progress, the work review concluded that modifying the pump was not maintenance, a responsibility of the Westinghouse shops, but more like new construction, reserved for Kaiser Engineers at their shop in North Richland. With this work "turn down" in hand, Tony dispatched a truck to the 200 Areas to haul the pump back to Kaiser. Kaiser got the pump on June 12 and finished tearing it apart.

Tony had a Barrett, Haentjens, and Co. technical representative right there helping with the tear down. At about 10 PM on a Saturday night, they found a scratch on a mechanical seal. They were working 80-hour weeks and every problem seemed to appear late at night on a weekend! They feared having to order a new seal, putting the pump behind schedule, and incurring Jack Lentsch's wrath at the next morning meeting. But the Haentjens man told them not to worry, just polish out the scratch with emery paper. This simple expedient kept the pump on schedule and they survived the schedule meeting unscathed.

168 | Part Two: The Mixer Pump Era

> The pump modifications were designed and fabricated in parallel. They were using the new Westinghouse policy that allowed "prototypes" to be designed by marking up drawings to tell the shop "we want it to look something like this" without the lengthy formal review and approval procedures normally demanded for new drawings. This way, the markups were converted to formal drawings only after all the pieces came together correctly. Craig Shaw at first estimated that the pump modifications might require 20 drawings. He was low by about a factor of ten!

No matter how well founded, new requirements flowing out of Los Alamos' evolving safety assessment were seen as a constant distraction by the pump team. The design and fabrication might be 90 percent done when Los Alamos would raise a new requirement that meant tearing the pump down and rebuilding it all over again! One of the more complicated and disruptive changes that called for extraordinary ingenuity was to protect the tank against the very real possibility of the pump dropping from the crane during installation or removal. If this happened, the kinetic energy of the falling pump had to be dissipated somehow before it collapsed the tank dome or punctured the tank bottom. The riser pipe was not strong enough to hold the weight of the pump, let alone a drop, so a stout steel frame was built to distribute the load onto the reinforced concrete pump pit. A 16-inch-thick "washer" of special aluminum honeycomb material was placed on the load frame to absorb the impact load. Tony Benegas still has a chunk of the honeycomb on his desk. Finally, they fastened a flat plate on the very bottom. Pushing the flat plate through the viscous waste would help slow the falling pump, so it wouldn't hit the tank bottom so hard.

Then there were the seal plates. Los Alamos prudently wanted to maintain the tank vacuum during pump installation or removal to keep radioactive contamination and flammable gases inside the tank dome where they belonged. But because the ventilation fans couldn't hold a vacuum with the big 42-inch riser open, the area around the pump had to be sealed along its entire length. To make the seal, Tony's crew attached steel-backed rubber wiper rings to the pump column every 2 to 3 feet. At least one of these sealing plates would be inside the riser to keep it air-tight while the pump went in or out. The plates also protected workers in and around the pump pit from the shine of radiation up through the riser and, incidentally, made the pump assembly look much more massive than it really was.

The "wasteberg bumpers" were another safety feature demanded by Los Alamos. The alarming in-tank video of the big burp of Event E, showing thermocouple trees swaying and bending, created a reasonably justifiable worry that the pump itself might be damaged by the impact of a "wasteberg." Wasteberg impact was really a bit hypothetical. The video showed pipes bending in a rush of waste during a burp and showed large islands or "wastebergs" of apparently solid waste after the flow stopped. However, there was no direct evidence that the impact of a moving island caused the damage. It might have been just the overwhelming hydraulic force of the rushing slurry. Nevertheless, it would be embarrassing, after all the money and effort, if a small burp during installation were to wreck the pump before it had a chance to prove itself. So Kaiser built a cage of steel pipes around the pump column in the range of elevation where wastebergs might be lurking. It wasn't very difficult to design, but it came very late and added to the already high level of strife and turmoil.

HERE'S THE PUMP!

What did the mixer pump end up looking like after all these things were done to it? Today, the SY-101 mixer pump is standing dormant in the tank, mostly concealed by the waste. Only the large pump column with several grimy sealing rings remains visible when the camera pans around the dome. Before it was installed, the pump was seen most often lying horizontal and probably appeared even longer than it was. It was a frighteningly big, repulsively ugly, and dauntingly complicated-looking machine. It appeared to be something meant for dirty work deep in a mine. Actually, a hot, damp, muddy mine is not that much different from a waste tank and the pump would probably have been quite effective there, too.

The entire pump assembly is over 60 feet long from the base plate under the nozzles to the control box above ground. About 44 feet of it hang in the tank. The business end of the pump has two diametrically opposed 2.6-inch-diameter nozzles, with centerlines 22 inches above the tank bottom. With the pump running at full speed, 1,000 revolutions per minute, a thick slurry similar to wet concrete and 1.6 times the density of water rushes out these nozzles at over 60 feet per second (about 40 miles per hour). The two nozzles together move 2,200 gallons of waste per minute. At this rate the entire 1.1 million gallons of waste in the tank would flow through the nozzles in about 8 hours.

Tony Benegas at the bottom of the pump hanging vertical at MASF. Nozzles, seal plate, and bottom drag disk are shown.
Courtesy Carl Hanson.

Sixteen-inch diameter impeller.
Courtesy Tony Benegas.

The nozzles are fed by 4-inch downcomer pipes extending a little over 19 feet up to the pump discharge ports. Two 4-inch "dummy legs" running parallel to the downcomers form a strong, stiff four-pipe cage. The outside of the cage is about 20 inches across in the lower 12 feet, expanding to about 38 inches along the top 7 feet as the pipes bend outward to join the discharge port. This area at first looks quite complex, but a closer inspection shows each pipe and plate logically fastened together to form a relatively simple, but very strong, structure.

Pump volute and downcomers in Kaiser shop. Pump inlet is through the 10-inch pipe in line with the jack behind the "dummy leg" in the foreground.
Courtesy Tony Benegas.

The downcomers join the discharge legs just below the pump itself. This is the spiral-shaped "volute," inside which the impeller spins to make the fluid move. The pump suction is a downward-opening 16-inch-diameter hole directly under the impeller. Spinning at 1,000 rpm inside the volute, the impeller slings the waste slurry up into an annular passage around the outside of the motor to cool it before turning back downward to enter the downcomers leading the nozzles. The 150-horsepower, 480-volt electric motor is canned in a mineral oil that tolerates radiation reasonably well. The pump needed good oil because, once in the tank, there was no way to change it.

The pump, motor, and piping hangs on 27 feet of 16-inch pipe called the support column. Wiring runs from the junction boxes above ground down through the column to the motor, controls, and instruments. A constant flow of dry nitrogen purge gas keeps any moisture or flammable gas from accumulating there. The aluminum honeycomb crush plate is fastened around the column about halfway up. The column also carries a series of bumper plates with nylon pads to keep the pump column centered in the riser as it descends the last 12 feet into the tank.

The pump column must be rigidly fixed to the riser, essentially becoming a structural part of the tank. But the radiation dose rate in

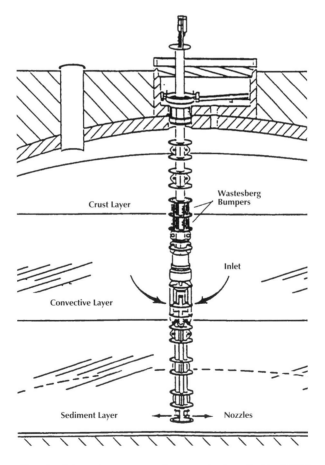

Sketch of the mixer pump as it was to be installed in SY101.

the pump pit is too high for workers to do lengthy tasks like welding or turning nuts onto bolts. So the pump is locked in remotely by an expanding ring of hydraulic brake shoes attached just below the crush plate. Under hydraulic pressure, the brake shoes push out against the riser walls and hold the pump as tightly and as accurately as a ring of bolts. It also makes removing the pump much easier, should it ever be necessary.

TESTING THE PUMP

The pump modifications were essentially complete by the end of July 1992, but one vital question remained to be answered. Would it all work as intended? Was some fatal design flaw buried in the tangle of pipes and wires? If there was, they had to find out and fix it before the pump went into the tank. The pump would have to demonstrate its ability to pump a liquid through all its pipes and passages. The control system needed to prove that it could start and stop the motor when called upon. The schedule for this testing was as relentless as it had been for the modifications. All knew that the tank had its own schedule for Event G that would hopefully open the window for mitigation.

First of all, where can you test a 150-horsepower, 2,000-gallon-per-minute pump enmeshed in a 60-foot-long column of pipes, plates, tubes, and wires? In May, before the Westinghouse 200 Areas shops turned down the pump modification work, Carl Hanson planned to test the pump in their pump pit, even though a hole would have had to be cut into the shop roof so that the tall pump would fit. When the turndown came and the work went to Kaiser Engineers, a different test site was necessary. Tony found an ideal location at the Fast Flux Test Facility. It was a massive, climate-controlled warehouse affair called the Maintenance and Storage Facility (MASF, appropriately pronounced "massif"). It was wide, tall, and comfortable and came with its own crew of engineers who had nothing much else to do with Fast Flux Test Facility shut down.

But the Fast Flux Test Facility crew were still in denial, believing their liquid metal reactor still had a mission. They were used to a methodical, step-by-step operation and were appalled, maybe even frightened, by the extremely tight schedule and the design-as-you-build, build-as-you-test style of the SY-101 mitigation team. Accordingly, they decided they could not support the pump tests and

174 | Part Two: The Mixer Pump Era

began to walk away. But "no" was exactly the wrong answer. Steve Marchetti recalibrated their attitude and wrenched them firmly into the reality of SY-101 mitigation.

Though Marchetti had to compel the Fast Flux Test Facility crew to take on the testing, they quickly proved their value. They forced more structure and formality on the job that served the later mixing tests in the tank extremely well. If it weren't for the pump installation exercises at the MASF, the actual installation would not have been nearly so smooth and quick.

By June 1, 1992, a formal plan was issued for the pump tests at the MASF. The tests would check the basic operation of the pump and the function of the instruments and controls with the pump submerged in water. Two major sessions would expose the pump to situations identical to that inside SY-101 minus the radiation. Basic operation of the pump, instruments, and controls was to be confirmed during August after which the pump would go back to the Kaiser shop briefly for some repairs and additional modifications. A longer series of tests would be done in September.

By early August, the pump was at the MASF having its wires hooked up. Doug Larsen, who had come to Hanford in the spring of 1992 from a background of industrial flammability and explosion prevention, was placed in charge of the testing. Doug wisely got right down to basics and asked, "OK, how do you turn it on?" He first asked Jack Lentsch, who referred him to Tony Benegas, who told him to call Tom Lopez, who passed him off to Ray Merriman, who suggested that Jack would know someone who knew! *Nobody* knew how to start the pump! Everyone was anxiously tracing out the circuit logic that was supposed to turn the pump *off* in response to the growing list of alarms and controls. They all assumed that someone else had taken care of the start-up procedure. It turned out that only one person, a contractor from H&N Electric in Pasco, knew how to start the pump! It was turned on the first time early in the second week of August.

About 8 PM one evening, everything had been hooked up and the crew was sitting around the MASF reviewing notes and talking over procedures. After a while there was nothing else left to do. Doug looked around and, getting a nod from each person, concluded, "I guess we're probably ready to start the pump." Working from the start-up crib notes from H&N Electric, they closed the main circuit breaker and hit the start button. All the lights and dials on the panel came to life, but nothing happened with the pump. Nothing at all! Nobody knew what the pump would act like when it started. A few climbed down and bent over the pump listening, but they didn't hear anything. They eventually concluded the pump had not started.

Doug crawled all over the pump tracing the wiring trying to find why nothing happened. By about 11 PM, he had found two safety disconnects downstream from the main breaker that were not mentioned in the crib notes. Both were still open, as they should be. No wonder the pump didn't make any noise! About midnight, after closing the two disconnects, Doug repeated the start procedure. This time they were rewarded with an accelerating throbbing hum of power. According to Doug, "It was like a powerful magnet or some other giant thing powering up." There was now no doubt the pump was running!

A crisis emerged immediately after that first run. To do the prescribed cold tests, where the pump was placed in a water tank, the operators had to monitor several pressures, temperatures, and flow rates to satisfy safety controls and to make sure the pump was running according to plan. But the gauges weren't working right! They began to give strange readings and drop off line one by one.

This problem was critical to the schedule and attracted all kinds of "help." Tom Lopez opened a junction box and tried to find the faulty wiring, while numerous managers and senior engineers crowded around, pointing and offering suggestions. Tony saw what was happening, elbowed his way into the melee, and cleared the area for Tom. Firmly but politely, Tony said, "We appreciate your help, it is very useful, but we can only stick so many heads in this box at one time. So please move back behind the barrier." This command hurt some feelings, and Tony still winces at it, but it saved Lopez' sanity and allowed him to make some progress.

Even with room to work, Tom could find no quick answers, so they called out PNNL instrument expert Mike White to run some diagnostic tests. He had not yet been exposed to the SY-101 teamwork and was amazed to find none of the usual Westinghouse versus PNNL rivalry during the operation. Almost no one even realized he was from PNNL. Regardless of the good feelings, Mike couldn't find the problem either. By now all the instruments had died. In desperation, Jack ordered the pump pulled out of the MASF and sent back to the Kaiser shop, where it laid it on its side for an urgent diagnostic deconstruction.

When Kaiser pulled the covers off some of the electric junction boxes that had been submerged in the test tank, water poured out! Now they knew exactly why the instruments weren't working! Because of the tight schedule, the original caulking on the junction box covers had not had time to dry before getting dunked in the water tank. To make sure it would work next time, they added gaskets on the junction box covers and carefully re-caulked. Despite the schedule, they waited to be sure the caulking dried really well. Kaiser had to work through the Labor Day holiday to do all this, but they received hearty thanks from Westinghouse and DOE Richland for the trouble.

In the meantime, another testing enterprise was happening in the 200 Areas. The Cold Test Facility had been set up in the sagebrush between the 200 East and 200 West Areas earlier that summer to practice raising the pump vertical from its cradle and lowering it into a mockup tank. A dummy pump was built that had the same dimensions and weight as the real thing but not the full detail. The dummy pump fit into a dummy riser that had the correct dimensions and general configuration. There operators could walk through procedures with real hardware to confirm the correct choreography. More important, the crane operators and rigging crew could practice swinging the massive load to precisely the right position. They could practice emergency pump removal too, just in case something didn't go according to plan. Things like the "flexible receiver," a big plastic sack meant to encase the pump if it had to be pulled out after being dipped in the waste, could be tested to see if it fit. The Cold Test Facility saw service before each of the three windows planned for pump installation: G, H and I. The steel scaffolding still stands as one of the many monuments to SY-101 mitigation. All it lacks is a historical plaque!

When the pump got back out to the MASF, it ran fine and most of the instruments were giving good readings. But, to Tony's disgust, the flow meters still wouldn't work. These were new, technologically advanced devices that used the Doppler shift of ultrasonic waves bouncing off tiny particles in the fluid to measure the flow speed. The Doppler effect is what makes the sound of an airplane's engine drop in pitch as it flies overhead. The flow meters only worked when nitrogen bubbles were injected. Apparently this gave the ultrasonic waves something to "see." They were originally intended to measure sewage flow in treatment plants, where the sound waves would be bouncing off bigger chunks with lots of bubbles.

Even after a factory rep came out to re-tune the flow meters for a slurry, they never gave a good measure of the flow. At Jack Lentsch's suggestion, proven pressure-based flow meters from the Barton company were installed on each nozzle. The "Bartons" provided a reliable if prosaic flow measurement for cold testing and remained useful throughout the life of the pump.

> Though the ultrasonic flow meters were not removed and their output continued to be recorded, they were never more than a rough indication of flow. However, by some interesting quirk of physics, their reading appeared to be sensitive to the amount of gas in the slurry passing through the pump. This set off much conjecture during the initial tank mixing tests and the first year of nominal operation as the flow meter indications rose and fell with changing tank conditions. We never did explain the effect or correlate it to tank data in any satisfying way.

Despite the various setbacks, the team now had the entire system operating. DOE Richland was pleased. On August 26, 1992, mitigation project monitor Ron Gerton sent a formal letter to Westinghouse President Anderson expressing pleasure with the progress of the mixer pump cold tests. He was especially encouraged by the creation of the Westinghouse, Los Alamos, PNNL, Kaiser Engineers team that was working so hard to get the testing completed in time for installation.

Because there had been so much trouble with instruments, Doug Larsen *insisted*, in the face of crushing schedule pressure, that the new DACS trailer that Los Alamos had procured from the Nevada nuclear test site be set up at the MASF for the second cold test. There they could hook up every gauge and gizmo pretty much the way they would be at the tank, and wring out the entire system in enclosed, air-conditioned comfort. Any trouble discovered at the MASF could be repaired much more efficiently than in the tank farm. When the pump actually went in the tank, the Los Alamos safety assessment allowed only 48 hours before it had to start up. If some instrument or control circuit went bad at the tank, fixing it would be an emergency with the whole world looking on with great interest. In the worst case, delaying the start up until after the next burp could subject the pump to damage, thereby subjecting the project to much embarrassment, and Westinghouse to contract curtailment. Thanks to Doug's resolve on the cold test, there were no serious surprises, and the crew started the motor well within the required 48 hours.

Cold testing did reveal one unexpected shortcoming in the pump's capability. On September 17, 1992, a report of initial cold test performance results stated that the pump motor oil temperature rose higher than expected! Based on the trend of increasing oil temperature during the longer tests with water, it was apparent that the pump would not be able to run continuously at its maximum speed of 1,200 revolutions per minute (rpm) at the current waste temperature. The motor speed would have to be backed way off to

Testing the Flexible Receiver, intended to bag a contaminated pump in case of emergency removal, at the Cold Test Facility.
Courtesy Carl Hanson.

Mixer pump being lowered into the test tank at MASF. Motor at bottom, support and brake shoes above last seal plate.
Courtesy Carl Hanson.

keep the oil temperature below the limit of 200 degrees F (later reduced to 195 degrees F). Even at 880 rpm the oil temperature hit the limit after many hours in the cold test. Apparently, it was pumping cooler water during full-speed shop-testing at Barrett and Haentjens.

Running the pump in the hot gooey waste a year later quickly showed that, first of all, the pump could not even be run at 1,200 rpm. The electric current in the motor hit the limit of 210 amps before reaching that speed. The maximum practical speed was actually 1,000 rpm, at which the pump had only 25 minutes before the oil temperature hit 195 degrees F. It took at least a day to cool back down for another run, limiting the pump schedule to three or four 25-minute runs per week. At lower speed, the pump could run a little longer—40 minutes at 920 rpm and an hour at 750 rpm. But lower speeds did not do enough mixing to be useful, even if longer runs were possible.

Even with these limitations, the pump proved more than capable of mitigating the tank. Running only three times per week for 25 minutes at 1,000 rpm, it kept the waste mixed and prevented SY-101 from burping for 6 years. This is only 0.7 percent of continuous operation.

Cold testing was finally completed on September 19, 1992, and the pump was ready to go out to the 200 West Area and make its grand entrance into the tank with hundreds of people in attendance. Instead it waited in the MASF for the drama of Window G to play out. Then it had to wait through the winter while the embarrassment of Window H concluded. The winter of 1992-93 was cold and snowy. Fortunately, the MASF was nice and warm.

THE SAFETY ANALYSIS

Los Alamos National Laboratory began working on the mixer pump safety analysis, which I shall call the "SA" as it was then, immediately after getting the job at the Fiasco in Pasco meeting in February 1992. Harold Sullivan drafted 15 to 20 good people for the SA who worked every bit as hard as the crew designing and building the pump. Of the whole Los Alamos crew, three individuals became most familiar to the Hanford team. Jack Edwards was the SA project manager who carefully protected Los Alamos' scope of work and acted as the primary spokesman in presentations to the TAP and DOE. Bob White came to Hanford full time as the technical leader of the analysis and creation of the actual document. Kemal Pasamehmetoglu was the third horse in the Los Alamos "troika" that pulled the SA into shape. He actually did a great deal of the analysis and developed the concepts and strategies for the controls. Kemal would trade off with Jack Edwards, working alternating two-week shifts at Hanford.

Bob White's full beard, broad shoulders, and powerful belly would have served him well in a renegade motorcycle gang. Bob, in fact, enjoyed riding his big bike around New Mexico, but the fearsome exterior concealed a gentle nature with a heart of gold. Though Bob could be tenacious and firm when seeking knowledge or defending a point of safety, he was as kind and helpful a colleague as one could find anywhere.

He was a well-trained worrywart and a self-styled "safety weenie" from the Tennessee Valley Authority nuclear power plants, where he acquired his southern drawl. Working through the fallout from the cable tray fires at the Browns Ferry plant in the 1970s made Bob very leery of radioactive contamination.

Bob did most of the writing for the SA. According to Bob, many of his contributors were either foreign speakers whose English was pretty rough or scientists who wrote in scientific jargon that nobody could understand. Some just plain "couldn't write worth beans." Many were "prima donnas" who were hard to guide and not always in agreement with each other. Bob had to not only herd these people around so that they were saying the same thing, but also translate their writing into good technical English, which he did very well indeed.

The SA was a huge responsibility for Los Alamos, and they took it very seriously. They had to find a way to guarantee that the mixer pump could be installed and run without hurting people or spilling radioactive contamination. They faced the same challenge as Bob Marusich had stated in 1990, "You had to be conservative because you didn't *know* anything, but you couldn't be too conservative or you couldn't *do* anything." Los Alamos quickly rediscovered the truth of Bob's statement.

Standard SA methods showed it was too dangerous to do anything in the tank. But nobody knew enough to discount a possibility or back off on conservatism. The only recourse was to assume that the tank domespace was always flammable and use design features and work controls to prevent ignition, if possible, or to withstand a burn if not. This necessity led Los Alamos to propose what seemed like an endless series of controls and design modifications to prevent or mitigate what many thought were absurdly improbable accident scenarios. It was a continuing challenge to Jack Edwards and Jack Lentsch to keep an amicable working relationship. Sometimes Edwards had to stand firm and insist on a control, and sometimes Lentsch had to stand firm and refuse one. Somehow both were able to find a way to accommodate these conflicts within their technical and professional integrity.

One of the more plausible accident scenarios Los Alamos dealt with was the "pump drop" in which the pump dropped from the crane and punched a hole in the tank bottom. Several mixer pump design changes to counteract this possibility were described earlier. But it also made sense to try and prevent a drop in the first place. The first step was to confront the crane and rigging crew with it to see if they had any good ideas. Tony Benegas brought in expert riggers from the Pasco-based Lampson Company, who did really big and difficult crane jobs all over the world, to consult with the Hanford people. The Westinghouse rigging guys were really upset, even insulted, that anyone would even consider a pump drop. Riggers are a very small, tight community where everyone knows the other. The Lampson folks were in a tight spot. They didn't want to criticize fellow riggers, but they had to agree that pump drop was not impossible. After that realization sunk in, the group got down to work and began brainstorming things that could be done to beef up the rigging and anticipate the unexpected. This led to the "critical lift procedures" that guide all Hanford crane and rigging operations to this day.

One of the more extreme and improbable scenarios was the "pump-ejection accident." In this hypothetical event, a big gas release during installation happens to get ignited. The resulting pressure pulse blasts the pump out the riser like a Poseidon missile launch from a submarine. After shooting high in the air, the 20-ton pump falls right back onto the tank and collapses the dome if the gas burn hasn't already done so. To prevent this horrific possibility, Los Alamos proposed to hook a long, thick cable to the pump to swing it away from the tank during its hypothetical flight. Jack Lentsch simply couldn't stomach such a bizarre idea and quickly quenched the proposal.

But a great many other worry-driven modifications did get built in. The seal plates, wasteberg bumpers, and impact limiters have already been described with the story of building the pump. Another was the "Giant Nickel," a blow-off riser cover named for *The Giant Nickel*, a bi-weekly paper of classified ads published in Kennewick and distributed in supermarket doorways. The blow-off riser cover was just a big flat steel disk with a counterweight that would flop open in a tank burn, presumably letting enough of the hot gas escape through the 42-inch riser to keep the dome pressure below the failure limit. The Giant Nickel was finally installed in Window H, February 1993. By early 1994, it evolved into the "multi-port riser" that mounted the blow-off panels sideways to allow other equipment to occupy the big riser. A year later, better burn calculations showed that tank pressurization was too fast for blow-off panels to do any good. The Giant Nickel still lies rusting on the gravel in the SY farm.

In the hurry to stay on schedule, dumb things sometimes happen. The "Giant Nickel" blow-off plate had to be bolted to a concrete pad. The engineer doing the design specified a huge concrete pad 5 feet thick weighing many tons. It was much thicker than necessary and far too heavy. As the concrete was being poured, the engineer went out to see how it looked, only to be confronted by a skeptical Dan Niebuhr.

Dan asked if he had calculated the weight of the huge pad and whether it would be too heavy for the tank dome. The engineer explained that, because the concrete only replaced an equal volume of dirt removed, the net weight gain was small. He had calculated the density of concrete from the densities of water, sand, and Portland cement using his handbooks. Dan called Central Concrete to get the correct density from the people who mixed it for a living. The answer was much higher than the calculation! The pad really was far too heavy! But it was also too late, the pour was done. They had to go back and recalculate the dome load and move some other equipment to make sure it wouldn't collapse.

The pump column seal plates, already described, also seemed like a good idea, but actually created a severe hazard of their own.

When the pump went in, many of the seal plates submerged in the waste. If the pump were removed, truckloads of radioactive waste piled on the plates would be a huge source of contamination. True, a high-pressure water spray was provided to decontaminate the pump and would normally wash waste off the plates as it came out. But pump removal was also subject to the pump ejection accident where a flammable gas burn shot the contaminated pump out of the tank before the sprays could do their work. The large mass of waste carried out of the tank on the plates made this accident by far the worst of all the accident scenarios in the SA and placed the most restrictive controls on pump operation. But by the time these unexpected consequences came to light, it was too late to remove the plates.

Los Alamos was also worried about pieces of hardware breaking off and causing sparks that might ignite the gas in a burp induced by the mixer pump or a big earthquake. They also suggested that the mixer pump jets might make the long pipes hanging in the waste swing back and forth like a "sweeper" log in a river current, eventually breaking by fatigue. To prevent these hazards they wanted to have strain gauges glued on each item and shut off the pump automatically if the strain got too high. A strain gauge is a group of three little patches of fine wire arranged in a triangle. When the material they are glued to stretches under stress, the tiny wires in the strain gauge stretch along with it. The resulting change in electrical resistance can be read on a voltmeter.

The strain gauges were an endless source of grief. Occasionally the strain gauges showed the proper indications, such as an MIT bending gently away from the pump when the jet was aimed at it. But more often they would show high readings or even alarms for no good reason at all. On these occasions, Jack had to call an emergency safety review meeting to consider the evidence which most often showed that water from precipitation or condensation somehow got in and shorted out the wiring. Nevertheless, controls and alarms based on strains stayed in force for years. In about 1996, they were finally eliminated as much from pure weariness as from common sense.

NEW GIZMOS

Working on the SA and having to make so many sweeping assumptions fostered a longing for more knowledge about the waste, what it was doing, and what the pump might do to it. Accordingly, consistent with their scientific bent, Los Alamos constantly pushed the project to put in new instruments. The MIT, one of the most useful instruments ever, was proposed, designed, and built by Los Alamos. The first was installed in SY-101 in Window G and a second in Window I. Several more were put in other burping tanks. Other Los Alamos brainchildren included improved gas monitoring methods, the velocity-density-temperature trees (VDTTs), and calculating gas volume from barometric pressure effects.

Los Alamos super-scientist Herb Fry introduced improved gas monitoring capabilities for SY-101 that revolutionized gas concentration measurement in all tanks at Hanford. Herb was an expert from the nuclear weapons testing group at Los Alamos. Up to 1992, hydrogen concentrations had been measured with relatively insensitive Whitaker electrochemical cells and a Teledyne thermal conductivity probe in the vent header. An online mass spectrometer of doubtful validity and low reliability measured hydrogen and other gases. New metal oxide semiconductor hydrogen monitors were tried in mid-1992 but never worked because of interference from other gases.

Gas concentration measurements were absolutely vital for the mixer pump tests. Fry pushed gas chromatographs (GC) that could measure hydrogen, methane, and nitrous oxide concentrations to within a few parts per million, and the Fourier transform infrared spectrometer, or FTIR, that did the same thing for ammonia and nitrous oxide. They eventually proved to be excellent instruments, but there were "growing pains." The FTIR software often crashed, requiring someone to notice the loss of data and restart the system. The GC automatically calibrated itself by sucking gas from a standard bottle at intervals. But the software recorded the calibrations as normal data that analysts had to somehow sort out. The GC power supply shut down frequently, and it took a week to stabilize after an outage. It was a very rare thing to have all the instruments operating simultaneously!

The FTIR was extremely useful. It generated a spectrum of data that theoretically showed the concentrations of all gases that could be detected with infrared waves. The huge spectrum was recorded and stored at each reading, though the software in the equipment on SY-101 was programmed to read only ammonia and nitrous oxide. Herb Fry used the full spectrum on one occasion to search for particularly toxic "nasty gases" to satisfy one of Bob White's worries.

Much later PNNL scientist Sam Bryan took a look at the spectra to see what might be coming out of the waste during dilution. In neither case was anything noteworthy discovered, which was probably noteworthy in itself.

The FTIR gave the first good data on ammonia concentrations during the last big gas release, Event I. The measured ammonia concentrations were quite a bit higher than assumed in the SA. The resulting alarm required workers in the SY farm to be on self-contained breathing apparatus during pump installation and access roads to be blocked within a half-mile radius. This was the birth of the "ammonia issue" that plagued Hanford until mid-1995 and still echoes to this day!

Because of their demonstrated importance in SY-101, Westinghouse created a sort of in-house industry to develop, install, and care for gas-monitoring instruments on other tanks. The Standard Hydrogen Monitoring Systems or SHMS (pronounced "shims"), used Hanford-wide for the next 10 years, grew from the systems first set up on SY-101. Roger Bauer set up a "SHMS-R-Us" shop in the 300 Area with benches and plumbing to do calibrations with bottles of precise standard gas mixtures of the gases of interest. Tom Schneider and Eric Straalsund strove tirelessly to make SHMS work and keep them working.

One of the most ambitious designs was the VDTT built by Kaiser Engineers at Los Alamos's instigation. By monitoring the waste velocity, density, and temperature continuously during a burp with sensors spaced every 3 feet up the tree, they could confirm Rudy Allemann's gob theory or discover a new burp mechanism. When the pump went in, the VDTTs would be able to monitor mixing and evaluate its success or failure. The initial SA actually had controls based on VDTT data for waste density and velocity.

Unfortunately, the VDTTs did not live up to their expectations. Not all the sensors were installed in the first place because there was not enough room in the 3-inch pipe for all the wires! The velocity sensors were never turned on. They were designed to be used in fresh air, thrust into rivers and creeks on a stick by squinty-eyed ecologists to measure the flow. They could not be certified sufficiently spark-free for a hydrogen atmosphere.

Two VDTTs were installed during Window G and gave some inconclusive information during the small Event H burp in February 1993. But even by Event H the sensors had begun failing, and by

Event I, they were all essentially dead. They never saw what the pump could do. The VDTTs failed because the velocity sensors used aluminum to seal their sapphire windows from the waste. Unfortunately, the aluminum seals dissolved in the sodium hydroxide-laden waste, destroying all the electronics inside. In the rush to Window G, no tests were done to confirm that the design materials would withstand the waste.

> The best piece of information the VDTTs supplied was an approximate waste level indicated by the white scale rings painted every 2 inches on their exterior. They had become multi-million-dollar dipsticks! But it was not such a joke as it first seemed. A few years later, dipstick measurements of the waste level would be of great value in assessing the urgency of SY-101's surface level rise problem.

Los Alamos also initiated a valuable and far-reaching measurement concept that didn't even involve hardware. At the May 21, 1992, TAP meeting they suggested that because gas bubbles would change volume with pressure, it might be possible to calculate the total gas volume in the waste by measuring the change in waste level that accompanied variations in barometric pressure. The information was free because all tanks had a level measurement and the Hanford Meteorological Station supplied barometric pressure data. PNNL statistician Barry Wise and Dan Reynolds both immediately tried this and, in Jack Lentsch's June monthly report, stated that they found a correlation between pressure and waste level. Initial calculations indicated that SY-101 might contain as much as 18,000 cubic feet of gas. This was quite a bit more than most thought could possibly be there. The frequent large burps with the gas re-accumulating between them made any retained gas volume estimate a moving target. After the pump went in, the circulating gas bubbles during and after mixing gave the calculation too high an uncertainty to be useful.

> Barry Wise thought one reason the barometric pressure calculations were so difficult might be the randomness of the natural meteorological pressure changes. He believed that he could produce a much more precise gas volume estimate if the

188 | Part Two: The Mixer Pump Era

> tank domespace were pressurized and pulled back to a vacuum in some kind of controlled fashion. Understandably, nobody was willing to deliberately pressurize the tank against the strict safety controls to maintain a vacuum. Besides, Steve Eisenhawer of Los Alamos realized that the barometric pressure changes in a typical weather system are much greater than could ever be provided by tweaking the ventilation system. However, Barry's idea would have allowed measurements in the summer when storms were generally absent and natural pressure changes were minimal.

Though the barometric pressure effect method was never very useful in SY-101, it became a standard for estimating trapped gas volume in tanks that did not burp or burped infrequently. In 1994, another PNNL statistician, Paul Whitney, applied the barometric pressure effect method to show that a waste level drop did not necessarily mean a tank leak requiring emergency pumping. If the level drop happened in a period of high pressure and went back up again during a period of low pressure, it was just gas expanding and contracting. In 1995, Paul turned the method into a science by applying sophisticated statistical techniques to correlate the waste level and barometric pressure to actually measure the trapped gas volume. In 1996, using an improved version of the method and better data, he could even estimate the waste strength! About this time, direct in situ measurements of gas fractions in the five burping double-shell tanks confirmed that the barometric pressure effect estimates were pretty good. It became the mainline method for estimating gas volume at Hanford and was soon exported to the Savannah River Site and even the United Kingdom!

> Sometimes a good thing gets pushed a little too far. In 1996 Westinghouse applied the barometric pressure effect method in an attempt to find the trapped gas volume in all the waste tanks. They discovered that some tanks apparently had a lot of gas. One thing led to another and, after much laborious calculation and soul-wrenching decisions, Westinghouse formally proposed that 32 tanks be added to the flammable gas watch list! This sent up a lot of red flags with DOE Richland and other Hanford folks. So, before this proposal was sent beyond the Site's boundaries, Don Vieth of DOE Richland convened an expert panel to review the

method and the data. The panel had Westinghouse, Los Alamos, and PNNL scientists and engineers, including me, along with some TAP members.

Vieth tried to kill the watch list proposal by discrediting the barometric pressure effect method as too simple to be of any value. That strategy failed, though he accomplished his primary objective. The review panel could not deny the basic fact of the barometric pressure effect on waste level or that the level change resulted from gas in the waste, but they found problems with the application. The level readings in many tanks were too imprecise and infrequent to correlate accurately with barometric pressure. In many old single-shell tanks, a level gauge near the warmer and softer center might show more gas than one on the hard crusty waste near the wall. At the same time, Westinghouse had applied a series of conservative assumptions to the gas volume and flammability calculations that over-stated the hazard beyond reason. Thus the watch list proposal died, as it had to. The barometric pressure effect method survived, though with a better understanding of its limitations.

Los Alamos also supplied a unique kind of central location to monitor the data being recorded in the tank and to control the mixer pump. They had found an excess control trailer at the Nevada Test Site that was built to support underground nuclear weapons testing. The trailer was strong enough to withstand being rolled across the desert and sealed tightly enough to keep the inside clean in the middle of a puff of radioactive dust. If you examined the window sills closely, you could see the stress cracks left over from the shocks of the underground explosions. It was ideal for use next to a big tank full of radioactive waste that might explode at any time. Jack Lentsch and Carl Hanson hoped that the DACS trailer would be entirely "plug and play." If they could simply plug the pump into a portable control trailer, they could avoid the frustration of building one at Hanford.

Carl Hanson and George Vargo traveled to Los Alamos on February 28, 1992, shortly after the Fiasco in Pasco, to meet with Charlie Hatcher and his crew to kick off the DACS task. Los Alamos had two complaints right off. The schedule was accelerated far more than they expected, and Westinghouse had not as yet told them what they expected the DACS to do. Like the Fast Flux Test Facility staff had been earlier, Los Alamos people were not used to the intensity of the mitigation project schedule and had to be "recalibrated." No doubt George Vargo turned his used-car salesman enthusiasm all the way up, and Carl applied his most gentle but irresistible persuasiveness. By the end of the meeting, they along with Charlie Hatcher had drafted functional criteria for the DACS and agreed on a very fast schedule to make it ready.

A few days later on March 3, 1992, a full crew of Westinghouse staff visited Los Alamos and covered all of the technical details with Charlie and his crew. The project wanted the basic control trailer operational at Hanford in October, with fully functional DACS capability by March 1993 in time for Window H. When DOE demanded the pump be operational in Window G, the DACS fit-up had to be accelerated to match. The Window G schedule showed the DACS being installed and the pump plugged in by September 17. Los Alamos met the schedule, and the trailer became part of the scenery of the SY farm. In January 1993, the DACS was one of several features Westinghouse chose to show to Oregon Public Broadcasting to emphasize the positive achievements of SY-101 mitigation. Hiroshi Hoida was the point man for getting the DACS trailer delivered on time and within budget. He wound up making 26 trips to Hanford to get the work done.

> The Los Alamos people were issued clip-on film badges to record the radiation dose they might get at Hanford. The badges soon showed a much higher dose than onsite workers were getting! The radiation safety staff wanted to know where they were going and what they were doing to get the extra radiation exposure. It turned out that the badges got a high dose on the plane flying back and forth, and the background radiation was higher at Los Alamos, located 7,000 feet above sea level. Eventually, their badges were taken away.

Today, the DACS trailer would look quite primitive, but it was fairly modern in 1992. It was bright and shiny and had a very substantial look one would associate with weapons testing. You entered from a narrow dirt path between the fence and the trailer, up metal stairs, and through one of two doors. The control room on the east end was a large curved console of computer screens and terminals. One moderate-size window looked through the chain-link fence to

TOP: DACS trailer being placed next to the SY Farm fence during Window G, September 1992. Control room section is on the right. BOTTOM: Control room and operator in the DACS trailer shortly after installation.

192 | Part Two: The Mixer Pump Era

SY-101. Someone replaced the trackballs at the terminals with billiard balls matching the number of the workstation. The rest of the building had a central floor-to-ceiling rack packed with computers, modems, printers, instrument controllers, and any other piece of electrical equipment that needed to be there. On the north wall there were shelves, cabinets, and drawers for spares, manuals, tools, and anything needed for repairs or safety response. The west end had a little work bench meant for a technician to work on whatever might need fixing. There was no restroom or any other amenities. It wasn't much, but it was home.

> How can a 1-ounce mouse shut down a 10-ton mixer pump? It could happen at Hanford. Mice have a phenomenal ability to get into tightly sealed buildings. Hanford mice and the DACS trailer are no exception. This would have been only a nuisance except that the seeds and bugs the mice ate had been growing in or crawling on contaminated soil. Thus the mice and their leavings were radioactive. After a while the DACS trailer became sufficiently contaminated to be a concern, and Tank Farms workers had to do a thorough cleaning and rodent control exercise. While no mixer pump runs were delayed, this issue did make the front page of the *Tri-City Herald*. This problem might be the worst barrier to restarting the mixer pump today.

WHAT DID THE SA ACTUALLY SAY?

Jack Edwards first outlined the draft SA to a TAP meeting August 19, 1992. It was just about done, and the mitigation team had agreed on the basic concepts and control framework. They hoped the TAP would also. The primary hazard the SA had to deal with was the potentially large volume of flammable gas that the tank could release at any time. Los Alamos had to assume that a burp could happen at any time, although tank history showed that a big one was less likely right after a preceding one. Nobody knew enough as yet to say what would or would not trigger a gas release. Any waste disturbance might do it, from punching a hole in the waste to installing the pump, down to even small vibrations transmitted to the waste down the pump column. Running the pump was the stimulus most likely to cause a burp. After all, mixing was supposed to release gas. For this reason, the SA controls described tank conditions necessary to start a pump run and events that required a pump run to stop.

Los Alamos reasonably assumed that a longer pump run at higher speed would be more likely to cause a larger gas release. At the same time, tank history told them that the higher the waste level, the more probable and the larger a burp might be. To combine these two concepts, Los Alamos depended on simulations with Don Trent's TEMPEST computer program, to find the bounding gas release volume for any given recipe of waste level, pump speed, and pump operating duration. The maximum allowable waste level for a specific pump run was that which caused the maximum allowable gas release as calculated by TEMPEST. Los Alamos set the maximum allowable gas release, or "maximum safe burp," as 13,100 cubic feet. The "maximum expected burp" of 10,500 cubic feet, corresponding to the biggest "natural" burps recorded in Events E and G. A simplified but conservative calculation showed that a 13,100-cubic-foot gas burn would damage the tank, but that the damage would be "acceptable," but a 20 percent larger burn would begin to tear the dome.

In 1994, hoping to relax the waste level controls, Los Alamos engineers Jack Travis and Ed Rodriguez did a very sophisticated and detailed computer simulation of the maximum burp accident. Their simulation tied the tank pressure pulse from a burn to the dynamic action of the tank structure and the soil surrounding it. It took a very long time, even using Los Alamos' monster supercomputers, and produced some unexpected results. First, the simulated pressure spike from the gas burn was too fast for the "Giant Nickel" blow-off riser cover or the blow-off panels on the new-fangled multi-port riser to relieve effectively. This result would have saved the project several million dollars, had the calculation been done 2 years earlier.

Second, the same pressure spike suddenly inflated the tank dome and tossed a huge load of overlying soil into the air. The impact of all this weight falling back on the dome caused more damage than the pressurization. This surprise required not only reducing the maximum allowable burp to 8,650 cubic feet, but it also led to a new definition of "allowable." A burn of 8,650 cubic feet of gas, and the concurrent soil re-impact, would crack

> the concrete, but the steel rebar, the dome structure, and steel shell would stay intact. After the burn, the tank would be weakened but still leak-tight. However, just a 7 percent bigger burn of 9,230 cubic feet would toss the gravel high enough to collapse the dome when it fell back.

Placing TEMPEST in the position of supplying actual numbers for a formal SA had always made Don Trent and Tom Michener very nervous, besides putting them under excruciating pressure to produce results. TEMPEST had some fundamental limitations. The most compromising of these was that gas never left the waste, bubbles just moved around. So, to simulate a gas release, they had to assume that bubbles rising above a set elevation called the "accounting plane" were released. Not only that, but the waste properties were unknown and had to be assumed. Tom said, "Sometimes we [assumed it was like] peanut butter, sometimes mayonnaise, sometimes yogurt." Under these conditions TEMPEST could do a series of simulations with different inputs that revealed a lot of *qualitative* information about tank behavior. But Los Alamos wanted hard *quantitative* predictions. So when Kemal Pasamehmetoglu wanted to know how big a burp would be if the waste had 10 percent gas instead of 8 percent, Tom was aghast. TEMPEST calculations just couldn't distinguish such small differences. But Kemal was writing the results down as fact, believing that as long as the calculations were conservative, they'd be OK.

Given the shortcomings of the calculations on which the burp controls were based, everyone had to admit that gas releases might happen despite the waste level limits and pump operating restrictions. Accordingly, Los Alamos set up another set of controls aimed at preventing the released gas from igniting. The key flammability control was the maximum domespace hydrogen concentration, initially set at 0.75 percent (7,500 parts per million), less than one-fourth of the minimum flammable hydrogen concentration. When the hydrogen concentration exceeded this maximum, pump runs, equipment installation, or any other work in the tank or around an open riser had to stop. Additional controls also specified a minimum ventilation flow rate and tank vacuum, and required electrical bonding of tools, riser covers, and instruments to prevent static charge buildup. These controls eventually evolved into the Ignition Control Set now applied to all tanks as part of the tank farm safety authorization basis.

Though there were many unknowns, historical experience gave some idea of what tank behavior to expect. But there was no good theory or data to forecast what the tank might do if conditions departed from historic norms. This prompted the Los Alamos team, and Bob White in particular, to set controls limiting waste temperature, for example, to a range between the historical minimum and maximum temperatures recorded in the tank. If kept within historic bounds, the tank should burp within historic bounds. After the mixer pump began operating regularly, similar worries arose about a long-term change in particle size and the increased uniformity of the waste. This prompted a control requiring the hydrogen concentration to be greater than 10 parts per million before starting the pump, as Los Alamos feared that a low concentration might indicate unexpected gas retention.

Besides possible erosion of the tank wall or floor by the mixer pump jet, or metal fatigue in suspended hardware that might be swinging back and forth in the mixing-induced current, one of the most contentious concerns was removing too much gas from the tank! This seemed contrary to the very reason the mixer pump was put in the tank. Nevertheless, Bob White demanded a control to stop running the mixer pump, except for short "bumps" to keep the nozzles clear, when the waste level went below 400 inches!

The first test of this control came in early 1994 when the waste level fell to 398 inches, and the control threatened to delay full-scale testing. The SA-mandated Test Review Group (TRG), of which Bob was a member, agreed to relax the limit down to 398 inches. The tank obliged by raising its level back to 399, so everyone was happy.

Later everyone apparently realized that years of mixing really had changed the waste, maybe irreversibly, but almost certainly for the better. The search for potentially hazardous long-term effects was over. But nobody caught the one really hazardous long-term effect of mixing. The hazard was the runaway crust growth that forced the second SY-101 emergency remediation project into being in 1998.

The SA instituted the SY-101 mitigation Test Review Group or TRG. This group had the authority to interpret data and make

196 | Part Two: The Mixer Pump Era

decisions about operations and some of the safety controls. The TRG was intended to be broad enough to represent all important viewpoints but small enough to act decisively. The TRG thus had more of a scientific flavor than the Joint Test Group that more closely represented management and operations. The TRG did not replace the test group, who still opened and closed the window and scheduled the work therein. But the TRG focused on the mixer pump itself and other safety-related equipment on the tank. Once the mixer pump was installed and running, there were no more burps, hence no more windows, and no more Joint Test Group. The TRG was essentially in charge of pump operation and most everything else done to the tank and kept control up through the Surface Level Rise Remediation Project until the tank went off the flammable gas watch list in 2001.

On October 2, 1992, Jack Lentsch sent out a memo forming the first TRG. For the record, the initial TRG membership list was:

- Project Management—Jack Lentsch (chairman), alternate Nick Kirch. Jack and Nick remained in this position until the mixer pump was turned over to operations in 1996 and Doug Larsen became chairman. In 1999, Fred Schmorde took over until the TRG was dissolved in 2001.
- Mitigation Engineering—Carl Hanson (vice chairman), alternate Tom Burke
- Facilities Operations—Doug Hamrick, alternate Gary Dunford
- System Engineering—John Schofield, alternate Dave Reberger
- Environmental Safety and Quality—Mohammed Islam, alternate A. Zaman
- Los Alamos SA—Harold Sullivan, alternates Bob White and Jack Edwards. From 1994 to 1996, Los Alamos was represented by Kemal Pasamehmetoglu and Bob White. Los Alamos left the scene in 1996.
- PNNL—Max Kreiter, alternate Rudy Allemann, later Chuck Stewart and Frank Panisko. Like Los Alamos, PNNL left the TRG in 1996, but resumed duty in 1998 when the surface level rise problem needed attention.
- DOE Richland—John Gray. Craig Groendyke took over from Gray in early 1994 and served to the end.

The initial version of the SA was ready when the pump was but only at the last minute. Event G, that opened Window G in which the pump was to be installed, happened on September 3. The Westinghouse Safety and Environment Advisory Council approved the SA on September 19. This was only a couple of weeks before the pump would have been installed, had not extracting broken hardware occupied the entire window. But the first version was just the beginning. Los Alamos kept refining the SA, correcting inadequacies and operational absurdities, refining analyses with new data and insights gained from old data, and sometimes even finding simpler, less oppressive ways to do things.

About the time the first version was ready for review, there was a big meeting in Washington, D.C., where the SA was to get its first airing. The upper ranks of the Defense Board and DOE would be there, expecting a good show. The Los Alamos contingent was scheduled to leave New Mexico Sunday at 5 PM and give the presentations Monday morning. Bob White had a $1\frac{1}{2}$ hour talk, and Jack Edwards would give a 2-hour presentation.

On Friday, two days before the meeting, somebody found a problem in the gas burn calculation, and they had to re-do all the analyses from combustion up through radiation dose. By some miraculous effort, it was done by Saturday afternoon! Late on Saturday night, Jack Edwards called to say he had the flu, and Bob would have to give his presentation. Bob dutifully got out of bed and went to Jack's office to get his slides, but found they had not been updated for the new burn analysis! Bob had to work all day Sunday re-doing Jack's slides. He was still revising slides on the plane to Washington, D.C.

Bob arrived in D.C. at 11 PM on Sunday and got to bed about midnight. About 1 AM Monday, he woke with an attack of tachycardia, a rapid heart beat with low blood pressure that happens when one is under stress and drinks too much coffee. Bob took a cab to the local emergency room, got a shot, and was back in bed by 4 AM for a 6 AM wakeup call. He had to endure pre-pre-meeting and pre-meeting dry-runs and then do both Jack Edwards' and his presentations. There were a lot of questions, and critics raised the issue of a possible transition from a gas burn to an explosion or detonation, a potential show stopper. Bob admitted, "It was just about the worst day in my life." He survived, but he took a couple of days off when he got home.

On March 31, 1995, long after the mixer pump became a fact of life, the last version of *A Safety Assessment for Proposed Pump Mixing*

Operations to Mitigate Episodic Gas Releases in Tank 241-SY-101, Hanford Site, Richland, Washington, LA-UR-92-3196, Revision 14, came out in two thick volumes. Many of the controls developed for the special purpose of mitigating flammable gas releases in this specific tank eventually became part of the safety authorization basis that now applies in all the tank farms. Some examples of now standard requirements that were new in 1992 include critical lift procedures for crane operations where a particularly heavy, awkward, or fragile load must be positioned accurately, inspections to confirm tank structural integrity and vital functions before work begins, requirements for gas monitoring and purging equipment where flammable gas might accumulate with nitrogen, and formal ignition controls to prevent stray sparks.

The full set of controls was far more complex and involved than this brief narrative can describe. The last revision had 14 pages of controls for running the mixer pump, 14 pages of controls covering work on and in the tank, and 5 pages for installing and removing the pump. This might appear excessive, but each line was subjected to the most critical review by a totally unsympathetic and skeptical crew of engineers and operators from Westinghouse and PNNL. DOE and Westinghouse management were just as paranoid about anything that would make work more difficult to start or easier to stop as Los Alamos was about leaving any hazard uncontrolled. The fact that most of the controls survived was a credit to the fearsome technical basis Los Alamos built to back them up. The Los Alamos team pursued their mission to make mitigation safe in the face of loud ridicule, strenuous rebuttal, and outright rejection. But they also understood the exigencies of schedule and budget and kept striving to make the SA more practical and less restrictive. Their wholehearted participation in the harsh necessities leading up to Window G fostered a solid respect for their integrity and talent that survives to this day.

1992–1993

A Window for Mitigation

With the safety assessment (SA) approved and the mixer pump tested and waiting, the project applied all its energy toward installing the pump in the first available window. But there wasn't time in Window G and Event H wasn't big enough. The whole period was a war of safety worries versus practical engineering and a conflict of bureaucratic desire to find fault and place blame versus the need to show accomplishment. Everything finally came together in Window I on July 3, 1993. In hindsight, Window I was the probably the soonest the pump could have gone in anyway.

NEW WINDOW CRITERIA

The window criteria had been a source of irritation to John Tseng from the first. He thought Westinghouse should be able to figure out a better method for opening windows. What he really wanted was to avoid having windows fail to open after smaller burps and to have them stay open longer so that more work could be done. At a February 11, 1992, project review meeting, after getting a briefing on the results of the Fiasco in Pasco, Tseng asked for revision of the

200 | *Part Two: The Mixer Pump Era*

window document. In his view, after spending millions and getting a lot more data, the window document should have a more scientific basis.

The revision request went to Nick Kirch. He disagreed with Tseng that new data, or any revised evaluation method, would be useful. Nick believed that the problem was not the method, but the over-conservatism of the SA that constrained the method. The window criteria could not be relaxed in any major way unless the safety documents were also revised to accept the risk of higher amounts of gas in the tank. In a memo to Jerry Johnson, he wisely recommended "that safety documents...refine assumptions regarding available gas volume versus measured tank parameters. This will allow mitigation to proceed with a different, perhaps 'always open', window."

Tseng flatly rejected Nick's approach as totally non-responsive. He firmly believed there should be enough information to redefine the windows to permit a much wider range of time and activities. More ominously, Steve Marchetti apparently agreed and suggested that Los Alamos review the window document! But Nick believed that Hanford staff had the engineering and scientific capability to evaluate and interpret Hanford-generated data. Apparently they convinced the right people because, even though they got a lot of "help" from Los Alamos, and later PNNL, on the window document, it remained firmly in Westinghouse's, namely Dan Reynolds', suite of responsibilities throughout the project.

Work on the window criteria wasn't easy! Not only were Tseng's demands difficult to meet, but they conflicted with recommendations coming from Los Alamos and the TAP. It got so bad by April 5, 1992, that John Propson, Nick Kirch's boss, sent a memo to Harry Harmon warning of the difficulties. Propson's memo lists the following conflicting recommendations:

Tseng: Provide a more flexible criteria and evaluation process, differentiate between waste intrusive and non-intrusive windows, relax criteria for opening and closing windows using level and temperature

TAP (Kazimi): Current criteria are reasonable

TAP (Campbell): More conservatism is in order, recommend a 6-inch level drop, the hydrogen limit should be much lower than 0.8 percent

> *Los Alamos:* There should be no difference between intrusive and non-intrusive windows, the 5-inch level drop is conservative, but window should close for ANY hydrogen concentration above the background
>
> Propson concluded that "it appears we are in a no-win situation. We will, however, continue to produce the best possible window document which will provide as much flexibility as we can justify while maintaining a position of safety first."

The new window document wasn't ready for Event F, but it didn't matter. Although the level drop measured by the radar gauge exceeded the required 5 inches, the waste temperatures measured by the bottom four thermocouples did not change, and those at the top did not change enough. Thus the window criteria were not satisfied, and Window F did not open. It was probably just as well. Everyone was busy enough working on the pump and the SA.

Dan finally got revision 3 of the window document done by the end of July with another quick revision 4 in early August to accomplish some minor touch-ups. Whether it made John Tseng happy is not known, but it was a masterpiece of distilling value from conflicting advice. As a result of Dan's common sense, it really was more flexible and easier to apply, but it did not lengthen the window or make them much easier to open.

The revised document kept the 5-inch waste level drop as the primary criterion. But the measurement was simplified by computing the drop from the "nominal" reading over a 2-hour period prior to the event, and using *any* available level gauge. The temperature criteria were made more flexible by requiring a 4-degree drop on low-elevation thermocouples *or* a 5-degree rise on high-elevation thermocouples. This allowed opening a window should the sediment immediately surrounding the thermocouple tree happen not to move. An additional type of criterion was added in requiring a peak hydrogen concentration of at least 0.25 percent (2,500 parts per million) during the burp. The difference between intrusive and non-intrusive windows was eliminated, but the window width varied with burp size. The window closed in 20 days if the level drop was less than 6 inches and 30 days if over 6 inches. The Joint Test Group kept the primary responsibility for applying and interpreting the criteria.

Dan also added a rather involved statistical calculation of the risk of a large burp as a function of time to beef up the technical basis that Tseng had been deriding. He calculated the probability of a burp during various window intervals by fitting a special statistical function to the recorded times between burps since 1984. This function was the Weibull distribution, named for the Swedish mathematician who invented it. Dan's calculation showed that the risk of a big burp at the end of a 40-day window was two orders of magnitude higher than at the end of a 20-day window, indicating that windows just could not be widened very much. This was definitely *not* what Tseng wanted, and Weibull became a "four-letter word" in the mitigation project.

> On April 20, 1992, at 9:35 PM, a small gas release, Event F, occurred in SY-101. The Food Instrument Corporation (FIC) level gauge was still broken. The radar gauge dropped 7 inches over 6 hours. The manual tape showed a net drop of 4.5 inches. The hydrogen concentration peaked at 1.48 percent. The domespace did not pressurize. The waste temperature on the bottom four thermocouples did not change. The event was not very vigorous on the in-tank video. Event F occurred 138 days after Event E, the longest time between burps recorded since the air lancing period of 1984–85!

The window criteria were revised one more time before the mixer pump went in. Revision 5A came out May 13, 1993. There were three major changes in the criteria. First, the difference between

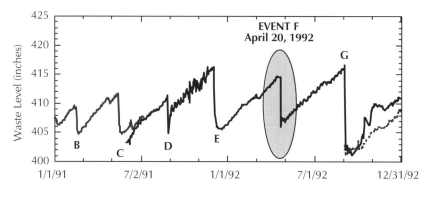

SY-101 Waste Level History: *1991–1992*

intrusive and non-intrusive windows was restored. Both windows were 20 days long for level drops of less than 7.5 inches. For larger drops, the intrusive window was 30 days long, while non-intrusive work could continue as long as 48 days for a huge 10-inch level drop. Second, the minimum level drop requirement increased to 7 inches, and the level drop was to be calculated as the *average* of that measured by the FIC and manual tape. The radar gauge readings had been discredited by then. Third, a provision was made for non-intrusive work for brief periods after a window closed if somebody could show the risk of the work was lower than the accumulated risk during the actual window. The risk basis now used a log-normal distribution to fit the time between burps instead of the Weibull distribution.

On June 22, 1993, John Tseng officially approved revision 5A of the window criteria in a memo to DOE Richland. This came just in time, four days before Event I, SY-101's final burp! To accommodate the review comments that Tseng attached to the approval memo, Dan hurriedly put out revision 6 on June 28, 1993, two days after Event I. It was a good thing that this event was big enough by anyone's calculation and was not a barrier to installing the pump!

NOT ENOUGH MONEY!

Because the SY-101 mitigation project had to go without a formal budget through all of 1992, the mitigation options evaluation and Fiasco in Pasco had to be covered by sapping funds from other projects. But when the real work began, Harmon and Marchetti soon had trouble matching the inflow to the outflow, even with John Tseng funding Los Alamos's work directly. On April 14, 1992, the Westinghouse Change Control Board, to which managers had to come, hat in hand, for money, was told that the project had $28 million worth of work started, but only a little over $17 million to cover it.

The first response was to cut costs where possible. In an April 21 staff meeting, John Deichman listed possible cuts that would save not quite $1 million. For example, stopping the PNNL Science Panel would save $150,000, and $350,000 of work scheduled for Window G could be deferred until the next window after October 1. These cuts were not enough. A week later, installation of several MITs in other tanks was deferred until the next fiscal year, saving about $3.5 million.

Even with this fairly large cut, Harry Harmon was forced to admit to Ron Gerton at DOE Richland on May 21 that the project would be out of money by June 5. He suggested dropping planned upgrades for the 222-S core sample analysis laboratory to save $3.8 million and stopping work on a new hard saltcake sampler for another $1.6 million. Harmon also suggested three different scheduling options for pump and test chamber installation to save a few hundred thousand dollars more. The most costly was the most ambitious. For $4.3 million, the pump would be installed in Window G, then expected to open in late July, and the test chamber in January/February 1993. The least cost, $3.6 million, would be to install *both* the pump and chamber in January/February 1993.

DOE's reply came quickly. Relaying demands from Headquarters, Gerton chose the high-stress, high-cost option! He asked Jack Lentsch for his evaluation of feasibility and limitations of accelerating the schedule to install the pump in Window G! Jack replied that it could be done, but there would be some risks and compromises. Pump modifications were behind schedule and might not be completed by mid-July as required. Los Alamos could not deliver the DACS trailer that soon, so there would be no automated pump shutdown or test sequencing, and no trailer or control room. The test plan was on schedule, but procedures and personnel were not yet integrated with DACS. DOE Headquarters would have to accelerate their review of safety and environmental documents and approve them in time for Window G.

Undaunted by Jack's assessment, Gerton ordered full speed ahead on a collision course with Window G! All the budget cuts and re-adjustments were apparently sufficient to keep the project going through the rest of 1992. But it was fortunate that the tank was following its own longer burp schedule and staged Event G in September instead of August as expected. The pump and SA just would not have been ready any earlier.

Funding was still extremely tight for the 1993 fiscal year beginning October 1, 1992, and the test chamber, among other items, fell from the schedule. Though it was a scientifically attractive concept, nobody could show how tests in the chamber would represent the behavior of the whole tank. The chamber might allow accurate measurement of changes in things like viscosity and particle size that were hard to do in the tank, but nobody knew how much these properties had to change to stop burps. And, because the waste in the long, skinny chamber would not burp in any way like the tank, it was not possible to tell if a test had stopped them. The test chamber was also extremely hard to design. Its large diameter and long length made it super-sensitive to seismic loads and waste impact loads during a burp.

The ambivalence and difficulties dampened everyone's enthusiasm for the test chamber and tight funding provided an excuse to slow and then stop work on it by the end of September 1992. There was no serious work on it after that, though the concept stayed alive until the mixer pump ended the burps in the fall of 1993. Then there was no need to test anything else.

There were some personal casualties of the budget stresses. John Deichman, who had resisted the deep cuts of 1992 and the low budget for 1993, was replaced as Tank Safety Program manager by John Fulton about June 1. On July 28, Fulton announced at his weekly staff meeting that Mike Siano from Westinghouse Pittsburgh would replace Steve Marchetti in August. Steve left Hanford at the end of September to become Vice President of Parsons Engineering in Pasadena, California, still far from his home in South Carolina. How much the mitigation money woes, or Westinghouse management's reaction to them, affected his decision we know not. We do know that, except for Fulton's announcement, Siano was pretty much invisible to the project.

WINDOW G HEROICS

It was not clear at first which of the anticipated windows would see the pump installation. The initial plan aimed at Window H, then expected in October 1992. But the combination of pressure by DOE to accelerate the schedule and the realization after Window F that the windows were going to come later, changed the goal to Window G, now expected some time in August. So the pump fabrication work, pump testing, SA approval, procedure writing, field crew training, and the final readiness review all had to come together by about Labor Day at the latest. The schedule would have mitigated the tank before Thanksgiving and delivered DOE a very welcome Christmas present!

But these earnest hopes and dreams did not come true. Later events showed it was probably just as well they didn't. Neverthe-

206 | Part Two: The Mixer Pump Era

less, the 45-day Window G campaign may have been the severest test of perseverance and ingenuity under fire that Hanford has been put through since the Manhattan Project. More work was accomplished than in any prior window. The battle was physical (it was hot, dirty, hard work in cramped quarters in full suits), technical (unexpected emergencies kept coming), and organizational (not everyone was working towards the same goal). It was a sort of rite of passage that the work crews, supervisors, engineers, and managers had to endure to gain the hard discipline and mental toughness needed to put the pump in when the time finally came.

> Working conditions in the SY farm were really terrible. All work was done in whites, in tents with fresh air. "Whites" are loose, white-hooded coveralls with no open seams or zippers that might admit contamination. You also wear a face mask, rubber gloves, and boots, all sealed to the suit with masking tape. To undress you carefully peel all this off into special containers to keep radioactive dirt from getting onto anything. Fresh air is supplied to the mask from a long hose. Workers could not eat, drink, or relieve themselves while in whites. It is difficult to imagine working for hours in a full, tightly sealed suit in 100 degree heat without water or toilet. Dan Niebuhr says it took careful planning, but you sort of got used to it. Los Alamos suggested that workers wear full encapsulation suits, a rubber balloon with self-contained breathing apparatus inside. It was the highest level of protection available in industry, but deathly uncomfortable to work in. The suggestion was not adopted.
>
> Before doing any work, they put a plastic isolation tent around the riser as a vapor boundary to confine contamination and toxic gases. Fearing that nasty stuff would burst out of the tank during a burn, Los Alamos required everyone to be in fresh air within 30 feet from any riser. Between tasks when they were not working, the operators in the tent had to get back by the wall of the tent and kneel or lie down to get out of the "shine" radiation from the riser. This was not normally allowed because lying down in the dirt contaminated the suits, but in this case it was necessary because the risk from the shine was worse than the risk from the dirt.

It was one thing to talk and plan about installing the pump, but quite another to actually do it. As late as August some were not

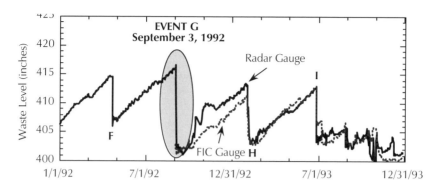

SY-101 Waste Level History: *1992–1993*

mentally ready for it. It was obvious that worker safety and radiation protection people were not familiar with the work ethic of SY farm operations. They had to balance contamination risk versus mitigation of SY-101's safety problem. Jerry Johnson records that "several people were really wound up about it." But eventually they were "recalibrated," as so many who served the mitigation project had to be. By August 25, 1992, the Joint Test Group reported that all Window G work packages were done and out for review.

> On September 3, 1992, at 2:45 PM, Event G, one of the largest recorded gas releases, happened in SY-101. By 4:30 PM, the FIC level dropped 8.7 inches and the radar gauge showed 10.9-inch drop. The FIC eventually reached 9.8 inches early on September 5. The hydrogen concentration peaked at 5.1 percent, over the flammability threshold of 4 percent. The domespace pressurized. The waste temperatures dropped all the way to the tank bottom. The in-tank video showed massive hydraulics. Event G occurred 136 days after Event F.

Event G was another monster, every bit as violent and fierce as Event E, 9 months earlier. Though its lights had "browned" in the radiation and the view was dim, the in-tank camera showed huge, upwelling surges and breaking waves sloshing across the tank. All the hanging hardware swayed back and forth, and an air lance bent.

208 | Part Two: The Mixer Pump Era

It might even have been possible for a person in the SY farm to feel the earth move and maybe even hear pipes banging in the risers.

At least there was no doubt the window should be opened. The release was big enough by anyone's criteria, and window activities began to gather momentum. At the September 9 morning mitigation meeting, the schedule announced for the 30-day window would install the mixer pump on September 23 with plenty of time to spare before the window closed October 4.

> Their traveling life could lead the Los Alamos crew into some interesting situations. Before each "scheduled burp," it seemed that all the bureaucrats and consultants in the country came to Hanford to hear and be heard. Kemal and Jack Edwards also had to participate in these events. They arrived at the airport late one Saturday night, probably September 5, 1992, just after Event G, to find all the rooms in the Tri-Cities already taken.
>
> A considerate man at the airport finally found them a place in Pasco. There were no signs on it or anything. It was apparently one of these places that rent rooms by the hour. It was hot, and the air conditioning was not effective. The doors wouldn't lock either. It was so hot that Kemal and Jack wanted to go for a walk. The manager advised them not to go walking around in that part of town at midnight. Edwards asked where they could get a cold beer. The manager looked at them squinty-eyed and said, well there's a bar over there, but the way you guys are dressed I don't think you'd get out. At that point they gave up and waited until dawn when they could drive out to Jack's air conditioned war room in 2750E Building and get some sleep.

The alarming discovery that determined the future of Window G came just as work was starting on the morning of September 10. Between burps, the video camera stares at a fixed angle that only shows about a quarter of the tank. After the burp, an operator goes out and scans around the tank, looking at the entire waste surface. That morning the camera operator couldn't find one of the old air lances that had been out of view during the burp. It had broken off and fallen into the waste! Only a stub remained in the riser.

The serious implications of the broken lance began to sink in as the news spread. Would the mixer pump jet bang the 30 feet of 2-inch

LEFT: Bent air lance in Riser 17C (southeast part of the tank) shortly after Event G.
RIGHT: Bent air lance in Riser 1B and stub in Riser 14A (northwest part of the tank) shortly after Event G.

pipe around and punch a hole in the tank? Or could it form a sort of dam that would hinder mixing? Could it have fallen right under the riser where the pump was supposed to go? All these worrisome questions led to the obvious conclusion that the rest of the obsolete hardware had to be removed so no more could break off.

Actually, the air lances were going to be pulled out in Window G anyway. Nothing could be done about the broken one. So the main question was whether or not to pull the bent thermocouple tree. A consultant asked to evaluate the thermocouple tree situation weighed in Friday, September 18, with the conclusion that it still worked, it was not structurally compromised, and did not obstruct anything else. Leave it alone! Despite this helpful advice, the question had already been decided three days earlier when Ron Gerton relayed instructions from John Tseng to remove it. Tseng also wanted resolution of the potential hazard of the broken air lance to be included in the SA and submitted to DOE Headquarters Monday morning, September 21.

Jack realized that pulling the severely bent thermocouple tree might present unforeseen problems and quickly floated a proposal to extend the window beyond the planned 30 days. DOE Headquarters gave this idea a chilly reception, but the waste level on the radar gauge was still dropping, so there was some reason to hope. Regardless of the extra work needed for the thermocouple tree, Jack fully expected to install the mixer pump. On Wednesday, September 30, 1992, he e-mailed the mitigation team that the bent thermocouple

210 | Part Two: The Mixer Pump Era

tree would be removed Friday, October 2, using the same tools and procedures that had already pulled the bent lances earlier. He wanted to start pump installation Sunday, October 4, and finish it on Wednesday. In the meantime, a window extension was being processed to make time for it. But the two days Jack allotted for pulling the thermocouple tree were not enough. The window extension was required all right, but not for the pump!

The next two weeks saw the climax of frustration, the disappointment of the unexpected, and the triumph of overcoming the unexpected. Let's follow each step in the effort day by day in journal format.

Friday, October 2, 1992: John Tseng approved the "justification for continued operations" for extending Window G work beyond the current 30-day limit to 40 days with a possible further extension to 45 days pending a review of the data. But he cautioned that "we expect that every reasonable effort will be made to complete all tank window work by...October 7, 1992. In *no* case, however, shall the window remain open beyond a post-event level increase of 3 inches." Good news, but Tseng's approval didn't come cheap! He required DOE Richland staff to meet and review tank data twice a day to assess whether the window should stay open and report the recommendation to Headquarters daily, *including weekends*! Of course this also required Westinghouse staff to collect and plot up the data and present it to DOE. Then again, everyone was already working weekends anyway.

A justification for continued operations is the Westinghouse administrative tool one uses when either the current SA and documentation has proved inadequate or non-existent, but work has to continue. Window G was a case where Event G was so large and the work so pressing that Westinghouse felt a good justification could be made to continue Window G operations beyond the normal 30-day period allowed by the current window criteria document. The primary justification was that the waste level was still lower at the end of the 30-day window than it had been right after most of the previous burps. Los Alamos endorsed the justification for continued operations as being bounded by the SA and attached a statistical study to the justification showing that there was actually a 50-day window after Event G with a *zero* probability of a gas release.

Sunday, October 4, 1992: At the daily meetings to extend Window G, the data showed that the 3-inch level rise limit was still 2 inches away. The window was extended 48 hours. The field crew would start pulling the bent thermocouple tree today and install the VDTT Monday.

Monday, October 5, 1992: Harry Harmon sent an e-mail to all staff announcing Friday's approval by Tseng of the justification for continued operations to let the window remain open until a level rise of 3 inches or 40 days. The mixer pump was still scheduled to be installed October 7.

Tuesday, October 6, 1992: The tragedy was announced at Jack's 7:00 AM mitigation meeting. About 25 feet of the thermocouple tree had been pulled out when it broke off at the riser lip! Fortunately, the bottom 30 feet were still hanging precariously by a few thermocouple wires. It was a delicate situation. Pump installation had to be pushed back to October 9 to finish October 12 with the first run on October 14.

Jack called a special meeting to brainstorm how to fix the broken thermocouple tree. It broke under a pull of 9,000 pounds force. This was equivalent to only 5,000 pounds per square inch stress while the tensile limit of the pipe should have been 25,000 pounds per square inch. There must have been a crack or other defect at the break. Pictures showed the pipe hanging by its wires with some of it still in the waste. The group agonized over whether to lock it in place, remove it somehow, or just cut the wires and drop it back into tank. They decided it had to be pulled and began looking for a way to do it.

The daily window meeting voted to extend the window 24 hours. The VDTT readiness review was still not complete. Some reviewer had a new issue about structural fatigue cycles. Apparently it would break after only 15 cycles of bending three to eight degrees.

Wednesday, October 7, 1992: At the daily window meeting, the team reported that they were looking for a tool to grab the broken piece of the thermocouple tree. A wasteberg was apparently holding the lower end against the tank wall. The window was extended 24 hours.

In response to a formal complaint from Roger Christensen of DOE Richland, the team decided to stop overtime for 24 hours or maybe 48 hours. Things were going too fast and people were overworked. Christensen had e-mailed John Fulton earlier the same day that he had found some rigging with mandatory inspections far overdue while checking progress on SY-101 work. The field supervisor first argued that overdue inspections weren't important, but he had riggers remove discrepant rigging. But THEN the supervisor "stood on a riser and lectured me in front of his work crew about too much safety oversight, etc." He claimed to be suffering from 130 hours overtime in 13 weeks.

Thursday, October 8, 1992: The broken thermocouple tree was going to blow the original 40-day extended window. Jack had to ask for more time and gathered his team to meet with Westinghouse's Safety and Environment Advisory Council, hoping to get their approval. He admitted that he had already sent the request to DOE Headquarters to get Duffy's approval on Friday. If he had waited, the window would have closed Sunday before Duffy returned from the weekend. The council was very upset that he had sent the letter off before they approved it, and the meeting became more heated from there.

There was much contention over the tank data and the methods being used to measure surface level. Dan Reynolds tried to calm things down by reminding them that there had never been a gas release at the current level of 405 inches no matter how long since the last burp. The Safety and Environment Advisory Council accused the team of using over-zealous technical calculations to push the extension. However, Rudy Allemann showed TEMPEST results, which most still believed, predicting that any gas release now would be far less than the maximum allowable burp stated in the SA. In the end, the council grudgingly agreed to extend the window to 45 days or a 3-inch level rise. But they were obviously uncomfortable, and some of them didn't buy it at all. The clear message to the project was, don't come back again!

Later that day, Jack called a meeting in the war room in the 2750E Building on recovering the broken thermocouple tree. Someone had contacted an old Oklahoma oil man who suggested what he called a "sucker" that they used to retrieve broken drilling bits from oil wells. They planned to insert the sucker into the broken pipe, expand it to grip the pipe tightly and just pull it on out. A water jet from another riser would break up the wasteberg pushing it against the wall. The sucker would be tested on a broken pipe mockup down at MASF, and the tree would be pulled over the weekend. Jack was forced to delay pump installation to October 15 with the first run on October 20.

Friday, October 9, 1992: Meanwhile, the VDTTs were finally ready, but mixer pump installation now seemed less than certain. Should they be installed now or wait? Because the VDTTs were designed to evaluate pump performance, some wondered what good they would be with no mixer pump to evaluate. Others countered that VDTTs would provide a measure of the waste density, "which has been somewhat of a mystery to date." Knowing the waste velocity during the next burp would also be very valuable. Besides, if VDTTs broke by wasteberg impact, new ones could be installed in time for pump installation in the next window at a cost of only $100,000 per probe. The team concluded that the VDTTs should be installed because neither safety nor structural issues preclude it, and they could provide valuable data.

Monday, October 12, 1992: Jack sent out a memo stating the "drop dead date" for pump installation was Saturday, October 17. It had to start no later than 7:00 AM on that day.

Tuesday, October 13, 1992: The first trial of the "sucker" didn't work! It gripped the broken pipe securely and pulled some of it up into the riser. But the tree broke again at a weld lower down. This time there was only a quarter inch of metal keeping the lower piece from falling into the waste, an even more delicate situation than when it hung by thermocouple wires.

They needed more time. Friday was the last day of the 40-day window, so they had to go for the full 45 days. The tank data looked good. The waste level was rising slowly and would only just reach the 3-inch limit at 45 days. So Westinghouse sent a letter to Ron Gerton, formally notifying DOE Richland of the decision. Window G would be closed at midnight Wednesday, October 19, 1992.

Friday, October 16, 1992: The field crew finally pulled the rest of the broken thermocouple tree out of the tank! It was a sweet victory but also a defeat. Harry Harmon was forced to concede that there was no time to install the mixer pump. Pulling the thermocouple tree ended up costing $3.9 million! Jack was right to remember it as "the thermocouple tree from hell."

Harry formalized the surrender in a letter to John Anttonen, the Tank Waste Remediation System (TWRS) program manager at DOE Richland. He told Anttonen that "several activities still must be completed before the mixer pump can be installed. Completion of these activities will push the schedule for pump installation beyond the 45-day limit for Window G closure which would occur at 2359 hours October 19, 1992...The likelihood of a gas release event approaches 0.01 [one-in-one hundred] for a window extension past 45 days. The safety of our workers is our top priority and we are not willing to place them at risk with a probability in this range. **Therefore, it is the judgment of Westinghouse that continued tank intrusive work to install the mixer pump should not proceed during Window G.**"

Even without the pump, there was still a lot accomplished in Window G. The old bent air lances and thermocouple tree were removed with all the attendant troubles narrated above. One MIT was installed in Riser 17B, and two VDTTs were placed in Risers 1B and 14A. The DACS trailer from Los Alamos found its permanent

LEFT: Broken thermocouple tree in Riser 4A after second attempt to extract it with the oil field "sucker." RIGHT: Hydraulic jack used to pull bent air lances and the broken thermocouple tree.

Project photos from *Window G Activities*

A Window for Mitigation: 1992–1993 | 215

site just outside the SY farm fence. The camera pan and tilt mechanism was repaired but replacement of the dimming lights had to wait for the next window.

Besides these specific tasks, the team also completed other big items necessary for the mixer pump. The initial version of the mixer pump SA was approved, mixer pump cold testing was completed, and the test plan for the first phase of mitigation testing was written. Less tangible, but just as important, the installation crew got to practice installing the mockup pump many times at the cold test facility and a few times with the real pump at MASF. Despite this progress, the tank still had a few more surprises waiting to trouble the project in Window H and Window I.

WINDOW G AFTERMATH

Jack Lentsch was extremely frustrated because the thermocouple tree used up his window. To this day, Jack is sure that, had the bent tree been left in the tank as the consultant advised, they COULD HAVE PUT THE PUMP IN! On the day of defeat, Westinghouse President Tom Anderson scratched a note to Harry Harmon on the mixer pump readiness review coversheet saying, "Harry—I want to make sure that DOE-RL/HQ knows where we were at this point—i.e. there were unresolved issues and we didn't just 'give up.' Secondly, we need to have an aggressive plan to move forward and make sure we're ready well before the next window."

Dave Pepson, the DOE Headquarters project engineer who reported directly to John Tseng, and DOE Richland project coordinator John Gray were very upset at the failure to install the mixer pump in Window G. The weekend after the window closed, they tried to get a conference call with the mitigation project office, but nobody was there. They continued talking to each other, unaware that their conversation was being recorded on the answering machine! Tony Benegas, Carl Hanson, and George Vargo got the whole unabridged DOE opinion of the project's performance when the played the tape back on Monday. Jack "righteously" refused to listen. But he already knew what they thought.

216 | Part Two: The Mixer Pump Era

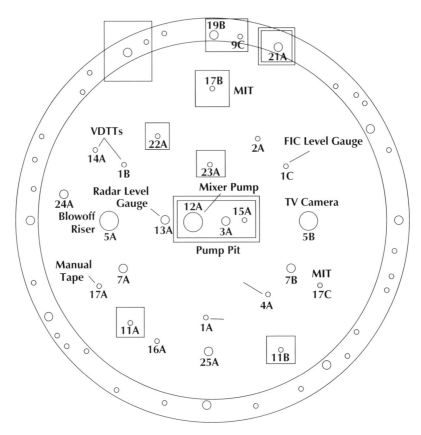

Map of primary tank risers on SY-101 showing what was done during Window G in October 1992. North is up.

John Tseng and Leo Duffy probably would have liked to fire Westinghouse for failing to get the pump in. Duffy said in a November 4, 1992, memo to DOE Richland, "I find that I have mixed emotions. On one hand I commend the hard work and dedication, extraordinary efforts...[that] resulted in substantial progress being made toward pump installation. On the other hand, I am very disappointed in the management of the pump installation effort . . . [Failure] was largely due to our inability to manage multiple tasks of equal importance and to alert management in DOE and Westinghouse about the need for additional assistance." The memo went on to demand a review of what happened and a set of plans to make sure the pump would go in next window. The message was clear: do it next time or else!

Tseng called a high-level meeting October 29, 1992, to uncover "lessons learned" in Window G and maybe to decide who would get fired. Besides Tseng, Roger Christensen, Mike Payne, Jack Lentsch, Carl Hanson, and others were there. On the positive side, they commended the contractor team of Westinghouse, PNNL, Los Alamos, and Kaiser for working effectively together as a team. They also recognized that DOE Headquarters and DOE Richland worked as a team more than in past windows. They lauded the demonstrated commitment and dedication of the people at all levels and the extensive use of pre-job training of field crews that enabled them to complete work quickly. MASF cold testing was a good example. Even Tseng had to admit that the team used a prompt, professional approach to the unexpected broken thermocouple tree. They felt TAP input was valuable in getting new technology deployed, mostly their insistence that it be used rather than advice about using it.

It all seemed to boil down to two problems. The first was that no single individual was responsible to coordinate all the window and non-window work. There was a different manager and separate schedule for mixer pump activities and for other window work. Sometimes the two schedules conflicted. The remedy was to assign George Mishko, senior Westinghouse operations manager, to be Jack's lieutenant, with overall authority to schedule and direct all Window H activities. According to Jack, Mishko had a jovial disposition, but he could be "mean as hell" when needed. He did a good job and made tank farm operations full members of the mitigation team.

The second, and probably the fundamental, problem was simply having too much work to do with too few people and too little time. Procedures and work packages had to be re-done several times because inexperienced people were pulled in to write them. A continual stream of late design changes and equipment modifications caused delays. Equipment and parts were not available in the SY farm when needed for a job. The readiness review seemed to be aimed at finding fault and causing delay instead of confirming readiness. However, the source of this problem, in effect, supplied the cure. So much was done in Window G that there was less to do in Window H. With the mixer pump completed and tested and the

218 | Part Two: The Mixer Pump Era

SA approved, there would be more spare time and more people available to do it.

The book closed on Window G on November 16, 1992, when Leo Duffy himself visited Hanford. The theme of his visit was to encourage Hanford to rise as a "Phoenix from the ashes" of 40 years of plutonium production. On his two-day visit, Duffy toured the site, visited the SY-101 DACS trailer, touted new technology for cleanup, and had dinner with community leaders. But he also attended a special award ceremony recognizing all the people for their hard work during Window G. Duffy probably should have paraphrased Churchill and admitted that Window G was "not the end . . . not even the beginning of the end, but, perhaps, the end of the beginning."

As Yogi Berra would have said, Window G wasn't over until it was over. It actually took a few more months to finish with the air lances. After pulling them out of the tank into long, sealed containers, the crew laid the containers down in the farm and roped off a high-radiation area around them. Although each of the three whole lances was rinsed internally with about 50 gallons of water and washed on the outside with another 100 to 400 gallons, they still contained some radioactive waste. The containers should have kept the waste in, but the design for their end caps did not specify gaskets to complete the seal. By the time Window G closed, the containers had leaked waste and contaminated the ground under them.

In November, Westinghouse and DOE jointly decided they needed to triple-rinse the lances to reduce the radiation dose, so they would not have to be classified as "mixed waste." Westinghouse designed and built long cylinders called "overpacks" to hold the lances and their leaky receivers, intending to flush them thoroughly, and ship them to the Central Waste Complex for storage.

The job started January 13, 1993. The procedure required the lances and their receivers to be shoved into the overpacks and then flushed and shipped. Unfortunately, it was very cold and there was a foot of snow on the ground. Some snow got pushed into the overpacks with the lances. Worse, the work plan did not instruct the crew to bolt the end caps onto the overpacks, even though the cap and bolts were right there! Instead of questioning the procedure, they stuffed the hole with rags and taped a plastic bag around it. They did this four times before someone caught the error and got the caps on. But it was too late. The snow melted inside the bags and the liquid leaked onto the ground, spreading contamination to about one-third of an acre.

In the spring of 1994, Dan Niebuhr and a crew finally used glove bags to pull the lances out, cut them up, and pack the pieces into standard burial boxes. The ultimate cost of disposing of the contaminated lances was $5.3 million! The state of Washington also fined Westinghouse heavily for it.

THE SA GROWS AND MATURES

Although the first version of the SA had now been approved, Los Alamos did not stop worrying about things. In fact, they had more time to think about safety issues with the press of Window G behind them. There were a lot of loose ends that there had been no time to tidy up. In early November, Bob White sent two long lists of urgent anxieties to Jack Lentsch for action. Many were good ideas, some really were major safety issues, and others were downright unreasonable. The list was long, but Westinghouse had to respond. On December 8, 1992, after due consideration at the morning mitigation meeting, Jack sent Bob his disposition of the safety issues. Some of the more important issues and the Westinghouse responses are described below to show what Bob was worried about and how the issues were dealt with.

Hydrodynamic oscillation: Bob worried that the long skinny pipes hanging in the waste started whipping back and forth at their natural frequency in the pump jet current, eventually fatiguing the metal and fracturing the pipes. To prevent this, Bob wanted to set limits on the fluctuating strains measured on the MIT and VDTTs and shut down the pump if they were exceeded. Jack was not convinced this scenario was real and consented only to set limits on steady strains, not fluctuations. Fluid dynamic oscillation remained a background worry until early 1994 when the pump jet was aimed directly at the MIT for several runs. Jack was right. There was absolutely no sign of an oscillation.

Blow-off riser cover: The SA calculations of the maximum allowable burp and corresponding mixer pump operation controls assumed that a big 42-inch riser would open and relieve the pressure. However, no such a pressure relief feature existed as yet.

Bob reminded Jack that either he had to get one built pronto, or the SA would have to further limit pump operation. Jack hurried to assign an engineer to get going on the design. This resulted in the "Giant Nickel" blow-off riser cover, with its over-heavy concrete foundation, that went on the tank during Window H.

Earthquake protection: The mixer pump might create a shower of sparks if an earthquake broke it while it was running. At the same time, the shaking would probably release some extra gas and the tank dome might be flammable. Bob wanted to prevent these sparks with an automatic seismic shutdown switch. Jack met the suggestion with extreme resistance because such a switch would require the DACS trailer and electrical system to be formally qualified to withstand an earthquake. There would be no seismic shutdown on the pump. A little later, Westinghouse closed the issue anyway by calculating that the mixer pump would be able to withstand an earthquake without causing sparks.

Post-earthquake fire protection: Automatic Halon gas fire extinguishers in the DACS trailer were designed to suppress fires after an earthquake. But the resulting Halon-filled atmosphere would also probably disable the operators, so they couldn't shut down the pump. This prompted Bob to request an automatic shut down when the Halon system activated. This caused Jack the same concerns about seismic qualification. Instead he agreed to make it an administrative control. If the Halon system fired during a pump run, the gasping DACS operators would have to maintain consciousness long enough hit the off switch!

Waste loading: A "lollipop" of waste sometimes builds up on things that hang in the tank. It was possible that one might build up on the pump column over time. So Bob wanted to know what maximum deadweight load would be acceptable without overloading the dome or damaging the pump's load frame. He was probably thinking of a control in the SA requiring waste buildups to be flushed off. Westinghouse engineers gave him a number, but no flush control made it into the SA. And the pump has not yet collected a waste lollipop.

Backup batteries: The DACS could run on its backup batteries even with normal utility power available. But there was no way for operators to know it was doing so. So Bob asked for a warning

A Window for Mitigation: 1992–1993 | 221

light in the control room to prevent discharging the emergency batteries unawares. Jack agreed to put an enunciator on the panel that would indicate switching to the backup battery and loss of utility power.

Resolving this first series of safety issues did not mean Bob White stopped worrying! After all, he was trained to worry, he was good at it, and he got paid to do it. There were always new safety issues and rumors of safety issues. As with the November lists, some were actually serious hazards and created more controls in the SA. Many could be discounted. But they all commanded somebody's time and attention.

One of the most onerous of the first category was the belief that a big burp that pressurized the dome would jet flammable gas out leaks in riser caps and flanges. If this happened, a spark outside the tank near a riser could theoretically ignite a burn that might propagate back into the riser and blow the tank. Accordingly, the SA applied all the in-tank ignition controls outside the tank within some radius around every riser. This was bad enough for SY-101, but it was an intolerable nuisance on the other burping tanks. It took several years before Hanford developed enough evidence and confidence in the evidence to relax this burdensome control. None of the other tanks was nearly as bad as SY-101 had been, and the probability for big gas releases just wasn't there.

Examples of some issues that were resolved before they infected the SA have already been mentioned. They included Bob's fear of "nasty gases," which was allayed by data from the FTIR, the potential for the broken air lance to penetrate the primary tank, and the hydrodynamic oscillation issue.

Bob White had asked for an inspection of the annular space between the inner and outer shell of the double-shell tank, looking for corrosion and cracks that might weaken the tank or make it leak. When he found out it was at the bottom of the project priority list he threatened to go all the way to John Tseng to get it done! The threat worked and the inspection was one of Bob's Christmas presents. The inspection video covered about 18 percent of primary tank and 30 percent of secondary liner.

On Friday, December 18, 1992, a group at Los Alamos reviewed the video tape. It seemed to show serious corrosion and some cracks on the primary tank! Someone saw a gray material and some white streaks and exclaimed that it "looks like waste!" Thinking there might already be a serious tank leak, Los Alamos sent an urgent message to DOE Headquarters without telling anyone at Hanford. Then they all left for a two-week holiday and were not available to explain themselves when DOE directed their own urgent questions and accusations at Hanford on Monday.

Responding to the emergency, Jack Lentsch hurriedly gathered a group together on Tuesday, December 29, 1992, to look at the video. The panel included international metallurgy expert Dr. Spence Bush, former Westinghouse whistle-blower Casey Ruud, and many others. Los Alamos engineer Ed Rodriguez, who apparently didn't take vacation or was called back, represented the New Mexico contingent.

On closer inspection, the experts found that the apparent flaws were not cracks, but mill scale from when the steel plates were manufactured. The gray material on the floor was concrete leakage and the white streaks were construction chalk marks! They saw some minor corrosion, but it was not as excessive as Los Alamos had thought. In short, the tank was OK! Ed Rodriguez agreed to transmit the panel's findings, and probably his own embarrassment, back to Los Alamos when they returned to work after the New Year's holiday.

The TAP was also reviewing the SA and reported their evaluation at a January 7, 1993, meeting. It was not supportive. They had never really bought into the concept of using TEMPEST simulation results to set mixer pump controls. Actually, Don Trent and Tom Michener would have much rather used the results only for insights and clues about what factors influenced tank behavior and how they did it. But TEMPEST was the only tool available to show what the pump might do to the tank, and Kemal had no choice but to use it in the SA. At this TAP meeting, Rudy Allemann showed results from a different computer model called HULL, designed to study deformation of materials like missile warheads on impact. HULL predicted that the mixer pump jet would tunnel under the sediment and eventually create a trench. TEMPEST showed some tendency in this direction but could not resolve it in detail. The difference in results confused and frustrated the TAP. They complained that "the striking difference between HULL and TEMPEST results do not add to our confidence in ability to predict behavior. Further modeling prior to use of pump will be of limited value." They were also getting fed up with the continual ratcheting of controls. They said, "The SA appears to be treated as an operations manual with unrealistic control limits not directly related to safety."

This tone of comment from the TAP was becoming typical. Actually, the mitigation team was no longer subject to the full TAP, which, under the leadership of Dr. Kazimi, usually tried to provide sensible and constructive advice. Since mid-1992, the SY-101 mitigation project had been enduring scrutiny from a smaller, narrower group designated as the "Chemical Reactions SubTAP." Some of them genuinely tried to understand the tank and help the project, but they were more and more often overruled or out-shouted by the others. The project's relationship with the SubTAP just kept going downhill from there!

WINDOW H—TOO SMALL

On October 30, 1992, Nancy Wilkins from Nick Kirch's group presented her predictions of when Event H might occur to the Joint Test Group, who wanted to know how much time they had to get everything ready. Nancy had a dilemma. The median time between events was 105 days, putting Event H at December 17, 1992. But, the longer times between Events E and F and between F and G, averaged 135 days, which would push it back to January 15, 1993. Event H actually happened February 2, 1993, 152 days after Event G. This is the longest inter-burp period in SY-101's history. The 144 days between Events H and I is next longest.

Even with Westinghouse President Tom Anderson's and Harry Harmon's resolve to prepare well in advance for Window H, things weren't going well. DOE Richland noticed, and Ron Gerton sent a letter to Anderson on November 25, 1992 (the 25th was a Wednesday and the Thanksgiving holiday started Thursday), expressing his concern about the lack of progress. Duffy had re-emphasized the need for mitigation to get top priority in his recent Hanford visit. Gerton felt that not enough staff and dollars were being expended to get the work done. His letter concluded, "I encourage you and your staff to energetically resolve these issues."

224 | Part Two: The Mixer Pump Era

In response to Gerton's concern, Harry appointed George Mishko as full-time Window H manager on December 1, 1992. John Tseng also got serious. Just before Christmas, he called a meeting with the TAP and his task force. Harry sent Westinghouse manager Cherri DeFigh-Price. Cherri reported back that "unless we have a strong 'natural causes' excuse, we had better get the pump in." And, to make sure Hanford got enough funding, Tseng cut the Los Alamos budget to about half what they asked for, keeping the SA fully funded, of course.

> Preparations for Window H were continually plagued by instrument problems, especially gas monitoring. On December 10, 1992, Jerry Johnson sent an e-mail to Carl Hanson, expressing his exasperation with the performance of the GCs that were supposed to measure hydrogen in SY-101's ventilation header. "The two GCs on the SY-101 vent header have been plagued with problems believed to be caused by physical arrangement of piping etc. If this has not been corrected then there is absolutely NO value in getting them ready for the next venting. They have not worked for the past 3 burps, and I do not have any confidence they will work for the next GRE [gas release event]." They weren't fixed, and the only hydrogen data reported for Event H came from the Whitaker electrochemical cells in the domespace.

Since before Window G, Los Alamos had been complaining to DOE that Westinghouse was not supporting them in reviewing the data. DOE Headquarters had the same complaint and demanded action. Accordingly, Nick Kirch agreed to start providing *daily* tank data reports starting Monday, January 18, 1993. The report would include all available tank data plus a list of instrument problems with dates when the problem would be fixed. Nick optimistically promised that the "plan is to have all required instruments except those on pump, and strain gauges on VDTTs and MITs operational by January 28."

> On February 2, 1993, at 2:00 AM, Event H occurred in SY-101. By 6:00 AM, the FIC level dropped 3.7 inches, the manual tape 9.5 inches, and the radar gauge 5.8 inches. The hydrogen concentration peaked at 2.8 percent. The domespace did not

pressurize though the vent flow spiked to 1,170 cubic feet per minute from the nominal 500 cubic feet per minute. The waste temperature below 100 inches did not drop. The video showed a "significant disturbance of the waste surface," though nothing like Event G. Event H occurred 152 days after Event G.

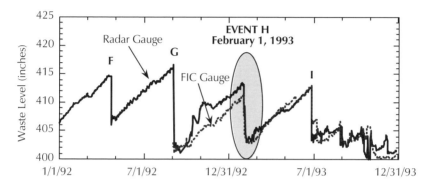

SY-101 Waste Level History: *1992–1993*

Nick Kirch had a dilemma. It was a good gas release with an adequate level drop. But the dome didn't pressurize and, most troubling, more than 8 feet of sediment on the bottom did not participate, according to data from the new MIT. Nevertheless, after a tense evaluation of all the data, Nick sent a memo to the project on Wednesday, February 3, 1993, declaring that the event met the criteria for opening Window H. The 30-day window would open at 5:35 PM on February 3 and close at midnight March 5. But Nick also recommended a Joint Test Group review to hedge his bet. The group met the same day and, noting a 6.85-inch waste level drop from the FIC, they voted to open the 30-day window. A grateful Harry Harmon told them to "GO FOR IT!" Jack wanted to have the pump in the tank by Thursday, February 18.

But trouble began almost immediately, though at first it was only an irritation. On Sunday, February 7, there was an emergency meeting about the potential radiation dose to workers if the "Giant Nickel" blow-off riser opened during a gas release event. Bob Marusich was there representing the Safety and Environment Advisory Council. It was clear to Bob that the worker safety representative didn't have much tank farm experience and didn't understand the exigencies of SY-101. It was really pretty silly. The blow-off riser was there to prevent dome failure following a burn, which would give a truly fatal radiation dose to workers. Compared to that, the dose from the blow-off riser lifting was inconsequential. Besides, Bob calculated that the probability of lifting the blow-off riser and dosing the workers during mixer pump installation would be less than one in a million. Apparently common sense prevailed and work resumed.

On February 9, DOE Richland Manager John Wagoner gave the approval from DOE Richland for Westinghouse to start Window H work on SY-101. He authorized installing the pressure relief riser, removing and repairing the video camera and lights, removing the old slurry distributor, which was the last piece of original hardware that had to go to make room for the pump, and installing the mixer pump load frame. All this was contingent on successful completion of the operational readiness review. Very notably, Wagoner did *not* authorize installing the mixer pump itself!

Doubts about how wide Window H really was had been festering and now erupted in a series of emergency meetings. Never mind it was a long holiday weekend. On Sunday, February 14, the overriding issue at the 7:00 AM mitigation meeting was the safety envelope for Window H. Earlier in the week, Rudy asked Don Trent and Tom Michener to run TEMPEST to see what inserting the pump into 125 inches of almost neutrally buoyant sediment would do. Don and Tom worked late, very late, and reported that the resulting gas release would reach the maximum allowable volume of 13,100 cubic feet! But the maximum allowable burp was still theoretically allowable, and questions arose on how to interpret the SA. Rudy was sent back to try to predict the gas release by extrapolating the results of water lancing and VDTT insertion.

Given DOE Headquarters' great expectations of a successful Window H pump installation, and the equally great consequences of a big gas release, Harry called a summit meeting on Monday, February 15, the President's Day holiday. All the important people attended, including Harmon, Lentsch, Mike Payne, John Fulton, and senior scientist Harry Babad from Westinghouse, Mike Knotek, Director of PNNL's new Molecular Science Research Center, along with all concerned from DOE Richland, Los Alamos, PNNL, Westinghouse, and elsewhere. Kemal Pasamehmetoglu, Bob White, and Jack Edwards from Los Alamos were most concerned about potential damage to the pump and worried that the dreaded pump ejection accident might actually occur. PNNL generally agreed with Los Alamos that there was really a lot of gas still in the tank, but knew the calculations were very conservative and the pump could probably go in without incident.

While steadfastly advising against installation, Los Alamos strongly urged that a hole be water lanced in the waste for the pump in the event that management decided to do it anyway. This would have the benefit of inducing any big potential gas release without having the expensive pump damaged or even ejected. It sounded like a plausible way out of the Window H dilemma. Carl Hanson was assigned to get a plan together to do water lancing for pump insertion. The rest of the crew went home to enjoy what little was left of their holiday weekend, and maybe think some more about the decision that would have to come soon.

The same discussion went on with the SubTAP on Tuesday, February 16, only this time it was even more heated! The TAP asked whether the pump was ready, and the answer from Westinghouse was an emphatic YES! But Jack Edwards firmly re-stated the Los Alamos position that, with over 100 inches of unturned waste, "the current tank conditions are outside the SA" for pump insertion and operation. When asked the basis for this recommendation, he said that TEMPEST results predicted a big rollover under these conditions. This prompted one outspoken TAP member to exclaim, If you don't put the pump in, at least throw a version of TEMPEST in the tank so it will know how it's supposed to behave! Many would have probably rather thrown him the in the tank to instruct the tank personally!

A tall, slim Turkish gentleman of striking good looks, Kemal Pasamehmetoglu might have done as well in the movies as he did in science. A brilliant, conscientious nuclear engineer, Kemal could create an onerous safety control backed by a formidable technical basis in a moment and state it in a way that all had to agree it was necessary.

At the President's Day meeting everyone came in blue jeans and flannel shirts

> so you couldn't tell managers and vice presidents from tank farm operators. Kemal stood up and told them NO, there was too much gas in the tank to install the pump. Naturally he was asked to explain his position in some detail. One person kept pushing Kemal by asking the same question again and again. Finally, annoyed at having his holiday messed up, being singled out of a hundred people to stand up for four hours trying to explain his position, Kemal angrily told the man to stop asking questions, "ask somebody else if you don't believe me." Afterwards, Jack Lentsch cornered Kemal and asked, do you know who you were talking to? That was HARRY HARMON! Thoroughly a gentleman, Harry later apologized to Kemal, explaining he had to be absolutely sure that Los Alamos had a firm conviction because he would have to explain the NO to Tseng. That was not a pleasant prospect at all!

Because the tank conditions were outside the safety envelope that Los Alamos had calculated in the SA, there was no way to justify installing the pump. Harry and Jack could not figure any way to get around it and called it quits sometime that week. On Monday, February 22, 1993, Jack sent the formal letter to John Wagoner, delaying the mixer pump yet again. The letter said installation was postponed "because of safety concerns expressed by . . . [PNNL] and Los Alamos" about the apparently large volume of waste that had not released its gas. On the same day, Harry Harmon also described the postponement as a safety issue via e-mail to Westinghouse staff. He explained that Westinghouse was not going to do anything unsafe. "Technical experts from PNNL and Los Alamos have been working together for last two weeks to evaluate the safety implications of the residual gas. The safety assessments tell us that the potential gas release from installation is beyond the limits of the safety criteria."

Delaying something so important to DOE and Westinghouse management as the mixer pump set off a sort of crisis of credibility for the SA and especially TEMPEST. Jack Lentsch recalls that after the Window H fiasco, "everyone hated TEMPEST since it was the cause of not making the window." TEMPEST was also challenged internally. At Jerry Johnson's weekly hydrogen meeting on March 2, 1993, Dan Reynolds reminded everyone that the huge gas release predicted by TEMPEST for pump installation would create a level drop of more than 10 inches. This would have lowered the waste level below its original fill level! This was definitely unrealistic, but the blame lay more with the extremely high gas content ascribed to the unturned sediment that was input to TEMPEST.

The whole embarrassment came to a head in an April 26, 1993, letter from Roger Christensen, DOE Richland, to the Westinghouse President. The letter exhorted Westinghouse to exercise better management of analytical conservatisms on the mitigation project. In fact, it was a very reasonable and timely request. Christensen noted accurately that conservatisms may appear prudent individually, but "when acting synergistically with other conservatisms, render the project increasingly difficult to execute by setting a margin of safety beyond reasonable limits. Safety is paramount, but failure to act and mitigate SY-101 is also a safety risk." This exceptionally concise and articulate statement is possibly the first real recognition that the traditional approach of bounding safety analysis just does not work with flammable gas releases.

The traditional analysis method assumes bounding values "(i.e., it can't be any worse than this) for each parameter, deliberately compounding conservatisms like Christensen said. But if the consequences calculated with all the bounding inputs are still tolerable, one can be assured it's safe. The trouble is, most of the time, these bounding consequences are totally intolerable or require unworkable restrictions to make them tolerable, as was the case in Window H. A better solution is to use the best possible estimates of the input numbers but assign each a reasonable uncertainty. Using fairly simple statistical methods, all these uncertainties can be carried through the math to give an estimate of the overall uncertainty in the final answer. Then, only at the end, the managers and bureaucrats can decide how conservative they need to be to ensure safety, and choose an answer from its final uncertainty range. This method finally began to come into use about 1996, and it is now pretty much the Hanford standard.

In the meantime, Christensen assigned Westinghouse some specific actions to correct the condition. First, they had to identify the conservatisms used in key modeling activities, namely TEMPEST; analyze their cumulative effect; and ascertain the changes required where the effect was unreasonable. Second, they were to compare TEMPEST predictions of past gas release events with measurements. Finally, from the results of the analysis and comparisons, Westinghouse was to give PNNL directions on how to remove compounded conservatisms in the future.

230 | Part Two: The Mixer Pump Era

It is unclear just what these directions to PNNL may have been or whether they had any effect. Regardless, in late 1993, shortly after the mixer pump was installed, Kemal was able to find a satisfactory correlation between gas release volume and waste level prior to a burp that allowed him to eliminate TEMPEST from the SA altogether! But by then it didn't matter because there were no more burps to deal with!

THE PUMP IS *TOO BIG*!

The Window H embarrassment was quickly followed by another. Though the pump wasn't allowed to go in, the window was fully open for all the other work. The Giant Nickel blow-off riser was installed, the camera and lights were brightened, and the old slurry distributor was replaced with the mixer pump load frame. This was good.

Then, on Wednesday, February 24, the riser for the pump was sized with the "go-no-go" gauge plug, which is a large barrel-like device used to determine if the pump would fit through the riser. THE PUMP WAS TOO BIG! The gauge plug, built to the same 41.5-inch diameter as the outside of the pump, went in only 5 inches and stopped! The pump would not have fit had they tried to install it in Window G or H! Two days later, the *Tri-City Herald* ran the headline "101-SY Tank Pump Too Big." What an embarrassment! But what if they had tried and got the pump stuck tight part way in, still hanging from the crane! *That* would have been a real career-shortening embarrassment for a lot of people!

There had been early indications that the fit would be tight. Way back in Window G on October 6, 1992, Doug Lenkersdorfer reported to Jack Lentsch that, during testing at MASF, the hydraulic brake shoes on the pump column hung up on the riser lip on several installations. The shoes would not retract to less than 41.5 inches in diameter, leaving only half an inch to spare in the 42-inch riser, a quarter-inch on each side. That was tight! If the actual tank riser wasn't absolutely straight and round, or the tolerances stacked up in the wrong direction, the pump would probably hang up.

Just prior to running the gauge plug on February 18, 1993, John Galbraith, who was a quality assurance engineer during SY-101's construction in 1975, summarized what was then common construction practice that might affect the size and shape of the riser.

The drawings describing the risers did not specify dimensional tolerances. Though quality assurance engineers took verification measurements with a tape measure, they did not document them, and DOE did not want to spend the extra money to document the as-built dimensions. Besides, the construction method made accurate as-building very difficult. Welders made risers out of rolled steel plate, and the weld joints made the riser slightly oval. The welds were stress-relieved by heating and pressurizing the tank, distorting the risers a little more. At that time nobody worried about building to a precise dimension because all the equipment was supposed to be much smaller than the riser. Nobody had ever thought of inserting a 10-ton monster like the mixer pump that used every available millimeter of riser opening.

Jack Lentsch described the riser fit problem to the project in an e-mail the same day. The action plan was to first get an accurate measurement of the riser all the way down to see just how much they would have to trim off the pump. For this task they called up the Tank Riser Characterization Unit, a fancy automated robotic inspection device originally designed to measure the big risers for the now dead test chamber. The characterization unit was supposed to have run back in Window G just before the pump went in, but there was no time. By the way, the height of the 42-inch riser turned out to be less than shown on the original drawings, and the mixer pump load frame needed to be stretched a little to make up for it.

On Wednesday, March 3, Tony Benegas and Mike Ostrom sent some operators into the pump pit, a high radiation zone, to get some video of the riser in preparation for running the Tank Riser Characterization Unit. They were able to report back to Jack Lentsch that the riser was generally in good shape, but there were two welds around its circumference that protruded in about one-eighth of an inch. No vertical seam welds were visible, but there had to be at least one.

The Tank Riser Characterization Unit was ready that afternoon and Mike and Tony stayed in the DACS trailer far into the night running it up and down the riser. The data showed the riser to be slightly out of round in addition to the weld seams. The minimum inside diameter of the riser was 41.34 inches, about 9 inches down from top. Because the pump was supposed to be a maximum of 41.5 inches, a little over an eighth of an inch needed to be shaved off the brake shoes. Mike and Tony recommended that the nylon skid pads, designed to keep the pump centered in the riser, be removed. Without the pads, there would be metal-to-metal contact between pump and riser, but because there would only be at most one insertion and one removal, the wear would probably not be excessive. Nobody said anything about sparks. The characterization unit was never run again. It is out in the SY farm lying in the gravel with a tarp over it.

They couldn't go home yet! The pump designers had been assuming that they had a 41.5-inch diameter to work with up and down the entire pump column. Because the riser was now known to be a little smaller, Tony had to make sure there was nothing else anywhere on the pump sticking out just a tiny fraction of an inch greater than the newly measured minimum diameter. Mike and Tony ran over to MASF and carefully checked every dimension on the pump, slipping all kinds of quickly concocted gauge rings and test risers over the pump to make sure. They documented the exact dimensions of the pump in 44-inch sections all the way from top to bottom. When they were done, they knew exactly where the pump had to be shaved and by how much.

On March 4, 1993, Jack was called on the carpet to brief Tseng's deputy, Dave Pepson, a contingent from DOE Richland, and Defense Nuclear Facilities Safety Board staff on the riser fit problem. Jack explained that Window H was the first opportunity they had to accurately measure the riser. Even though it was part of the plan for installing the pump, other work consumed Window G before it could be used. He reminded them that as-built drawings did not exist, so they had to work from design drawings. He described the results of the riser inspection and their plan for slimming down the pump to fit. DOE was apparently satisfied, but demanded that Jack have TAP do a step-by-step review of the mixer pump test plan and procedures.

We do not know the results of the TAP review, or whether they even did one. Harmon and Anderson did their own high-level Westinghouse internal review later in March. They seemed to be most concerned with the brake shoes sticking and preventing pump removal and whether they might create sparks during installation. They wanted to know whether the brake shoes could be removed altogether. They also asked the obvious question about what the installation crew could do if a wasteberg under the pump kept it from going all the way in. There was another independent review of pump installation issues in April. The main issue from that review appeared to be that nobody had an accurate estimate of the dome loading and could absolutely guarantee the pump's added weight would not collapse it. Lampson Company crane and rigging experts reviewed the procedures and concluded that Westinghouse could do a safe lift.

Besides these reviews, many other people offered opinions and raised issues that had to be at least replied to before the pump went in. But there were no further doubts about fitting the pump into the tank. Everyone from Leo Duffy at DOE Headquarters to the Westinghouse tank farm workers felt that everything that could go wrong had already happened and the mixer pump was sure to go in during the next window, just a couple of months away. They were at least half right! What went wrong next was not the tank's fault. Fortunately, it didn't happen until after the pump was installed.

March–July 1993

The Pump Goes In

Even the tank worked late that Saturday night, June 26, 1993, when it let loose Event I at 7:08 PM. It was a big one, and Nick Kirch opened Window I a few days later. The pump was ready, the SA was approved, and everyone was on hand and rehearsed to install the pump. Even the calendar cooperated, providing the July 4 holiday to keep intervening officials and reviewers away. This happy conjunction of accomplishment and opportunity let the pump slip in right on schedule and exactly according to plan. The brake shoes locked at 1:32 PM on Saturday, July 3. The first short pump "bump" happened at 6:32 AM on Monday, July 5, and the tank was finally open for the business of mitigation.

BUSINESS AS USUAL

After the furor over Window H died down, it was back to business as usual during the countdown to the next window. As usual, the project was short of money. At the March 17 morning mitigation meeting, Jack announced that the project needed another $5.5 million. In fact, he was drafting a letter talking about stopping work if

235

he didn't get it. At the same time, DOE Richland was increasing the work scope by adding a second MIT to help fill in for the failing VDTTs and directing Los Alamos to put an ultrasonic probe in SY-101, this time on Westinghouse's money! Actually the demand for the ultrasonic probe came from Headquarters. They had not forgotten the alternative mitigation methods that were to have gone into the cancelled test chamber and wanted to see firm plans to get them tested anyway. Besides ultrasonics, the project needed to plan for heating, dilution, and Mark Hall's sonic probe. Despite DOE Headquarters' pressure, it was difficult to generate much enthusiasm right now with everyone primed for the pump.

As usual, Window H and the steady stream of pump and tank issues kept folks working hard. On May 3, 1993, Carl Hanson's boss sent his staff a memo, warning that many of them had overtime pay equal to more than 20 percent of their year-to-date earnings. His message was to stop it and take a break!

As usual, people kept trying to derail the project. Distracting criticisms, concerns, and issues kept flooding in from within and without. Most were on the silly side, probably from people without enough work to do. On May 24, 1993, Al Fisher from Westinghouse's quality assurance (QA) organization passed on a grumbling message from DOE Richland about a lack of devices to secure or capture nuts and bolts to keep them from coming loose and falling off into the waste when the mixer pump was running. They were also miffed because there had been no reply to the original complaint several months ago.

Jack replied to DOE that there were actually no formal requirements for locking nuts and bolts on the mixer pump. Nevertheless, most had been double-nutted or tack welded when the pump was at MASF and the rest were torqued to specified values. Al Fisher reported back that DOE had reacted to Jack's explanation in typical bureaucratic fashion, insinuating that the fastener issue was a symptom of basic flaws with Westinghouse's design, fabrication, and QA process. He warned Jack that DOE might ask for all the mixer pump QA paperwork, find some fault, and demand a full re-inspection of the pump right in the middle of Window I, maybe right in the middle of installation. Apparently wiser folks at DOE Richland accepted Jack's reply, and no QA inspection happened. One could speculate on the fate of the DOE Richland managers had they delayed mixer pump installation a third time over a few nuts and bolts!

As usual, Los Alamos kept thinking up new things to put in the SA that, reasonable or not, made extra work for everyone. On May 5 they conceived a need to spray off the pump while it was still in tank if it had to be removed. Los Alamos was concerned that "dinosaur eggs" of waste could build up on the pump, making it too hot for the workers to handle. It was a good suggestion, and Carl provided a high-pressure "water wand" to wash the pump off before it came out. The wand is still stored in the SY farm, awaiting use if needed.

> Sometimes troubles flowed back upstream. On May 27, Bob White complained to Jack Edwards on what Westinghouse thought was going into the next SA revision. Bob said that "Carl Hanson asked me if Rev 3 allowed sluicing out of the pump or use of the vibrators during normal or emergency removal. I told him 'no' and he seemed surprised . . . Sigh . . . no sooner than the ink is dry on one revision, then something else comes up. Somehow, we are not communicating well with project folk- their expectations of what we are working on versus what our understanding is are often two different things."

As usual, the SubTAP made its presence known. On June 11, Sub-TAP Chairman Billy Hudson faxed a letter to Dave Pepson at DOE Headquarters complaining about a fundamental difference in philosophy between the Los Alamos SA and the rather naive SubTAP view on pump installation safety requirements. In the SubTAP's view, Los Alamos intended the SA to minimize the possibility of a maximum allowable burp and burn, and to bound the consequences of any accident by a collection of analyzed accidents. This demanded a set of complex controls based on arduous analyses of doubtful validity. The SubTAP believed that the "position of keeping an accident within analyzed bounds (Los Alamos approach) is impractical and that avoiding an accident based on measured values (SubTAP approach) is not only practical but more conservative. . . . The required condition for pump insertion be that the hydrogen concentration remains less than 0.75 percent (7500 ppm [parts per million])." Sounds simple. But what if the tank didn't oblige the

SubTAP and the hydrogen rose above the limit, even beyond flammability, with the pump hanging halfway into the riser? It was clear to most everyone else that Los Alamos's careful forethought about the host of potential, admittedly improbable, dangers was prudent, if frustrating.

Some of the Hanford stakeholders had last-minute concerns. On June 8, Bob Cook, now in the employ of the Yakama Nation, expressed a concern that because there was ammonia in SY-101's exhaust gas, ammonium nitrate might build up around the pump or elsewhere, become confined, and explode on a shock or spark. This concern prompted the Yakama Nation to ask the Secretary of Energy to postpone pump installation. But nobody had found any ammonium nitrate in or around the tank. A few months later, PNNL chemists Larry Pedersen and Sam Bryan explained that ammonium nitrate precipitation is not possible in most tanks, including SY-101, because ammonium nitrate can only precipitate in an acid environment, and the waste is strongly basic. Also, ammonium nitrate is extremely soluble and stays dissolved in water if any is available.

As usual, people strolling around the tank farms without enough to do noticed things to write memos about. On May 25 someone from the Westinghouse Safety and Environment Advisory Council waved a red flag about a crane parked near SY-101 with its boom raised, advising that the crane was a lightning hazard that should be grounded. Someone took the time to check this out and tell the council that cranes were normally stored with boom raised, and they were not normally grounded. Grounding was not necessary because the SA forced SY-101 work to shut down if there was a lightning storm within 50 miles! Speaking of lightning, a series of collapsible lightning towers were set up around SY-101 back in 1992, despite experts' advice that they would have little value. The project team and tank farm managers decided to leave them down for pump installation so that they wouldn't get in the way. They were eventually removed altogether.

As usual, organizations continued to change. On June 9, 1993, Bill Alumkal was named to replace Harry Harmon as Executive Vice President and Manager of the Tank Waste Remediation System (TWRS). Harry was moved down and sideways on the organization chart to Vice President and Deputy Manager of TWRS for technology development and external interface. This change did not deter

INSTRUMENT IRRITATIONS

As usual, the tank tried to destroy every instrument that anyone put into it. Somebody was always working on the various instrumentation to keep it working or make it ready for installation in Window I. Gas monitoring was still a major headache; the sensors on the VDTTs were failing, and the radar level gauge readings were persistently confusing. Though these troubles were not new by any means, they were getting more and more irritating because mitigation testing required a "baseline" set of tank data from before the mixer pump was operated to compare with new data after mixing to see if it was worthwhile. Also, the mitigation test plans had lists of required instruments and specified shutdown and startup conditions based on various measurements. Obviously, the gizmos making the measurements needed to be working with some modicum of reliability. But, as of the spring of 1993, they were not.

As of March 23, Gas Monitoring Shack #1 (GMS-1) had only the Whitaker electrochemical cells operable to measure hydrogen. The low and high range metal-oxide semiconductor monitors were unusable because other gases were chemically interfering with hydrogen measurement. The mass spectrometer was operable, but with suspect validity. Things were a little better in GMS-2. Two of the three GCs were running, as was the FTIR, though it wasn't very reliable. In the two standard hydrogen monitoring system, or SHMS, the Whitakers were OK, but their metal-oxide semiconductor units were showing the same interference as in GMS-1.

Not only were the instruments themselves not working very well, but the numbers they were spitting out were difficult to use. Jeanne Lechelt, who had the job of gathering all the data together into spreadsheets, found that the data files from the low-range GC and the FTIR were corrupted when readings got too high, just when you needed the data most. A simple change in the data format software would fix this. When the data were not being corrupted by software error, they were being corrupted on purpose by an automatic calibration system built into the GC! Jeanne pleaded with instrument guru Eric Straalsund to figure out a way to sort out the automatic calibrations from the GC data. Unfortunately, Jeanne had to continue putting up with these irritations through Event I and into the mixer pump testing program.

The radar level gauge had not lived up to expectations since it was installed in Window C back in May 1991. While it gave reliable, stable readings between burps, it fluctuated wildly immediately after one, just when Nick Kirch needed good level data to evaluate the window criteria. Engineers scratched their heads, PNNL ran tests, and Jerry Johnson summarized the findings to Ron Gerton on May 12. The tests showed that the properties of the waste and shape of the surface altered the radar reflection that made the measurement. The gauge could not detect dry waste at all. Unfortunately, the manufacturer could not suggest any modifications that would help. Jerry had already told the TAP that radar would not be the primary level measurement. They agreed, but they kept agitating for a better level measurement than the FIC and manual tape. The search would eventually turn up a gauge that worked by buoyancy rather than electrical contact with the waste. It was also sold by Enraf-Nonius, the manufacturer of the radar gauge and became known simply as "the Enraf." After the pump began running and burps stopped, the radar gauge steadied out and its readings were close to the other devices. But, by then, its credibility was blown and nobody believed it anymore. Nevertheless, it stayed on the tank, sending data until it succumbed to radiation damage in the spring of 1995.

Another conspicuous case of new technology not meeting expectations was the two VDTTs. Their velocity sensors were never allowed to operate, and the pressure and temperature sensors began failing in December 1992, only three months after they went in the tank. By the end of January 1993, one VDTT had only three of seven pairs of its sensors working, and the other still had four pressure and five temperature sensors. At the end of March, they were down to two functional sensor pairs on one VDTT and four on the other. All of the sensors had failed by the time of Event I. On June 3, the mitigation project's monitor, John Gray, told Jack Lentsch that DOE did not want the dead VDTTs removed because it would take money away from other important things, and they still hang forlornly in the dark. They aren't even good dipsticks anymore because the white level marks were smeared over with waste during the level rise of 1998 and 1999.

In March, at the request of DOE, Jack agreed to install a second multifunction instrument tree, MIT, to partially make up for the failing VDTT. Los Alamos delivered it just in time to be installed in Window I after the pump went in. Having a second good, reliable temperature profile in the waste while the pump was being tested would have been extremely useful. But sometimes engineering judgment fails just when it is most critical.

The first MIT, installed in Window G, was wired to record the temperatures from each of its 22 thermocouples in increments of 1/10th of a degree Fahrenheit. Rudy Allemann and Dan Reynolds could use these precise measurements to detect faint trends and small changes that explained a lot about what was happening in the waste. But when the other MIT was plugged in, someone read that thermocouples in general were only accurate to within a couple of degrees, and set it up to record temperatures in increments one full degree Fahrenheit! Failure to understand the difference between accuracy and precision made the second MIT almost useless in explaining what the pump was doing. Jack Lentsch threatened at one morning meeting that "someone is going to loose their job over that MIT!" But the temperature scale was not corrected until 1994 after all the testing was done.

The new equipment was causing trauma in the tank farm for other reasons—other than just not working right. Dan Niebuhr was a shift supervisor for tank farm monitoring. His crew was responsible for recording tank data around the clock and was also supposed to take care of any emergencies that happened to the installed equipment. But they were tripping over the huge accumulation of hardware and nobody bothered to tell him what most of it was or what he was supposed to do to it. The work crews had little or no information about the function, the alarm response, emergency shutdown procedures, or what kind of noises the things were supposed to make. For example, the GC first had a blue flashing light on it, which normally meant "nuclear criticality!" Then someone changed it to a red light that normally indicated "fire!" Finally it got an amber light that usually called for "caution." Were they supposed to run for their lives or just be careful?

On May 28, 1993, Dan had had enough of this, and sent an e-mail to Carl Hanson demanding a change. Dan described the problem

242 | Part Two: The Mixer Pump Era

they were facing and how it could lead to serious trouble. He concluded by warning the engineers that "if we don't get the information we need in two weeks, we won't let you in the farm!" Carl saw the problem and instructed his team to collect what Dan needed as soon as possible. Dan constructed a training manual for the operators to give them some ownership of all the gizmos and to foster a desire to keep them running.

TEST PLANS AND TANK DATA

It was not enough just to get the mixer pump installed. The pump had to be run too, if the tank was to be mitigated. Just as important, somebody had to look at the data and make a convincing case that mitigation was or was not happening as expected. These efforts needed a test plan to describe what to do with the pump, and a data management plan to describe what the data would look like and where to get it.

Joe Brothers of PNNL wrote the data management plan. Joe had been working on the design of the Hanford-wide Tank Characterization Database that PNNL was putting together for Westinghouse. This system would house all the tank core sample analysis results as well as waste temperature and level histories. Rudy Allemann knew that data management was not one of the Westinghouse's strong points, and invited Joe to the Fiasco in Pasco in February 1992 to get him exposed to SY-101. Around the time of Window G, Joe volunteered to pull together a plan for data collection and management during pump installation and the testing to follow. Joe worked with Jeanne Lechelt and Roger Bauer's instrumentation crew to develop the plan. Though it wasn't activated in Window G, Joe's work did not go to waste. He got the honor of presenting it to a TAP meeting in mid-November 1992. They supported the plan, though they were somewhat skeptical of the broadband digital network Joe advocated.

Even if the pump wasn't in, data were still spewing out of the tank, and it had to be managed somehow. Jeanne Lechelt began operating according to the plan in her little office in 2750E. There was a data management meeting every week where data managers like Jeanne and Joe, analysts like Rudy and Nick, and instrument gurus like Eric Straalsund and Tom Schneider could figure out how to fix, foil, or work around all the debilitating instrument problems. During mixer pump testing in 1993 and 1994, these meetings became an institution where important people came to find out what was happening with the tank. Everybody supported everyone else, and a lot of work was started that improved data quality and reliability.

> About the time of Event H, George Vargo, who shepherded the DACS, decided he needed a software expert on the team. Joe Brothers suggested one of his group, Barry Gregory. Max Kreiter worked with Jack Lentsch to get George to accept his help. Barry was a no-nonsense Type A personality with a very commanding presence, who went straight for the guts of any problem despite any obfuscation or dissemblance. When he first looked around the DACS, he pointed out that the mixer pump control screens were not only ugly, but the screens forced the operators to page through several screens to do an emergency shutdown! That was unacceptable. An emergency shutdown button should have always been showing. Barry got the job of re-designing the screens, and Los Alamos programmed the change.
>
> Barry became involved with the tank data when the pump started running. He was responsible for the dynamic data management, while Jeanne did the initial raw data translation and final archiving. Barry and test engineer Mike Erhart wrote the data management bible describing each item completely, showing its position in the computer files, which instrument was producing it, and how often. Though unofficial, this document was one of the most useful and most used during mixer pump testing.

Like the SA, the test plan kept evolving as new issues and constraints came up and new knowledge or new assumptions about the waste developed. The mixer pump was to be tested in three phases. Phase A was a very tentative test of a few extremely short runs at slow speed to feel out the waste and make sure that a little disturbance would not induce a huge burp. Shortly afterward, Phase B would be the actual mitigation test with a series of increasingly aggressive pump runs, aimed at mixing the waste and releasing the gas as completely as possible. Finally, assuming Phase B proved mixing was effective, full-scale testing would try various pump run schedules to find out how it should be run to keep the tank mitigated in the long term.

The whole concept of Phase A was a matter of contention. The safety crew wanted the shortest, most gentle pump runs to prevent what they feared might be the "mother of all burps." This fear came

244 | Part Two: The Mixer Pump Era

from TEMPEST predictions that, if there was enough trapped gas, a little disturbance would produce a tank-wide burp. The pump experts and test engineers were convinced that these gentle little "bumps" would quickly plug the pump, maybe irreversibly. Their first concern proved correct, but, fortunately, not the second.

The final Phase A test plan was issued April 19, 1993. This plan and the follow-on Phase B test plan were worked out by Westinghouse's Larry Efferding and Walter Knecht under PNNL's lead. The plan called for the first run after the pump was installed to ramp up to 380 revolutions per minute in 4 seconds, hold that speed for 8 seconds, and ramp back down quickly to shut off. Regardless of the duration, this speed was so low that the flow meters would be unable to detect plugging, so the plan directed the operators to monitor temperature sensors in the piping. If the temperatures did not change as cooler supernatant liquid coursed through the pump, there was probably a plug. In that case, the motor speed could be increased, subject to TRG approval.

The 12-second runs were to be done daily until the TRG authorized the actual Phase A tests. The Phase A tests were actually to be at a slower speed, 340 revolutions per minute, but for a longer time. The first run was a ramp up to 340 revolutions per minute in 1 minute and quickly shut down. After a 12-hour wait to see if anything bad would happen, there were four more runs of 1, 2, 5, and 10 minutes at 340 revolutions per minute, each separated by the 12-hour wait. Needless to say, there would not be much mixing in Phase A.

The Phase B test plan essentially turned the pump loose, step-by-step, to work its will on the waste. The plan allowed pump speeds from 520 to 750 and then 920 revolutions per minute with durations of up to an hour with the purpose of actually mixing the waste. Full-scale testing was supposed to demonstrate the pump's ability to control the gas content in the tank and to disturb the waste out close to the tank wall. But let me defer further discussion of pump testing until after the pump is installed and we have something to test!

THE PUMP GOES IN!

On June 26, 1993, at 7:08 PM, Event I, the last of the "natural" gas release events, occurred in SY-101. By 10 PM, the FIC and manual tape levels dropped 3.6 inches and 6.5 inches, respectively. The FIC reached a maximum drop of 6.7 inches at

10:30 PM on June 28 and the manual tape hit 9.75 inches at 9:30 PM on June 29. The gauge marks on the VDTTs showed an 8-inch drop. The hydrogen concentration peaked at 3.3 percent on both the Whitakers and the GC. The FTIR data showed ammonia and nitrous oxide concentrations around 1,400 parts per million. The domespace pressurized to +1 inch of water and the ventilation rate spiked to 1,500 cubic feet per minute. The gas release volume was estimated at 7,400 cubic feet. The waste temperatures dropped above 24 inches from the bottom. The in-tank video showed violent waste motion. Event I occurred 144 days after Event H.

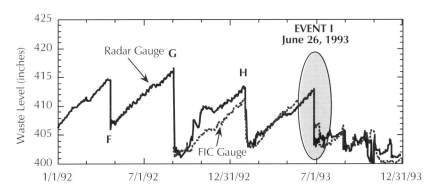

SY-101 Waste Level History: *1992–1993*

On the day of Event I, SY-101 seemed to take a perverse pleasure in changing people's plans. Jack Lentsch was walking his daughter down the aisle on her wedding. As the story goes, Jack got the news from George Mishko after the reception. He had not had a day off in eighteen months, and it was supposed to be a time for family celebration, but he came back to Hanford that Sunday anyway. Event I was also Nick Kirch's 10th wedding anniversary, and the burp doubtless disturbed his Saturday evening plans. Bob White was one of the few who was not very disturbed. He had just finished a big push to get the last SA revision out and was on a long vacation in Hawaii.

In spite of having to work the weekend instead of celebrating his anniversary, Nick Kirch was very relieved that this burp left no doubt about its meeting the window criteria. He recommended that

246 | Part Two: The Mixer Pump Era

Window I open at 6 AM, Wednesday, June 30, 1993. The Westing-house Plant Review Committee unanimously approved Nick's recommendation on Tuesday, June 29, and John Fulton sent a formal letter to Ron Gerton at DOE Richland that very day, declaring Window I open.

> At last, it was time to ship the pump out to the tank farm for installation, but there was one last delay. The long, 16-inch-diameter pump column through which all the motor power, control, and instrument wiring ran was supposed to be filled with nitrogen to at least the hydrostatic pressure of the waste at the depth of the pump motor. This was to prevent flammable gas buildup and keep waste from leaking in. But the pump column was leaking nitrogen faster than it could be supplied.
>
> At the morning mitigation meeting on Monday, June 28, right after Event I, Jack asked Tony when the pump would be shipped to 200 West. Tony said it wasn't ready and explained that "the nitrogen purge pressure inside is not holding. If we install it like this it will fill up with waste. It needs to be fixed." Tony and Troy Stokes worked all that day and into the night sandblasting metal and sealing leaks with a special commercial "goop." Tony's caulking job held for the first run and for all the years the pump was running. The nitrogen is still connected and the pump column is still holding pressure down there in the dark.

Jack wanted the pump in NOW! But there was a last major flap! The ammonia concentration measured during Event I was 1½ times what they had assumed in the SA. On June 29, 1993, right after Event I, the Yakama Nation sent another letter to John Wagoner at DOE Richland about the large amount of ammonia released in the burp. The letter warned that "ammonia was not considered in prior safety documentation. Don't install the pump until you figure this out." They were right about the SA ignoring ammonia, and Jack Edwards, Kemal Pasamehmetoglu, and the Los Alamos crew were feverishly working on that issue. Wagoner was as committed to getting the pump in as everyone else and tried to reassure the tribe as best he could while pushing the work as hard as he could. Bob White said, "ammonia was a new zinger that hadn't surfaced before. We were working on what to do about it until the day before installation."

Ammonia is only marginally flammable, but it has a vicious toxicity. At the July 2 pump installation meeting, Jack Edwards warned that ammonia was immediately dangerous to life and health at a concentration of only 500 parts per million. The worst accident for ammonia would be a large burp without a burn because ammonia would be consumed in a flame. To keep pump installation on schedule, Los Alamos decided to add an evaluation of ammonia effects as an appendix to the SA, rather than changing all the existing sections that dealt with gas composition, burn analysis, worker protection, and accident consequences. Los Alamos had a lot of work to do in a hurry!

The immediate concern was to protect the workers. The field crew would put on masks supplied with fresh air when they were working near open risers and would evacuate the area if the ammonia concentration exceeded 12.5 parts per million. Besides the workers, Jack Edwards argued for setting boundaries to keep nonessential people far away from the tank. They didn't know what was going to happen, and there wasn't time to calculate a safe distance precisely. So, guided by his best judgment, Edwards asked for a 300-meter (about 1/5-mile) radius—nicknamed the "Zone of Death"—around the tank farm. When Edwards announced this proposal, Jack Lentsch cringed. Many objected loudly to the idea that a cloud of hot gas could descend to ground level and go searching for people to smother. But Jack knew there was no way around it and asked each person in the room "will it keep the people safe?" Everyone had to agree that it would, and the Zone of Death was summarily invoked.

The Hanford Patrol ran the perimeter roadblocks, and operators stood duty as guards, checking people against a master list of who was authorized to be there. The news media were not allowed in. The Zone of Death originally included the entire S, SY, and SX farms. After a couple of months, it was relaxed to just the SY farm. About a year later, after the tank was fully mixed, the barriers came down, and operators could actually walk around in the farm during pump runs. However, the ammonia issue remained a plague up through 1995.

The time had come! At 2 PM on Thursday, July 1, in response to Fulton's notification that the window was now open, DOE Richland

Manager John Wagoner authorized Westinghouse President Tom Anderson to proceed with installation and initial operation of mixer pump in SY-101! Before the pump went in, Wagoner directed that the water lance be operated to clear a hole for the pump in the sediment and that the go-no-go gauge be run up and down the pump to check one last time that it would fit. He also wisely pre-authorized pump removal in case of emergency along with installation of water wands in a nearby riser to wash it off. Installation of the second MIT was also authorized.

The same day Carl Hanson e-mailed DOE monitor John Gray to report that the Federal Aviation Administration had established a temporary flight restriction within a 5-mile radius of SY-101 to 5,000 above the ground from Friday, July 2, to Tuesday, July 6, to prevent news media or other curiosity seekers from flying low to get pictures. The Hanford Patrol had arranged with the Benton County Sheriff to have their fixed-wing aircraft on standby at Kennewick airport with a deputy and pilot to persuade violating aircraft to leave if necessary. Evidently, not even a hang-glider came near the tank that week and the deputy had light duty.

The first few days of Window I were occupied with training, staging the pump and other equipment into the farm or nearby areas, getting all the final signatures on various required documents, and innumerable other details. On Friday, July 2, the video camera was repaired and given fresh light bulbs. It was in top shape to make a sharp, clear record of the drama to take place over the next few days.

The pump went in the tank over the 4th of July holiday weekend. The timing was very fortunate because it kept various meddlers away. Dave Pepson was the only DOE Headquarters representative in attendance, and he was not a meddler.

It so happens that a DOE Headquarters staffer had been doing a safety review of the mixer pump installation plan. Before leaving for the Independence Day holiday, the reviewer told Dave Pepson, "I have a few comments and I'll get them to you next week." Fortunately, by the time the reviewer got back to work, the pump was already in and had been run once! His comments may have been important, but the installation plan had been proven safe, at least once.

The tank farm was now ready, protected by the Zone of Death and equipped with a forest of lights on high masts, bright enough to make midnight like noon. There were video surveillance cameras all over the place. There was even a camera continuously monitoring the crane load and speed, the water flow meter, and a couple of other gauges so that operators could watch them from inside the DACS trailer. Video and radio communications were fed to monitors set up in the 2750E war room. Jack Edwards, Dave Pepson, Jack Lentsch, and Carl Hanson were there throughout the excitement. JR Biggs was shift supervisor for the installation working from the DACS trailer, Dan Niebuhr was out on the road working perimeter access control, Jim Lee ran the entire field operation, while Butch Hall spearheaded the work inside the farm. The weather was hot and sunny, and everybody got a sunburn.

The job everyone had been waiting for the last 12 months began at midnight on Saturday, July 3, 1993, with a thorough pre-job briefing. The briefing took a while and then the field crew had to get suited up with masks and turn on their fresh air before taking their positions in the farm. Work began a little after 4 AM. Here begins a detailed minute-by-minute chronology of pump installation recorded by Carl Hanson with descriptive notes and interesting stories included as they happened.

2400   Started pre-job briefing for 78 operators, engineers, and managers.

Around this time Carl Hanson recorded predictions of hydrogen concentrations or gas release volumes induced by the water lance. Here are everyone's predictions recorded for posterity. PNNL: Rudy Allemann 500 cubic feet, Barry Gregory 2,500 cubic feet, John Hudson 750 cubic feet, Zen Antoniak 600 cubic feet, Barry Wise 300 parts per million; Westinghouse: Nick Kirch 500 parts per million, Jack Lentsch 2,000 cubic feet, Larry Efferding 3,000 cubic feet, Mike McElroy 1,100 cubic feet, Guy Bear 5,000 parts per million, Steve Krogsrud 900 cubic feet, Dan Reynolds 700 cubic feet and 125 parts per million, John R. 630 cubic feet, Carl Hanson 1,000 cubic feet, Jack Edwards 800 cubic feet, Bill Alumkal 950 cubic feet, Rajiv Mulhan 1,300 cubic feet, George Mishko 750 parts per million, Mike Payne 110 parts per million, Harry Harmon 200 parts per million; DOE: Dave Pepson 1,400 parts per million, Gary Rosenwald 1,800 cubic feet, Ron Gerton 100 cubic feet, John Gray 250 parts per million.

250 | Part Two: The Mixer Pump Era

0418 Removed the riser shield plug for the water lance.
The shield plug was a large chunk of concrete weighing several tons designed to absorb radiation shining up into the riser. It had to be lifted off with a crane. JR Biggs remembers the two-crane pick for the lance. It was a hurry-up-and-wait thing where you were briefed on the work and went into the farm but waited hours for your job. People had been working 16-hour days getting ready for the pump, and they were tired. There was a crane operator sitting in his cab waiting with his head laid way back, very obviously napping. JR had to send someone over to tell him to at least keep his head up so it wouldn't look so bad on the video camera recording the operation.

0430 Picked up the water lance with two cranes and positioned it over the riser.
It was a "two-crane pick," where one crane grabbed the lance a little below midpoint, while the crane that would lower it into the tank hooked on the top. With two cranes, there was less chance of the long skinny pipe bending or damaging the lance by jostling it around in the dirt. Just as the cranes began to pick the lance up, Troy Stokes ran up to JR and yelled "Stop the lift!" Troy explained from a little notebook he carried that a mark he had made on the lance to indicate the bottom of the tank didn't allow for the height of the nozzle assembly. It would have hit the bottom before the marks lined up. JR stopped the lift, so Troy could paint new marks.

0455 Moved water lance into the pit.
The business end of the water lance sprayed water at a high pressure from many tiny nozzles threaded into two 40-inch lengths of 1-inch pipe. The two pipes were fastened across each other horizontally. Hence, its alternate name, the "cruciform lance." The X-shape spray section was bolted to 60 feet of 2-inch pipe. It fit through the riser with only a little more clearance than the pump had.

0457 Opened the plywood hatch over the riser.
After pulling out the concrete shield plug, the crew laid a plywood hatch over the gaping 42-inch hole to keep tools and workers from falling into the tank. It also cut down the shine radiation in the pit.

0503 Moved water lance into the riser.
The hour and 45 minutes from when the lance entered the riser to starting the spray seems like a long time. The crew needed the time to hook up the water hose and complete the startup procedure.

TOP: Shield plug being removed. The plywood cover has been placed over the riser inside the mixer pump load frame. BOTTOM: Cruciform water lance being removed from its container.

Courtesy Tony Benegas.

252 | Part Two: The Mixer Pump Era

0648 Sprayed water with lance placed just above the waste!

Everyone was expecting a pressurization when the lance agitated the waste near the tank bottom. But there was a huge dome pressure spike as soon as the water came on. When the fine aerosol spray of hot water first blasted into the dome at 3,000 pounds per square inch, it created an instant fog and rapidly heated the air. The expansion pulsed the pressure from -2 to +5 inches of water. A second later, the water vapor that had evaporated from the fine spray hit the tank wall and suddenly condensed, pulling a vacuum of -8 inches of water.

0651 Stopped water lance at the tank bottom, 50.28 feet from the top of the riser, 350 gallons of water used and still spraying. *Many feared a large burp during lancing. When the lance stopped, everyone thought it might be a problem or emergency. Maybe it was that big burp pushing up on the lance! It turned out that the lance just couldn't go any farther.*

0655 Turned off the water; a total of 500 gallons was used in lancing.
The water for the spray was supplied by a huge positive displacement pump on a truck. A big, stiff, high-pressure hose led across the farm to the lance pipe. When the pump started, the hose began pulsing and digging dirt. When they finished lancing, the hose had excavated itself right down into the soil!

0700 Gas monitor readings: hydrogen 93 parts per million, ammonia 350 parts per million.

0710 Order given to pull out water lance.

0715 Started moving lance out; 50 gallons of decontamination water used.
The lance mast and the lance itself were contaminated by the slime of waste they picked up during immersion. As with most equipment designed to poke into the waste, a high-pressure spray ring was built into the lance's riser housing to wash off the clinging, muddy waste.

0725 Gas monitor readings: hydrogen 109 parts per million, ammonia 349 parts per million.
This was the highest hydrogen concentration measured during lancing. Mike Payne's prediction of 110 parts per million was the closest.

0739 Turned off decontamination water; a total of 300 gallons was used.

0805 Removed water lance and piping from the tank.
In the next 2 hours and 18 minutes, the crane swung the water lance away from the tank and walked it to the northeast corner of the farm. There it re-joined with the other crane, and they laid it down in a shielded revetment. Then the big crane crawled into the farm near the pump cradle to get ready for lifting the pump.

1023 Started raising the mixer pump in cradle.
The cradle lifted the pump vertical with hydraulic rams like a missile launch.

1033 Raised pump to vertical.
There was some difficulty disengaging one locking pin on the cradle base. They finished raising the pump to vertical with a 150-ton link belt crane, the biggest available on the site. The pump weighed only 20,000 pounds (10 tons), so there was plenty of lifting ability to spare. Then the crane had to crawl about 100 feet to the tank with the pump hanging. The crane had two big American flags flying from the top, a rigger tradition for a heavy, difficult lift. JR Biggs said it was a powerfully emotional sight that still gives him a chill

LEFT: Riggers in man-lift unbolting the mixer pump from its cradle. RIGHT: Mixer pump hanging from crane with two American flags flying. Pump motor is at the bottom of the picture.
Courtesy Tony Benegas.

to think about. There was some complicated plumbing on the water line for flushing the nozzles. As the crane crawled to the riser, the engineer that designed it was "freaking out" because he had found that the piping was too complicated for an accurate calculation of the flow. He wanted to redo the flush line plumbing on the move! The shift supervisor wisely wouldn't allow it.

1200 Removed three bolts holding the pump to the cradle.
This was a daring and athletic exercise. They had to work from a swaying man-lift 60 feet above the ground reaching over to the top of the pump. Even with safety lines, it would not have been a job for anyone with a tendency to acrophobia.

1218 Pump free from 3 bolts.
Now there was a flap over a lift restriction in the work package. They were not supposed to lift the pump more than 3 feet high to prevent damage in case it fell. But at the pump pit, the isolation tent and other necessary equipment was 7 feet high! They had a 10-ton pump dangling from the crane. It was not sensible to walk the crane back and lay the pump down on the cradle while someone rearranged the clutter around the riser. They had to do an emergency amendment to the SA on the spot to allow the higher lift.

1242 Halted crane and boomed the pump out over the riser.

LEFT: Mixer pump about half way into the tank. SY Farm ventilation exhaust on the right. RIGHT: Pump motor going in. Vertical pipes on the top third are the wasteberg bumpers. Wooden 2x6 being used to align the pump.
Courtesy Tony Benegas.

The Pump Goes In: March–July 1993 | 255

1252 Positioned pump over the pit.
This was a remarkable piece of precision rigging work! The crane operator, Ken Flowers, who had worked the Alaska pipeline, could not see the riser that the pump had to go into. He had to control the crane boom and swing to within fractions of an inch by radio instructions from an operator in the pit!

1302 Removed plywood pit cover.
The plywood hatch had been placed after lancing to protect the workers in and around the pit from the shine radiation from the riser.

1304 Pump enters riser.
One of the SA controls said there could be no dust devils in the area when the pump was being installed. Just as the pump was going into the riser, the surveillance camera showed a dust devil whirling right across the tank farm, as if taunting the watching managers in 2750E. Nobody said a word.

1305 Second disk in.
Rubber-ringed sealing disks were attached to the pump column about every 3 feet to seal the riser so that the tank could hold a vacuum. If nothing else, they provided a convenient reference to gauge the progress of installation.

1317 Turned on pump burrowing ring.
The burrowing ring sprayed jets of water at the base of the pump to help it push into the sediment. It was used even though the big water lance had just been run to excavate a hole for the pump and make sure there was nothing in the way. A burrowing ring was standard practice for installing saltwell screens and liquid observation wells through the stiff saltcake in single-shell tanks.

1318 Started moving down at 4 feet per minute.
This creeping motion, less than 1 inch per second, would have been barely perceptible. It was mandated by Los Alamos to reduce the probability of permanently jamming the pump in the riser if it should bind on something.

1320 The base of the pump touched down on the waste surface!
This extraordinarily impressive scene has become a Hanford trademark. Even years later, the local news media have run this video clip with practically all stories dealing with Hanford.

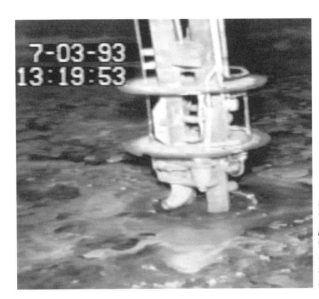

Frame from in-tank video showing mixer pump nozzles just entering the waste. Note the wet waste surface one week after Event I.

1327   Pump motor casing goes in; hydrogen 12 parts per million.

1330   Brake shoes go into the riser.

1332   Brake shoes locked, pump installed. JOB WELL DONE!!!

That night there was a celebration at the Gaslight Tavern in Richland. Everybody was extremely happy at getting the pump installed. It was said that Dave Pepson from DOE Headquarters may have been the happiest celebrant. Harry Harmon's daughter had arrived that day with her fiancé. It was also her birthday. The family went to dinner at the Blue Moon Restaurant in Kennewick. Harry told me, "It was the best day in my life."

On July 6, the *Tri-City Herald* carried the exuberant headline "101-SY PUMP INSTALLED" with a photo of the mixer pump with the two large American flags flying. On July 6, the *Herald* carried a more detailed article headlined "Officials praise tank pump crew," with photos of crane operator Ken Flowers and Vice President Harry Harmon describing the installation. Harmon extolled the crew, "This proves what Hanford workers can do. . . . It's a confidence builder for us and should be a confidence builder for the outside world." Assistant Energy Secretary Thomas Grumbly, Leo Duffy's replacement, described the importance of the pump installation to DOE in this quote from the July 9 *Energy Daily*. "The successful installation of this pump in Tank 101-SY is a major step in meeting our commitment to reduce the most urgent risks first, and to make real progress in addressing health and safety problems." In contrast, long-time Hanford critic Matt Wald described the event in a more sinister context in the Sunday, July 11, edition of the *New York Times*. His article, which also appeared in the *Seattle Times*, began "After months of frustrating delays, workers . . . have installed a mixing pump in a giant tank of nuclear bomb waste, the first step toward greatly reducing the risk of an explosion."

But even a less-than-enthusiastic response from eastern media did not dampen the celebratory mood at Hanford. On Thursday evening, July 15, Westinghouse Vice President Bill Alumkal hosted a buffet dinner at the Shilo Inn at Richland. Over 300 people involved with Window I work and SY-101 issues were recognized and honored. Westinghouse President Tom Anderson told them, "Thanks for a job well done." Alumkal presented Harry Harmon with a framed copy of the *Tri-City Herald* photo of Harry describing the progress of pump installation. Alumkal also singled out Jerry Johnson, who had led flammable gas studies from the start, and George Mishko, who actually orchestrated the installation, along with about a dozen other key players, for special recognition.

DOE Richland Manager John Wagoner congratulated the project team, "Your accomplishments have been recognized nationally" and relayed Assistant Secretary Grumbly's personal congratulations. In early September, new Secretary of Energy Hazel O'Leary came for a personal visit to bask a little in the success. Harry

Harry Harmon (left) congratulated by Bill Alumkal.
Courtesy *Hanford Reach*.

Harmon took her out to the SY farm and explained how the mixer pump was installed and what it was expected to do later when full-speed testing began.

The installation and first runs of the mixer pump signify the zenith of institutional and individual success over the whole history of SY-101. Never before and never after would so many years of struggle be capped off with such a dramatic, definitive physical accomplishment. Though final remediation of the tank in 2000 was a greater victory, it took months to perform while installing the pump took days. Instead of the tension and excitement of the first pump run, the success of remediation was simply not turning the pump back on. It was an anti-climax where pump installation was a climax. Getting the pump installed was truly grounds for a great celebration. However, typical of anything having to do with SY-101, it would be short-lived.

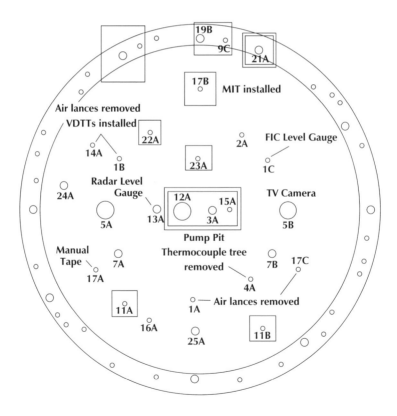

SY-101 riser map showing location of equipment and instructions after mixer pump installation in July 1993. North is up.

*Energy Secretary Hazel O'Leary visits SY Farm in September 1993.
Left to right, Bill Alumkal, Harry Harmon, Hazel O'Leary, and John Wagoner.*
Courtesy *Hanford Reach*.

*DOE Richland manager John Wagoner (left) and Energy Secretary Hazel O'Leary
(right) face news cameras at the SY Farm fence in September 1993.*
Courtesy John Wagoner.

1993–1995

Mitigation Happens

The thrill of victory soon gave way to the agony of defeat. The first low-speed tests not only did little to the waste, but they also plugged the pump. The flush system worked well, and the pump was quickly restored to physical health. But just before the full-speed mixing tests could get started, two fateful incidents precipitated the great 1993 stand down that shut down all operations in the tank farms for 2 months. When work could resume, the mixer pump was finally able to run at full power and thoroughly mix the tank through November and into December 1993. After a final test program ended in April 1994, pump runs evolved into a normal, humdrum operation, and SY-101 gradually faded from public view—the ultimate proof of its effectiveness. This is the final chapter of the mitigation story.

STARTING THE PUMP

The electricians jumped into action as soon as the pump was locked in and worked feverishly all through the swing and graveyard shifts of July 4 and 5 to get the pump plugged in for its first run. The

DACS operator commanded the first pump start at 6:43 AM on July 5, 1993. Mike Erhart was test engineer, and Doug Larsen was the test director. It ran 18 seconds at 380 revolutions per minute with little effect on the tank, but a lot of action in the DACS trailer and the war room in 2750E Building.

> Tension was high everywhere. Everyone was still worried that the pump might induce a huge burp, and everyone was watching to see what happened. A gaggle of senior Westinghouse and DOE observers filled the DACS trailer. Doug and Mike finally had to shoo all but two of them out. The Zone of Death was active and, except inside the DACS, everybody was supposed to be suited up on fresh air. But, in honor of their hard work on the wiring, somebody arranged to have some union electricians huddle in the little motor control shack outside the DACS, ostensibly to monitor the electrical equipment. During startup, Doug had to exit the DACS, walk over to the shack, visit a little with the electricians, ask one of them to close the main breaker, and return to DACS. Then the DACS operator clicked his mouse on the START button. The pump ran, the procedure worked smoothly, and the electricians were happy to witness the fruit of their labors.

The euphoria subsided a little after the first pump run, and everyone went back to work. The second MIT was installed Wednesday, July 7, and Kaiser finished hooking up all the bells and whistles on the pump on Friday, July 9. The first small step in mitigation, Phase A testing, started on Wednesday, July 14, at 8:40 PM with a 1-minute run at 340 revolutions per minute. The test was delayed until evening to wait for the National Weather Service to cancel a thunderstorm warning for the Hanford area.

There was more than just a thunderstorm on the horizon that evening. The next day at the July 16 morning mitigation meeting, Doug Larsen reported that the second Phase A test had started at 5:37 AM with a 1-minute ramp up to 340 revolutions per minute, then a 1-minute run at that speed. But the pump data were not quite right. It looked like there was a problem with flow through one downcomer leg—the first indication of pump plugging! On Monday, July 19, it got worse. The fourth Phase A test at 11:41 AM was a 6-minute run at 340 revolutions per minute. But the flow meter on one of the nozzles stayed on zero. Neither was there any temperature change there during the run. One of the nozzles or down comers was plugged! Early the next morning, July 20, Doug tried a quick 13-second pump bump at 510 revolutions per minute, but it wasn't enough to clear the nozzle.

Jack called an emergency TRG meeting to figure out what to do about the plugged nozzles. It was clear that the original bump schedule specified in the SA was too slow and too short to prevent clogging the nozzles. Because a plugged pump was much worse than potentially triggering a big burp, they decided to let the TRG set the motor revolutions per minute and bump duration to whatever was needed, and do multiple bumps with no waiting. The motor speed only had to stay below the limit of 1,020 revolutions per minute. The changes were approved by TRG and Westinghouse's internal Safety and Environment Advisory Council and reported to DOE Richland.

At 8 PM Doug tried out his new freedom with a 15-second bump at 720 revolutions per minute. That was enough to clear the clog and force flow through both nozzles. With this encouragement, Doug completed the fifth and last test of Phase A, a 6-minute run at 340 revolutions per minute, at 11:09 PM. Again both the temperature readings and flow meter showed one nozzle plugged! But again, at 6:24 AM the next morning, July 21, another 15-second run at 720 revolutions per minute cleared it.

The Phase A test results were encouraging if you looked at them right. Though the gentle pump runs did not affect the waste temperature profiles, they did produce small, but detectable, gas releases. This showed that there was still gas present, and the pump was releasing it. Mitigation by mixing might actually work! At the same time, the sediment was stable, and there was no huge, almost-buoyant gob lurking, ready to be triggered by a tiny disturbance. This was definitely good news. The experience also showed that the short, slow runs really would plug the pump nozzles as the engineers had warned, but a quick burst at higher speed would apparently clear them. A more thorough lesson in nozzle unplugging was next on the tank's agenda!

With Phase A completed, there was nothing to do but wait for the Phase B tests. While waiting, the pump had to be bumped daily to make sure it would keep running. On Thursday, July 22, the DACS crew ran another 15-second, 720 revolutions per minute bump. It came off uneventfully, but, maybe because someone worried about running the pump at the higher revolutions per minute without a formal test plan, the next day's bump was only 5 seconds. This plugged the pressure sensor tap on one of the nozzles. It was still plugged during the next 5-second bump on July 24. At 6 AM on Sunday, July 25, the third 5-second bump plugged *both* nozzles. They stayed plugged through a 5-second bump at 1,000 revolutions per minute at 9:46 AM and a 2-minute run at 1,000 revolutions per minute at 7:02 PM. Even 5 minutes at 1,000 revolutions per minute at 11:30 AM on Monday failed to clear the nozzles.

The pump was now plugged tight! The short little 5-second bursts might have been just enough for the jets to squirt a little plume of sludge straight up around the downcomer legs and into the inlet as some early TEMPEST predictions indicated might happen. However it happened, the project had a very embarrassing problem on its hands. What if they couldn't get the multi-million-dollar pump unplugged? Instead of mitigating the tank, it would become an awkward impediment to any future mitigation effort. All the labor and agony of the last 9 months would have been for nothing—or worse than nothing.

This was the time to exercise the procedures and equipment designed into the pump to clear just such a plug. It was a comedy of errors and a "make do" job in many ways. Tony Benegas called JR Biggs about noon, right after the last attempt to "bump" the clog out and announced that they had to flush the pump volute. It was a hot July afternoon. The volute flush was accomplished by connecting a small, positive displacement flush pump to fittings on top of the mixer pump. The pump pulled hot water from a 55-gallon tank on a truck. The little pump had a heavy duty extension cord to avoid an excessive voltage drop that a small cord might have produced. But, as usual, preventing a potential problem caused a real one.

The safety controls excluded operators from the farm during the flushing procedure. JR had planned to just plug in the pump and start it from the DACS control room. But the on-off switch at the end of the big extension cord could be operated only with power on the line. That meant someone had to go into the farm and press the ON button after the cord was plugged in. JR had to call Jack Lentsch as TRG chairman to get on-the-spot authorization to go into the farm to turn the pump on! Then they found that the big power cord was too short! Tony brought out a couple of six-foot power strips from an office. JR first refused to use them because they were totally contrary to the procedure, but nothing else was available.

Problems didn't end when they turned on the flush pump. The suction line to the water tank was a regular garden hose that was not reinforced to take a vacuum. The hot water and the hot sun weakened it further, and the line collapsed. The only pressure available to relieve the vacuum was the gravity head from the truck bed to the ground. JR had to pick up the line and hold it over his head to keep the flow going, but it worked. In only a few minutes, the temperature sensors in the nozzles showed the hot water had come all the way through the mixer pump from inlet to nozzle.

Right after JR finished the flush, Doug ran another 5-minute, 1,000 revolutions per minute bump. The instruments showed good flow through both nozzles, though the unequal reading on the pump column strain gauges showed one had more flow than the other. The hydrogen concentration peaked at 250 parts per million. Much more waste was getting disturbed this time! There was even a small upwelling of waste about 10 to 15 feet west of the pump in line with the jets. Two more 5-minute bumps happened July 27 and 28. The DACS crew kept on running two 5-minute bumps per day until August 22 when they returned to one per day. The strain gauge differential quickly disappeared and the hydrogen concentration peak after a bump decreased to about 100 parts per million.

Because the pump jets were held in the same position for all these bumps, the pump was apparently carving a trench in the waste where most of the accumulated gas was released, but leaving it to build up in the bulk of the waste on either side of the trench. None of the thermocouples on the two MITs, one about 90 degrees and the other about 30 degrees from the trench line, showed any change in temperature when the pump ran. Nevertheless, each pump bump stirred up the sediment and suspended a considerable cloud of tiny particles in the liquid. While these particles were settling out, they made the liquid effectively heavier, thereby making it easier for the gas-bearing sediment to become buoyant. This subtle, but important, effect would have some interesting consequences at the end of August. Meanwhile, the consequences of some unfortunate human events created one of the most sweeping changes ever to disturb the Hanford work culture.

THE GREAT 1993 STAND DOWN

"Stand down" is a military term signifying going off duty or ending one's period of duty, usually with the connotation of a change in the duty. In DOE parlance, a stand down means all but absolutely essential work ceases until some stated change is made. Hanford stand downs had, up to now, lasted for several days, maybe a week at most, to correct specific problems or investigate an accident. The great 1993 stand down was different in both its long duration and in its powerful influence on Hanford work.

Two completely separate events happening only a week apart precipitated the stand down. The first was an unplanned mixer pump startup in SY-101 on August 4, 1993. The second was a contamination incident during some pit work on dirty old Tank C-106 on August 10 where a worker lowered a rock into the waste on a rope and pulled it out again. Because the latter incident was much more sensational, the whole sequence was soon called simply "Rock-on-a-Rope."

It was a truly a seminal event. Everything else was time-referenced as before or after Rock-on-a-Rope. It was a great constriction, like an hourglass. All projects were halted and not all were restarted afterward. Even saltwell pumping of liquid waste out of the leaky old single-shell tanks got shut down and had to go through a painful restart period, even with consternation from the U.S. Environmental Protection Agency and Washington State Department of Ecology hurrying it along. Everything was different afterward. Here's how the two events happened.

Unplanned Pump Start

After the morning pump bump on Wednesday, August 4, 1993, acceptance testing began for new DACS software. The test procedures called for the DACS software to be run through its full suite of functions, including sending signals to start and stop the pump. But the pump motor circuit breaker was to be opened to make sure the pump did not actually start.

After lunch, the tank farm operations crew began their rounds, preparing for the afternoon pump bump. When they came to the pump motor circuit breaker, they closed it in accordance with the normal procedure, unaware that the DACS testing was still in progress. At that moment, the DACS happened to be commanding the pump to start. When the breaker closed, the pump motor dutifully began to spool up. The people at the DACS control panel noticed the pump motor current rising and immediately hit both the normal and emergency stop buttons. By then, the pump had run about 30 seconds, reaching a maximum speed of 533 revolutions per minute. Though many other, much more aggressive pump bumps had been done without hazard in the prior month, everyone was advised to leave the farm and acceptance testing was suspended. Eventually, the second pump bump was accomplished normally at 5:45 PM.

An "unusual occurrence report" was issued that evening. On Monday, August 9, 1993, there was a meeting of Westinghouse staff to formally investigate the inadvertent pump start. The corrective actions, most of which were already accomplished, dealt mainly with the formal lock and tag procedure, better defining ownership of the pump systems between tank farms and test engineering, and fostering better communication between them. The TRG later decided that tank farm operations owned everything from the motor circuit breaker down to the pump, while test engineering owned everything from the motor circuit breaker up through the variable frequency drive to the 480-volt power supply.

Had this been the only incident, these corrective actions might have satisfied DOE, and work could have proceeded with only the nominal sternly worded letter from John Wagoner. But the very next day, the second link in the chain of events was forged!

Rock-on-a-Rope

On Tuesday, August 10, 1993, two laborers and a health physics technician from Kaiser Engineers were told to enter a pit on Tank C-106 to see if a cover was installed over the pit drain. A pit is a covered, rectangular concrete box set below grade to house piping connections, pumps and other waste transfer paraphernalia. They were also told to look for an open riser to serve as a drain for the pit. The laborers took this to mean they were to search for an open path through a riser into the tank. Between 10:00 and 10:15 AM, they found two holes in a steel plate on the bottom of the pit and decided to see if either could be used as the desired drain path.

To find out if the path was clear, they taped a rock to a length of nylon cord and lowered it 25 to 28 feet into the larger hole. When cord went slack, they believed it had hit a blockage. So they pulled the rock out and laid it and the cord on the pit cover. But it wasn't a blockage! The rock and part of the cord had apparently gone all the way down into the waste. The workers' clothing and gloves were highly contaminated. The contaminated material was disposed of, and the workers were decontaminated by 10:45 AM.

This incident could not have been more ill timed. John Wagoner was escorting the new DOE Assistant Secretary of Environmental Management, Tom Grumbly, on his first visit to Hanford. They were just leaving the 2750E Building in the 200 East Area when a call came in from Dick French, president of Kaiser, describing the event. Grumbly was already disgruntled over Hanford problems and, as John says, "went ballistic!" On August 14, the *Bellingham Herald* quoted this reaction from an Associated Press story: "That is one of the more stupid activities I've heard about on a reservation. . . . This will not be business as usual," Grumbly said. "We expect management changes."

With two serious conduct of operations errors in quick succession, one of them witnessed by EM-1 himself, Bill Alumkal, the new Westinghouse Tank Waste Remediation System executive vice president, knew he had to take drastic action. At 11 AM on Thursday, August 12, he called an all-managers meeting to announce, among other items, "Safety is a condition of employment at Hanford!" In the days that followed, disciplinary action was taken with several employees, including managers, and some managers were reassigned to other jobs.

The fallout of Rock-on-a-Rope was tragic for those who had worked so hard installing the mixer pump. When the plant shut down, everyone was sent to town for training on how to do things right. About 30 operations staff spent months retraining and never got their old jobs back. One month they were heroes, and the next month they felt punished and insulted. They had nothing to do with Rock-on-a-Rope and felt unjustly treated. Later, when they were presented award certificates and plaques for SY-101 work, many refused them and some just threw them away!

Surviving the Stand Down

During the shutdown, the tank farm facilities were a ghost town. All the operators and supervisors were stuck in town for training. Alumkal had ordered the workers retrained, but there were no criteria to train them to. People spent several unproductive days just waiting around while the training material was being developed. The big change that retraining was trying to accomplish was that you can't do whatever you want, you have to follow the procedures. If it's not on your work package, you can't do it. The tank farm management team would say that this was not a new concept, and several years had been spent trying to instill this conduct of operations discipline. Obviously, everyone had not gotten the word.

> Tank farm supervisors were shown the depth of the change one bizarre evening after the shutdown. Bill Alumkal ordered all shift supervisors and shift managers out to the 242-AW Building for a meeting on a Friday at 6 PM. They all expected some kind of stern lecture on the new philosophy, conditions of employment, etc. They were amazed to find a catered steak dinner served by candle light! It was weird to be 30 miles out in the sagebrush on a deserted site, having a nice formal dinner in work clothes. After dessert, Alumkal got up to speak. He told the assembled supervisors, "We're going to start from zero and do one thing at a time. We are in charge, we have the power. We won't be doing very much for a while, but everything will be perfect! When we can do one job perfectly, we'll go to two jobs and do them perfectly."

DOE wanted to shut down the mixer pump along with everything else. But they had to keep the pump running, or it might be useless later. Jack Lentsch told DOE that the pump could plug if it wasn't run twice a day. DOE countered with: you unplugged it once before, can't you do it again? Jack responded that we probably can, but we can't guarantee it. Shutting the pump down would be negligent. That argument did the trick, and pump bumps continued. However, the first pump runs during the shutdown were "highly observed." DOE Headquarters, Westinghouse corporate, and everyone else came out to see. It took half a day to do the bump that first day because of the deliberate, painstaking pre-start procedures. They had to sign, date, and initial and have a second person do the same for each step. It took a long time to get through the procedure.

The stand down lasted into October. During the transition, each work package had to have an incredible marathon review and an oral board hearing before any work began. Rick Raymond, who later would lead the Surface Level Rise Remediation Project, was assigned as Engineering Manager as part of the stand down recovery team. It was 2 years before the work package system became fully operational again.

> For better or worse, nothing at Hanford was the same. The operational discipline was definitely an improvement. But the increasingly cumbersome system that projects have to use to get work done remains a burden to this day.

A FEW LAST BURPS!

The mixer pump was now in, but it had not yet run hard enough to have much effect. Only the trench in the sediment in line with the nozzles was being disturbed. Gas was still building up elsewhere. At the same time, by suspending sediment each bump, the pump was reducing the amount of gas required for buoyancy.

The gas buildup and increased buoyancy came together August 27, 1993. The now daily 5-minute pump bump began at 9:09 AM. About an hour later, there was a sudden level rise of 1.6 inches on

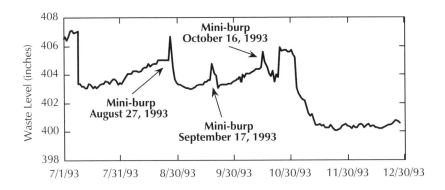

the FIC and 1 inch on the manual tape. The hydrogen concentration shot up to 7,200 parts per million and stayed above 5,000 parts per million for over half an hour. The FTIR showed a peak ammonia concentration of 1,250 parts per million. The tank dome pressurized to +0.3 inches of water. The in-tank video camera showed the big boils of liquid, but it was coming from an area that was not active in Event I, 2 months earlier. The total gas release was estimated at nearly 2,000 cubic feet. In pre-pump days, it would have been considered a pretty decent burp!

Apparently a big gob of gas-filled sediment had risen to the surface, releasing a lot of its gas, but keeping just enough to float on the surface for a while. The gas expanding from a hydrostatic pressure of about 2 atmospheres to 1 atmosphere at the surface made the level rise. By the next morning, the floating gob began to break up and sink. Some of the remaining gas released, but most of it was just recompressed in the sinking gob. By noon, the FIC and manual tape both showed a drop of 1 inch from the pre-burp level, eventually reaching about 2 inches by August 30. Had the pump not induced it early, this gas release would have been Event J that probably would have happened in late October or early November had the mixer pump not been operating. In fact, Rudy called it Event J in his weekly report. But that title was politically incorrect because the mixer pump was supposed to be preventing "events," and it became "the August 27 release" instead.

The August 27 release set the pattern for two smaller pump bump-induced "mini-burps" in September and October, as well as a major one at the beginning of Phase B testing. The first of these was on Saturday, September 17, shortly after the pump bump that started at 8:31 PM. The waste level jumped up 1.8 inches, and the hydrogen concentration spiked to just under 2,000 parts per million. The tank pressure increased a little but did not go positive while the ventilation flow rate went up to 870 cubic feet per minute. The level rise subsided about 24 hours later without releasing very much more gas. Another smaller burp happened on the morning of October 16, 1993, right after the morning pump bump at 6:05 AM. Again, the FIC registered a sudden rise of over an inch that subsided after about 24 hours. The hydrogen concentration jumped to 460 parts per million with a very slight dome pressure rise and a momentary ventilation flow spike to about 750 cubic feet per minute.

The End of TEMPEST

The small burps were far from dangerous, but they did keep the excitement level up during the stand down and sustained the pressure to get on with Phase B of the mitigation testing. They also gave the Los Alamos crew new data to use in the SA that allowed them finally to remove TEMPEST calculations as the basis for their level controls.

TEMPEST predictions of gas releases from the water lancing and initial pump runs had proven to be embarrassingly high. Rudy Allemann calculated that the water lance disturbed 140 cubic feet of waste and released 7 cubic feet of gas. TEMPEST had earlier predicted that water injection could cause a major burp. The gas releases from Phase A were barely detectable. TEMPEST predicted best estimate releases of 20 to 60 cubic feet for 1-minute and 10-minute tests, but it conservatively predicted a big burp of around 10,000 cubic feet! At their July 20, 1993, meeting, the SubTAP noted the disparity between TEMPEST predictions and observed releases and renewed their call for some method to make an in-tank measurement of gas volume. At their September 1 meeting, they stated that the August 27 mini-burp "emphasizes the unpredictable nature of tank and reinforces the need to be skeptical of model predictions."

Don Trent's and Tom Michener's worst fears had come home to roost. TEMPEST had been used in a way they never intended and had consistently advised against. But it was too late. Though it had given the team some key insights on waste behavior in the early days, TEMPEST had lost its credibility.

Kemal Pasamehmetoglu also recognized the disparity and realized that maybe TEMPEST predictions were too conservative to use in the SA for mixer pump controls. Instead, he had the insight to apply the data from the three small pump-induced gas releases directly. Kemal discovered that when the estimated gas release volume was plotted against the initial waste level, the mini-burp releases appeared to follow the same trend as the bigger, prepump burps. In early 1994, the simple level versus gas release plot went into the SA to replace the countless hours Don and Tom spent running TEMPEST predictions over the past year. However, it was TEMPEST predictions, conservative though they were, that allowed the mixer pump to be installed, and the Phase B mitigation tests to begin.

More Ammonia

The ammonia gas issue was introduced with Event I and temporarily accommodated with the Zone of Death. But that was just the beginning. More and more people began to believe it was a serious hazard. Norton McDuffie, former Oregon State University chemistry professor working for Jerry Johnson, was particularly alarmed by it. On September 3, 1993, he wrote, "The high ammonia production rates [gas release] from SY-101 (13,300 ppm [parts per million]) during the June 26 venting [Event I] justify a complete re-evaluation of our programs at Hanford."

Along with McDuffie, Bob White took the ammonia issue on as a sort of personal crusade. At a TRG meeting on September 21, he summarized a 4-page paper stating his perception of the ammonia issue versus what was already in the SA. Los Alamos calculated that the concentration of ammonia vapor in equilibrium with the liquid in SY-101 could approach 20 percent. Ammonia was flammable in the air at about 15 percent. While no one believed ammonia could evaporate fast enough to approach equilibrium with a ventilation system renewing the dome atmosphere every hour, they didn't have enough data to estimate just how far from equilibrium the tank might be.

Evaporation off the liquid surface was believed to be the dominant release mechanism, though a lot of ammonia also came out with the bubbles during the burps. This was why the ammonia concentration jumped up and stayed up quite a while after a burp that thoroughly wet the crust. The fear was that future aggressive mixer pump runs might circulate liquid to the surface, keeping the crust wet and increasing ammonia evaporation. This fear proved groundless, but it was obvious at the time that larger ammonia releases could potentially occur than Los Alamos had assumed in the SA. Accordingly, some SA changes were needed, including warnings of ammonia hazards, long-term monitoring of gas composition, and backup ammonia monitoring. Bob White recommended going ahead with Phase B with the current controls, adding the FTIR, or Fourier transform infrared gas monitoring system, to the minimum equipment list for ammonia monitoring. In the meantime, Los Alamos would prepare an extensive SA revision to address ammonia in time for full-scale testing.

Because the SA admittedly did not treat the ammonia hazard in its full magnitude, Westinghouse Nuclear Safety and Licensing thought it could be an official unreviewed safety question or USQ and dutifully brought the issue to the Plant Review Committee for a formal review. On October 15, 1993, the Plant Review Committee stated their conclusion that the new ammonia release mechanism and the error in SA calculations were, indeed, a USQ. But they also believed that the latest revision of the SA, expeditiously completed by Los Alamos, had resolved the question and Phase B mitigation testing could proceed. It was truly an open and shut case.

Even though the official USQ was closed as soon as it was declared, the ammonia issue persisted. Almost a year later on July 20, 1994, the SubTAP was still complaining that Westinghouse should investigate and eliminate the ammonia issue. If only that were possible! Proving a negative is nearly impossible and proving the improbability of huge ammonia releases was no exception.

> The ammonia issue developed into a controversy with two distinct sides. Dan Reynolds compared them to religious sects. Adherents to the "high ammonia" creed believed that the ammonia in solution was fully saturated and had their own data sets, models, and theories to support it. Those holding the "low ammonia" doctrine looked at the available ammonia measurements pragmatically—I don't see it, therefore I don't believe it! The seat of the high ammonia religion was Los Alamos, with a few believers in PNNL and DOE. Bill Kubic from Los Alamos might have been its high priest. Most advocates of the low ammonia faith came from Westinghouse and PNNL. However, each faction respected the other, and there was no violent reformation.

There was evidence on both sides. Waste core samples showed that the dissolved ammonia concentration was far from saturation. But the sample analyses were discounted because ammonia could easily escape the samples during the weeks of waiting in the hot cell. Likewise, laboratory mass spectrometer analysis of grab samples of tank headspace gas had low ammonia concentrations. But these findings were also discounted because ammonia could hide on the inside surface of the sample tubes and vessels without making it into the machine.

> Bob White brought up a bizarre twist to the ammonia hazard. In 1986 there was a disastrous carbon dioxide release from Lake Nyos in Cameroon that killed about 1,700 people. The bed of this deep lake had volcanic vents that saturated the lower layers with dissolved carbon dioxide gas at a pressure of 21 atmospheres, just like a carbonated beverage. At some point the water could hold no more and the gas began to form bubbles. Like opening a bottle of pop, this small disturbance grew quickly until the entire lake turned over, effervescing a huge volume of suffocating carbon dioxide.
>
> Despite the dubious comparison of a deep, cold, clear lake in Africa to a shallow, hot, murky Hanford waste tank, Bob thought a layer of ammonia-saturated liquid might exist at the bottom of waste tanks that, like the carbon dioxide in Lake Nyos, could effervesce into a catastrophic ammonia release. This hypothetical accident was called the Lake Nyos scenario. DOE and Westinghouse took the threat seriously and, on April 12, 1995, PNNL held a workshop to try to determine whether the Lake Nyos scenario was real. Everyone had to admit that, because ammonia is one of the most soluble of gases, there was a lot of it dissolved in the liquid in SY-101. But it had to reach saturation at a pressure above 1 atmosphere to set up a Lake Nyos-type release. Radioactive decay heat created convection currents that circulated liquid from top to bottom keeping ammonia from reaching saturation above 1 atmosphere. A higher concentration might build in a nonconvective sediment, but a solid sediment can't flow to start an effervescing plume. Also, a regularly burping tank loses excess ammonia in periodic releases. The meeting concluded that a Lake Nyos ammonia event was not plausible in SY-101 or any other tank.

There was never any "silver bullet" report or finding that could drive a stake through the heart of the ammonia issue. It just died away. New and better data showing low ammonia concentrations kept coming in. The new Retained Gas Sampler, which captured a core segment in a gas-tight container to keep ammonia from escaping, provided some of the best data. The weight of the data eventually smothered the sampling error objections. At the same time, understanding of ammonia release mechanisms improved enough to explain the rare cases where the concentrations were not low and

276 | Part Two: The Mixer Pump Era

to predict situations where extra care might be warranted. The accumulating data and knowledge changed the perception of ammonia from a dangerous accident waiting to happen into just another tank farm hazard, like radiation, that had to be treated with respect, but no longer with fear.

PHASE B TESTS—REAL MIXING

The Phase B or high-speed mixing tests would finally run the pump at full power to mix the waste as thoroughly as possible. In the process most of the gas that was not released in Event I or the "mini-burps," plus whatever had accumulated since, would be released. They had to be careful. The fear of the "mother of all burps" was still very real, and the mini-burps showed that the tank still held enough gas to be interesting. In September, a final set of 40 TEMPEST simulations predicted that the pump might start a burp if the sediment gas fraction exceeded 13 percent. Moreover, if the pump started a burp, shutting off the pump could not stop it!

Because of this fear, Phase B started gently. Larry Efferding's final revision of the Phase B test plan, issued in September 1993, started with a series of four 20-minute runs with the nozzles aimed 30 degrees clockwise from the trench excavated by the 9 hours of cumulative pump bumps on the east-west orientation since the very first run on July 4. The first run was a very gentle 20 minutes at 360 revolutions per minute. The second and third increased the speed to 510 and 720 revolutions per minute, and the fourth to almost full speed, 920 revolutions per minute. Five more sets of four 20-minute runs were planned, each set with the pump rotated another 30 degrees clockwise, ending back at 5 degrees, almost in the trench which may have been filled back in by the earlier runs.

The six nozzle orientation angles were chosen to aim the jets at important pieces of equipment to give as much information as possible about what the pump was doing. The first increment to 35 degrees clockwise from the east-west trench aimed one jet directly at the two VDTTs. Though the velocity, density, and temperature sensors were dead, the strain gauges up near the tank dome might show when the jet was pushing on them. Also, the jet opposite the VDTTs was aimed near the MIT in Riser 17C, where a waste temperature change would indicate mixing. The TRG later substituted a 28-degree orientation, which aimed the jet exactly at the MIT, for the original 35 degrees after Phase B, because temperature change was much more informative than the strain gauge readings. The 95-degree angle directed a jet at the other MIT in Riser 17B for another positive measure of jet penetration. Finally, the 155-degree angle was aligned with the FIC level gauge, located to the northeast, and the manual tape, located to the southwest, though these gauges were on the waste surface and not under the direct influence of the jet.

The initial series of twenty-four 20-minute runs were expected to mobilize the central core of the waste, probably releasing most of the gas. The next block of tests used 6 runs of 1 hour at 750 revolutions per minute in each of the six chosen angles. After this, a longer series of optional "tank sweeps," where the pump was run in each of the six orientations in the same day, would complete de-gassing and mixing.

The total pump run time for the entire Phase B test program was 56 hours. Adding the prescribed waiting time between tests, the program needed a minimum calendar time of 24 days, 15 hours, and 20 minutes to complete. Phase B was supposed to have started on August 3, according to Jack Lentsch's hope at the July 7 morning mitigation meeting. If this had happened, mitigation might have been demonstrated by early September. But the great 1993 stand down intervened, and the mitigation project had to submit to the discipline of the stand down transition team and accomplish the required crew training and reviews like any other project. Phase B could not start until the end of October.

One of the more troublesome factors was calibration of the mixer pump and tank instrumentation. Apparently, nobody had noticed that many of the administratively required calibrations were overdue. Before Phase B could start, everything had to be brought up to date. The initial schedule called for calibration of all expired instruments to be completed by the end of October. Devices needing calibration were lurking everywhere. For example, DACS relays, nitrogen purge devices on the camera, and analog-digital converters on the MIT thermocouples all had to be calibrated. The sheer volume of work threatened more delay, and there were some very spirited discussions about it at Jack's morning meetings.

After a few more delays, Phase B started at 5:17 AM on Thursday, October 21, 1993. The first run was to have been 20 minutes at 360 revolutions per minute with the nozzles aimed into "virgin"

278 | Part Two: The Mixer Pump Era

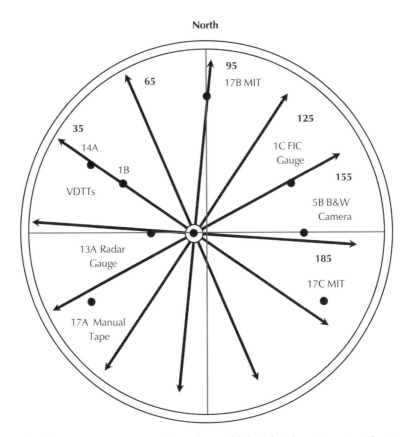

Sketch of the six mixer pump nozzle angles established for Phase B testing. The 35-degree orientation was changed to 28 degrees to pass through the 17C MIT during full-scale testing.

sediment at 35 degrees. But the sediment in that area was apparently ripe enough that the pump induced a small gas release after only 23 seconds. The DACS crews had been trained to jump on the STOP button at any indication of excitement in the tank, so the operator immediately killed the pump and called the TRG. The TRG wasn't as excited and authorized the test to restart. It ran its full 20 minutes at 7:45 AM without further incident. Likewise, the 510 revolutions per minute test went off without any excitement at 11:50 AM on Friday, October 22.

Then the tank decided to take control of testing. At 5:33 AM on Monday, October 25, halfway through the third 20-minute test at

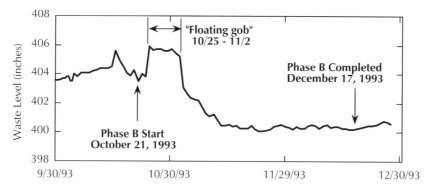

SY-101 Waste Level History: *Phase B Tests, 1993*

720 revolutions per minute, there was a sudden level rise and the hydrogen concentration climbed to about 2,100 parts per million. The test director quickly stopped the test as the vent flow jumped up. The FIC gained 2.3 inches and the manual tape showed a rise of 1.5 inches. This meant that a big chunk of sediment equivalent to maybe a quarter of the tank area had risen from the bottom and was floating at the surface.

This sudden rise pushed the average weekly level rise rate far above the maximum allowable level rise rate of 0.15 inches per day in the SA and testing had to stop. It didn't matter that the rise rate limit in the SA was conceived as a steady rise indicating rapid gas retention, not a sudden jump due to gas decompression in a rising gob of sediment. Jack had to wait until either the average rise rate dropped below the limit or until Los Alamos could revise the SA to clarify the limit.

Because a similar abrupt level rise had happened in the three mini-burps in August, September, and October, the TRG expected a level drop later that day or the next, after which they could get back to business without having to wait for an SA revision. But nothing happened that day or the day after. The gob just floated on the surface like a dead whale, holding up the waste level and the test schedule, much to Jack Lentsch's aggravation. He realized the tank was now in control, and all he could do was wait.

In the meantime, daily pump bumps resumed with the nozzles aimed back in the trench. Gas leaked out of the floating gob little by

280 | Part Two: The Mixer Pump Era

little, and the level crept lower. Finally, after more than a week of waiting, at 6:23 AM on Tuesday, November 2, they were able to try the 720 revolutions per minute run again. This time it lasted 13 minutes before a large gas release required the pump to shut down yet again. This time the hydrogen concentration reached 900 parts per million along with a ventilation flow spike to about 1,000 cubic feet per minute and a brief decrease in domespace vacuum. Though the gas release stopped the test, it indicated that the floating gob of waste had partially disintegrated and sunk. Phase B now could go ahead without a break. Well, not quite. Over the next week, the FIC level gauge was intermittently unusable. Because it was a required instrument, testing had to halt until it was finally fixed on November 10.

Even considering the delays and halted tests of the first three weeks of Phase B, everyone was encouraged at the pump's ability to stir up the waste. Even the SubTAP became a little excited. At a November 10 meeting in Kennewick, they acknowledged that "data from Phase B suggest that mixing will prevent burps" and went on to recommend that the existing pump be run for long-term mitigation. At the same time, they worried about the lack of evidence that the pump was actually mobilizing waste all the way to the wall. An undisturbed layer could be building at the wall that might someday yield a huge burp even though the central core of waste was perfectly mixed. There was no way to prove or disprove these fears. All Jack could do at present was to keep running the pump.

By November 15 the waste level had dropped about 2 inches from the start of Phase B. The radar gauge and manual tape were reading about 402 inches and the FIC sat at 400.5 inches. Not much gas was coming out anymore. Neither the full tank sweeps at 750 revolutions per minute nor the 35-minute runs at 920 rpm managed to raise the hydrogen concentration above 140 parts per million. Phase B testing finally became routine.

Rudy Allemann retired from PNNL on December 1, 1993, after working at maximum effort on SY-101 since 1990. He had assumed full responsibility for PNNL's contribution to Jack Lentsch's team in May after Max Kreiter, the first PNNL liaison to the mitigation project, retired. Rudy was perhaps most responsible for gathering together the sum total of the communal knowledge about SY-101's waste and weaving it into a general burp theory. Rudy's "gob theory" has been modified, but it still stands as the core explanation of why SY-101 was the way it was.

I returned from an educational leave at Washington State University on October 1, 1993, and Rudy handed the PNNL effort over to me. I will not soon forget the necessarily quick, eye-opening education that Rudy gave me in the 2 months we worked together. It was scary seeing careers hanging in the balance at Jack's morning meetings, being curtly but courteously told to "go away" so Rudy could write his weekly report, and being introduced to a team of hardened professionals who had very little time to waste with the "new guy." This team had been forged in the Fiasco in Pasco, exhausted in Event G, tried by installing the pump, and shamed by the Rock-on-a-Rope while I sat in a comfortable classroom in Pullman. But Jack gave me the benefit of the doubt long enough to get used to the powerfully swift flow of the work, and about the time Phase B ended I started having fun writing the report describing what happened.

The final test of Phase B ended at 5:33 AM on Friday, December 17, 57 days after the first test. The waste level had dropped about 4 inches, ending up at 400 inches on the FIC and about 401 inches on the radar gauge and manual tape. The actual test program used 28 hours of operating time, just half of the planned total. Of the optional tests, only 750 revolutions per minute tank sweeps and four runs of 35 minutes at 920 revolutions per minute were completed, two at 35 degrees and two at 95 degrees. But various holdups and problems stretched the calendar to more than twice the minimum time planned.

It's a good thing the tests stopped when they did. The 400-inch level was the *minimum* level allowed in the SA for mixer pump operation because Bob White and Kemal were afraid to de-gas the tank too much for fear of unknown long-term effects. This limit would become a plague during full-scale testing.

The pump's success was kind of a surprise to many. But, then again, in Steve Marchetti's words, "Everything that worked at Hanford was an amazement!" On the other hand, Nick Kirch and most of the project crew expected it to work. My formal PNNL report on Phase B test results, issued in March 1994, confidently proclaimed success. The final paragraph in the summary said, "Based on the objectives of the Phase A and B test plans, the behavior of the waste, and the performance of the mixer pump to date, we conclude that mixing has mitigated Tank 241-SY-101 by essentially eliminating the hazard of large flammable gas releases. Evidence continues to accumulate that the pump will maintain this state while it remains in operation." Following Phase B, pump bumps were no longer run daily, but about every other day. Also, instead of staying in the east-west trench, the pump was rotated 30 degrees each bump to cover each of the six angles used in Phase B. Effort now shifted to planning long-term tests to learn how the mixer pump should be run to keep the tank mitigated for the foreseeable future.

FULL-SCALE TESTS

There were a lot of opinions, ideas, and recommendations floating around about what the full-scale tests should do. As was his habit, Jack Lentsch decided it would be best to get everybody in the same room for a day or so and sort it out together. The meeting was December 14, 1993, at the Silver Cloud Inn in Kennewick near the south side of the Columbia Center Mall. The Silver Cloud has a unique meeting room that occupies the entire top floor. It is already up on a bit of a hill, and the big windows in the conference room give a beautiful view of the Columbia River near the confluence of the Yakima River. In the far distance, the dome of the Fast Flux Test Facility reactor is visible. Still further, though not in the line of sight, one can deduce the location of the SY farm in relation to the ridges of Gable and Rattlesnake Mountains.

The whole Westinghouse, Los Alamos, PNNL mitigation project team was there, along with numerous DOE Richland and TAP observers. The meeting first claimed success for the Phase B testing with the statement that "the Phase B tests demonstrated that with only limited pump operation, the tank level could be reduced to a historically low level of 400 inches and maintained at that level. The success criteria has (sic) been achieved by the mixer pump since none of the gas releases [since pump installation] have exceeded 25 percent of the LFL [lower flammability limit]." But they recognized that the pump had not yet run enough to demonstrate long-term success sufficiently to satisfy all of the skeptics. This is what the full-scale tests were for.

Besides the basic need to maintain the status quo for a longer time, there was much discussion about unknown long-term effects of the pump. Could mixing change the waste to eventually make the pump ineffective, or produce runaway gas retention and catastrophically huge gas releases should it fail? Was the pump jet slowly eating its way through the tank bottom or side wall? Could mixing cause dangerous new chemical reactions? Just how much of the sediment was still lying undisturbed around the edges or stuck on the tank wall waiting to burp? Each of these worries demanded some facet of full-scale tests to specifically address it. At the same time, almost everyone recognized that only a month or two of tests could not begin to answer the really long-term questions, though it was politically necessary to claim it would at least help. In fact, SY-101 had started a long-term experiment. The results would not be known for years, but it looked like we were off to a pretty good start.

By afternoon everyone settled down to work out the detailed test plan. The objectives were to 1) demonstrate that the mixer pump could control retained gas volume, 2) define a schedule of mixer pump runs that would keep the tank mitigated, and 3) measure the distance the pump jet penetrates into the sediment at full speed. Effective mitigation and control were further quantified as keeping the waste level between the lower limit of 400 inches and a threshold level of 402 inches by increasing or decreasing the amount of pump operation. The jet penetration measurement was needed to estimate how much of the waste might still be undisturbed and to confirm the design of the "permanent pump" that was eventually to replace the "test pump" in the tank.

The test plan ended up with four distinct phases: a "nominal" phase to try a candidate long-term schedule; a "regrowth" phase to see how quickly gas would accumulate if the pump were not run; a "knockdown" phase to demonstrate that gas could be removed quickly; and a "jet penetration" series. The first phase applied a candidate "nominal" pump schedule of one 60-minute run at 750 revolutions per minute every other day for three weeks. This was a first guess at a schedule that would keep the waste level between 400 and 402 inches and satisfy the second test objective. If the level rose or fell, the duration, speed, and frequency could be adjusted.

The regrowth phase applied minimal pump activity, only enough to keep the nozzles from plugging. There was just one 5-minute bump every two or three days at the same angle for an entire month. The waste level was expected to stay constant or even drop some initially as the sediment layer re-formed and then to rise steadily as gas accumulated. If it rose faster than it had between burps in prepump days, it might show that the mixing had badly exacerbated gas retention. If not, some of the SA controls based on worries about long-term effects might be relaxed, specifically the 400-inch minimum waste level.

The third phase ran three successive full tank sweeps to knock down the waste level and release the accumulated gas as quickly as possible. A tank sweep was six 30-minute runs at 750 revolutions per minute each day, with the pump rotated 30 degrees for each run. These tests addressed the first objective to demonstrate control of the waste level. If gas started accumulating on the nominal schedule, the pump could always be started up to get it out. The basic premise of the long-term operating plan was that running the pump more would make the waste level fall and not running it would allow it to rise. The failure of this fundamental premise was what precipitated the surface level rise emergency 5 years later.

The final tests were to measure jet penetration with a series of four maximum power tests where the pump was run as long as possible at high speed until the motor oil temperature reached the 190-degree limit. This amounted to two runs at 920 rpm for 45 minutes and two at 1,000 rpm for 25 minutes, all at 97 degrees. The 97-degree angle aimed one nozzle directly at the MIT in Riser 17B, so its closely spaced thermocouples could reveal the time and depth of jet penetration. The lower four thermocouples at 4, 16, 28, and 52 inches off the bottom were the most interesting.

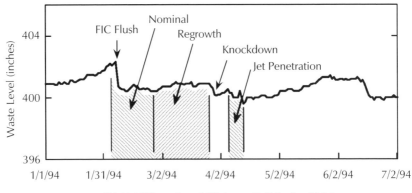

SY-101 Waste Level History: *Full Scale, 1994*

The first full-scale test began at 5:13 AM on Friday, February 4, 1994. It was a 60-minute run at 750 revolutions per minute at 35 degrees. Eight more nominal runs were made by the end of the three-week period on February 25. Each run released quite a bit of gas with hydrogen concentrations peaking at up to 250 parts per million. As expected, the waste level did not change much. The FIC stayed at around 400.5 inches while the radar gauge and manual tape hovered around 402 inches. Though everyone realized that three weeks was not long enough to confirm a pump schedule, it seemed plausible for long-term operation.

The regrowth period from February 26 through March 27 was one of the most instructive tests of waste behavior ever run on SY-101. The waste level rose less than an inch through the entire time, whereas in prepump days the level rise in a month between burps would have been more than 2 inches. More interesting, the temperature profiles on the two MITs showed that the solid particles stirred up by the pump settled back much more gradually than expected. These particles were really small! The sediment never really re-formed in any way equivalent to the deep prepump nonconvective layer. Instead, the temperature profiles showed two convective layers: an upper one about 100 inches deep of "clear" liquid and another below it with 150 inches of still-settling solids. Even a little bit of mixing would apparently last a long time.

The behavior of the hydrogen concentration following the pump bumps of the regrowth tests was a mystery that has never been adequately explained. Instead of the expected rapid increase in concentration to a peak followed by an exponential decrease to about the starting value, the hydrogen concentration *decreased* rapidly after a very small spike at the start of a bump. It was as if the 5-minute pump run froze every rising bubble in place and prevented its escape for about 12 hours. This "hydrogen suppression" phenomenon remained a source of curiosity until July 1994 when it inexplicably ceased. Nobody thought it was dangerous, but it was not very satisfying for the tank to be doing something we did not understand. It was also an operational irritation because the SA set a minimum hydrogen concentration of 10 parts per million to keep the tank within historical norms. If the concentration went below the limit, the TRG had to specifically approve a pump run. The concentration limit was soon changed to a weekly average, but it remained a headache until it was removed altogether in August 1994. It was almost as if the tank had read the SA and tried to challenge each control.

The two tank sweeps of the knockdown phase, following regrowth, were very effective at releasing gas. The hydrogen concentration exceeded 500 parts per million on each run and succeeded in getting a little more than the accumulated gas out in 2 days. The waste level on the FIC dropped about an inch, ending slightly lower than at the start of the regrowth period. Though the two tank sweeps released the gas, they did not mix the waste as completely as it had been prior to regrowth. The upper convective layer shown in the temperature profiles was reduced in thickness but was still there. The jet penetration runs would have to finish the mixing job.

The first jet penetration run began at 5:34 AM on Wednesday, April 6, 1994, with a 40-minute run at 920 revolutions per minute at 95 degrees. The last 20-minute run at 1,000 revolutions per minute ended at 5:27 AM the following Wednesday, concluding the full-scale test program. The gas releases from these first two tests were nominal, but the tank still had one last surprise. The first 1,000 revolution per minute run had to be aborted after only 9 minutes on high ventilation rate while the hydrogen concentration went to 600 parts per million. Apparently, there was still some waste out there with enough gas in it to do something that looked like a burp.

Special efforts were made to detect jet penetration at the tank wall. A special infrared video camera was lowered into the annulus in line with the pump jet to detect small temperature changes on the primary tank wall that might occur if the mixer pump jet hit it forcefully. Also, data loggers were attached to the set of thermocouples that were originally built into the tank bottom to monitor the heat treatment done to relieve welding stresses. Again, slight temperature changes registered by the bottom thermocouples in line with the jet would indicate excavation of waste off the tank bottom.

There was no detectable change in either the primary tank wall infrared signature or on any of the thermocouples attached to the tank bottom. But three of the bottom four thermocouples on the MIT in Riser 17B clearly showed the progress of the jet as it pushed through the sediment. On the first 920 revolution per minute (rpm) run, there was little change in temperature at the lower levels. The 52-inch thermocouple registered a sudden temperature rise of about half a degree Fahrenheit at the start of the second 920 rpm run, and the 28-inch thermocouple jumped one full degree at the start of the first 1,000 rpm run. Finally, about ten minutes into the last run, the thermocouple at 16 inches stepped up over three degrees! Although there was no response from the lowest thermocouple at 4 inches, these indications showed that the jet had penetrated below 16 inches at a radius of almost 30 feet from the pump centerline.

Like Dan Reynolds, Larry Efferding was a "fellow engineer," the highest grade a scientist or engineer could attain in Westinghouse. He had a rather stern, crusty demeanor and gray hair from a long career of hydraulics and pump calculations. Larry may not have been entirely comfortable working under PNNL direction and probably would have rather had it the other way. He wasn't about to take second place to any PNNL scientists. With a stream of plots and tables with artful combinations of data and creative calculations, Larry continually challenged our understanding of what the tank was doing. There were interesting possibilities everywhere, and none were easy to refute or to accept.

Larry had a theory for the lack of any temperature change on the infrared scan of the annulus wall during the jet penetration tests. He was convinced that a 3-foot-thick layer of solids on the wall was insulating it from any disturbances on the inside. My own theory was that the infrared scan was looking above the point where the jet would hit the wall and should not see any temperature change in a well-mixed tank. Besides, an adhering waste layer should make the annulus wall cooler towards the bottom, not uniform as the infrared scan showed.

Because I wrote the report on full-scale testing, Larry's layer theory was not included. Larry retired in 1994 and was not present when large volumes of waste were transferred out of SY-101 in January 2000, when much of the wall was revealed for the first time. When the in-tank camera panned to the wall, there, staring back in stark reproach, was a layer of waste about 3 feet thick! Larry, if you read this, please accept my humble apology!

Completion of the full-scale test program was somewhat of an anticlimax. While Phase B had proven the ability of the mixer pump to remove retained gas, the full-scale test had only showed more of the same. When the SubTAP got the results at a meeting on April 11, no resounding pronouncements or congratulations were forthcoming. Instead there was grousing that the jet penetration tests did not prove the jet actually reached clear out to the tank wall and even a recommendation to "initiate studies to determine the effectiveness of the pump." It was almost as if the full-scale tests had accomplished nothing at all. On the other hand, the focus of the SubTAP was shifting from SY-101 to other tanks and other issues. They admitted that the risk of another burp was low, and the mixer pump would probably keep it that way. So all that remained to do was keep on running the pump.

Returning to the hurricane metaphor I used in the beginning, we were now under blue sky in the eye of the storm. Even though large swells were still tossing in the ocean, the intense storm of mitigating the tank had temporarily abated. Throughout this trial we all realized that SY-101 was a personality. Though the mixer pump was in and running, the tank was still in control and seemed to know it. On June 21, 1996, DOE Richland closed the flammable gas USQ on SY-101 because the mixer pump eliminated spontaneous gas releases, and everyone turned away to work on other pressing needs. But it was a mistake to turn our backs on this tank. It was as if it was thinking, I'm going to stay quiet for a while to lull you into complacency, then I'm going to do something alarming and unexpected. SY-101 would be back in the headlines soon!

PART THREE
Surface Level Rise

September 1995 – Gradual waste level rise first officially noticed

February 26, 1998 – DOE declares SY-101 surface level rise USQ

October 1, 1998 – Surface level rise remediation project authorized

May 29, 1999 – SY-101 surface level peaks out at 434.28 inches, crust about 12 feet thick

May 1999 – Mechanical mitigation arm attempts to break up crust

August 1999 – High pressure, 42-inch-diameter water lance bores a hole through the crust

December 18, 1999 – First transfer and back dilution campaign begins

March 15, 2000 – Final transfer and dilution completed

April 3, 2000 – Last run of the SY-101 mixer pump

January 11, 2001 – SY-101 removed from the flammable gas watch list

August 13, 2001 – Flammable gas safety issue officially closed and all tanks removed from watch list

1995–1996

The Eye of the Storm

Though mitigation had become an almost universally accepted fact after full-scale testing in April 1994, there was a lot of unfinished business. But the work changed from a heroic battle against a force of nature to the thankless slog of cleanup and rebuilding. The mitigation project was still running the pump as a test, having yet to settle on a final long-term schedule. Then there were the other alternative mitigation methods in various stages of development whose fate had to be decided. It would take more than a year before the U.S. Department of Energy, or DOE, could bring itself to officially declare mitigation officially accomplished. The "glory days" were over, quenched by the great Rock-on-a-Rope correction and superseded by the mixer pump's success. It seemed as if mitigation was almost too easy. SY-101 still had enough mischief in it to fuel a short future period of glory, but Hanford had a few years to lumber off and turn its attention to other things.

EXCAVATION AND NORMAL OPERATION

The tank sweeps and jet penetration runs at the end of full-scale testing were almost too successful. The waste level had been flirting

292 | **Part Three: Surface Level Rise**

with the 400-inch limit imposed by the safety assessment or SA and now descended below it, preventing anything but pump bumps. Los Alamos didn't want to change it for fear that a more compact, stronger sediment would make larger natural gas releases if the pump quit. Consequently, the tank entered a second, sort of unplanned regrowth period of minimal pump activity. Towards the end of May the level began to rise again and the Test Review Group, or TRG, decided it would be wise to begin the "nominal" schedule of three 60-minute runs per week at 750 revolutions per minute that had been tried as the first part of the full-scale tests. At the same time Los Alamos was finally convinced to reduce the minimum waste level to 398 inches to give operations some flexibility. The minimum waste level limit was finally removed altogether in September 1994.

The definition of "nominal pump operation" would soon change. One of the main realizations from full-scale testing was that there was a layer of undisturbed waste at least a couple of feet thick and maybe deeper on the bottom of the tank. Los Alamos had calculated that this layer could generate a meaningful burp, even though the mixer pump was keeping the rest of the tank perfectly mixed. On the other hand, the jet penetration tests showed that the undisturbed layer could be swept out, at least to about a 30-foot radius, with a few high-speed pump runs aimed in the same direction. Accordingly, the TRG hatched a new strategy to remove the undisturbed layer around the whole tank, with a series of "excavation runs," beginning with one aimed at 28 degrees to hit the multifunction instrument tree, or MIT, in Riser 17C.

Excavation at 28 degrees began at 5:19 AM June 10 with a 25-minute run at 1,000 revolutions per minute. After the sixth run, July 6, the thermocouple 16 inches above the tank floor showed a response towards the end of the run. Because the temperature profiles now showed that the undisturbed sediment could be excavated down to below 16 inches with four to six runs, the TRG was confident the same thing could be accomplished at the other angles, even without an MIT in the line of fire to prove it. By September 10, 1994, four excavation runs had been done at 65 degrees, six at 125 degrees, and two each at 155 and 185 degrees. Gas releases were rewarding. One of the 65-degree runs produced a peak hydrogen concentration of 1,000 parts per million! This drove the level on the

Food Instrument Corporation, or FIC, gauge below 399 inches, the lowest recorded since the tank was filled in 1980.

The success of the excavation runs and the need to keep the maximum amount of waste mobilized led the TRG to the logical conclusion that excavation should be the nominal operation. Hence the long-term operating plan was amended to require three 25-minute runs per week at 1,000 revolutions per minute with the goal of keeping the waste level below 399.5 inches, but above 398 inches. This schedule began September 13 and continued through April 1995 when the TRG became concerned that the pump might not be mixing "scallops" of waste between the six nominal angles. Accordingly, the operating plan was revised once again to use six runs at new angles: 15, 50, 80, 110, 140, and 170 degrees, followed by six runs at the original angles. With an average of three runs per week, the pump jet would return to the same angle every fourth week. The first run at 15 degrees was made May 12. The peak hydrogen concentration increased to 300 parts per million for the first attack of the "scallops" before returning to the nominal 100 parts per million after the sediment within reach of the jet had been disturbed.

With this last change, mixer pump operation became truly routine. The only excitement came when bad weather, instrument breakdown, or some other problem foiled the normal Monday-Wednesday-Friday schedule, forcing weekend work to make the required six runs every two weeks. SY-101 was no longer perceived as an enemy threatening an imminent attack, but as an unrepentant and troublesome captive that had to be watched constantly, fed and exercised regularly, and disciplined occasionally.

FATE OF THE OTHER MITIGATION CONCEPTS

The mixer pump was clearly doing a good job of preventing burps and mixing the tank. But what about heating, dilution, the sonic probe, and Los Alamos's ultrasonic agitation scheme that had been chosen as alternative mitigation concepts at the Fiasco in Pasco back in 1992? Even though the test chamber had perished of budget starvation in late 1992, advocates of the various concepts had kept them alive all through 1993. But by the end of that year it was clear there was no need for them in SY-101 anymore. For a while the party line was that they might be useful in other tanks or as a pretreatment step on the front end of the future glass plant. But, because mixing was a necessary step in waste retrieval, the pump supplanted them in this arena also. It just took a little more time for DOE to take a deep breath and decide they were not necessary.

Mark Hall kept pushing his sonic probe mitigation concept through the fall of 1993 and into 1994. He and some Fast Flux Test Facility people with extra time on their hands set up some rather elaborate tests of commercial concrete vibrators in tubs and tubes of wet cement. Very little data came out of these tests, but they qualitatively showed that the sonic probe could shake up sediment over quite a distance. Mark finally produced a full conceptual design document on the sonic probe. It was a 5,000-pound steel pickle containing an electric motor driving a massive eccentric weight to make it vibrate at around 100 cycles per second. It probably would have freed bubbles from the sediment, but it would not have physically mixed the tank as the pump could. That was the fatal flaw. All it was good for was degassing and, because none of the other burping tanks had come anywhere close to flammability, DOE decided it was better to let them burp than spend lots more money on mitigation. The design document was the end of the sonic probe.

Pacific Northwest National Laboratory, or PNNL, was supposed to be working on the heating and dilution concepts with laboratory experiments on waste samples from Windows C and E. The samples were diluted with various amounts of water and sodium hydroxide solution and then run through tests at different temperatures to measure their properties. There were two problems on this front. First, PNNL was way behind schedule, and few data points were available for evaluating the concepts. Second, though some of the waste properties, at least in small samples, were measured as a function of temperature and dilution, there was as yet no way to positively relate the properties to gas retention and release mechanisms. To remedy this, PNNL scientist Phil Gauglitz was developing methods to test gas retention directly. But these tests showed that essentially any kind of sediment could retain a lot of gas unless it was almost liquid. There was as yet no clear criterion to establish the amount of dilution or heating required to prevent gas retention.

Meanwhile, Steve Agnew at Los Alamos continued to push his ultrasonic mitigation concept. But he could not reverse the fundamental fact that ultrasonic waves just cannot penetrate very far through bubble-filled sediment. It would have required tens of ultrasonic probes moving up and down through the waste to mitigate the tank. The other unexpected problem came from the Los Alamos SA! Bob White and company did not want to de-gas the tank too much or it would be outside the historical bounds where there was some knowledge of tank behavior. And, inside its tiny radius of influence, an ultrasonic probe was a very effective de-gasser. After Phase B ended, Jack Lentsch gave me the job of concluding that mitigation by ultrasonic vibration was just not practical and telling it to Los Alamos. They did not agree readily, but because John Tseng was no longer present to advocate it and the success of the mixer pump removed any need for it, the ultrasonic mitigation concept for SY-101 finally died for good.

The Chemical Reactions SubTAP, also noted that the success of the mixer pump was making all the other mitigation concepts less and less relevant. At the closeout session of their February 17, 1994, meeting, they stated that the "value of heating and dilution tests is now limited for SY-101, and their purpose needs to be reassessed." They also wrote, "Sonic vibration should be reconsidered since mixing is successful and the sonic probe does not provide for retrieval." In fact, the SubTAP was so satisfied with the mixer pump and what had been learned to that point that they concluded, "SY-101 has been studied enough to close USQ [unreviewed safety question] *now*. . . . independent of other tanks."

The other alternative mitigation concepts were finally put to rest in the spring of 1994 with a scheduled PNNL report comparing the mixer pump, sonic probe, heating, and dilution. It was supposed to have compared the results of research and testing of all concepts on an equal footing. But the test chamber that would have introduced the other options into the tank was never built and only the mixer pump was actually demonstrated to be effective. The other three remained pretty much just concepts. Dilution was deemed impractical because a little dilution was worse than none in terms of gas-release severity, and nobody could confidently say how much was enough. Heating was rejected because it accelerated gas generation by orders of magnitude and would be very slow to reverse if it proved detrimental. The sonic probe was most favorably rated but, because the mixer pump was already in and proven, it was not necessary to proceed further. Besides stating the obvious, the report probably served as an excuse to declare success and halt further work on the alternative concepts. After three years of hard work, people were getting very tired of mitigation. The report's

296 | Part Three: Surface Level Rise

conclusions plus the success of the mixer pump let them walk away with a clear conscience.

Moreover, previous talk about the "permanent pump" as a replacement to the current "test pump" evolved to discussion on designs for the "sister pump" as a backup to the current pump. It was necessary to keep running the mixer pump indefinitely, and there had to be another ready to go in when it failed. But the "test pump" was doing a fine job and there was no need to replace it unless it broke. The first spare pump was built and tested in MASF, in late 1994. It was an almost exact, but cleaned up, copy of the original pump. There were fewer gizmos attached and all but two of the conceptually obsolete riser seal plates were gone. New plumbing made it possible to change the motor oil, and new oil raised the temperature limit to 220 degrees Fahrenheit. Re-routing various instrument lines allowed removing the wasteberg bumpers. About a year later, a second spare pump was built. It also had the same basic design as the original and the improvements of the second spare, but used a more powerful 200-horsepower motor.

Los Alamos added a thorough analysis and detailed set of controls for pump removal and installation in the SA. If the old pump failed, the new one had to be running before an unacceptable volume of gas built back in. Los Alamos conservatively set the maximum waste level for removing the old pump at 401.8 inches and 403.3 inches to install the new one. If the old pump broke at 399.5 inches and the level rose at typical prepump rates, the crew would have only about 35 days to make the change. Fortunately, the old pump kept humming along and the preparations were never called to action. The two spare pumps are still stored in MASF just in case.

OFF THE FRONT PAGE AT LAST!

Though the test results and operational experience with the mixer pump clearly proclaimed SY-101's flammable gas release hazard mitigated, some of the more stubborn doubters remained unconvinced. But several new devices were being designed and built to measure the properties and gas content of the waste in the tank directly. This information would finally remove all doubt. The TAP had been pushing this work for years, and the Defense Nuclear Facilities Safety Board had just given it new impetus with their sweeping Recommendation 93-5 of July 1993 that demanded a better understanding of the mechanisms of gas retention and release. The ball rheometer and the void fraction instrument were the two main instruments in the works. They were being designed and built by Westinghouse, while PNNL did the conceptual design, test planning, and pilot-scale testing. Los Alamos did the safety analysis and mucked about with the fundamental physics. Work began in earnest in October 1993. Both instruments were designed, built, and deployed in a little over a year!

The falling ball viscometer, or ball rheometer, was designed to measure the viscosity, strength, and density of the waste in the tank. The TAP was getting tired of hearing the waste compared to honey, molasses, and peanut butter and wanted some real numbers. But the process of getting a small sample out of the tank and into a lab for traditional viscosity testing so disturbed the waste that the results were doubtful. Don Trent and Kemal Pasamehmetoglu badly needed this kind of data, and Max Kreiter squeezed enough money out of an already stretched budget to begin developing the concept. The ball rheometer was simply a 4-inch-diameter tungsten ball hung on a thin wire cable. Electronic load cells measured the tension on the cable very accurately as it dropped through the waste. The position of the ball was tracked as the wire wound off a drum. PNNL scientist and test engineer Chet Shepard described it very accurately as a "high-tech rock-on-a-rope." Though the ball was very simple and quick to run in a tank, it took two weeks or more to complete all the arcane data reduction calculations that converted ball speed and cable tension into viscosity, yield stress, and density as a function of elevation.

Originally conceived by Westinghouse engineer Troy Stokes, the void fraction instrument, or simply the void meter, measured the fraction of gas in a sample of waste about the size of a soup can by pressurizing it with dry nitrogen. The gas volume in the sample chamber was calculated from the accurately measured volumes, pressures, and temperatures in the system's vessels and piping before and after pressurization. The sample chamber had a sliding shroud that opened to fill it with waste and closed to seal it for pressurization. It was mounted on the end of a rotating arm about 3 feet long attached to a 60-foot vertical mast held up by a crane. The void meter went into a 4-inch riser with the arm set parallel to the vertical mast. When submerged in the waste, the arm rotated horizontally to begin sampling. While it took a crane and large field crew and a lot of planning to run the void meter, the data reduction was quick and simple. Jim Alzheimer was able to present the data to Jack Lentsch at the next 7:00 AM mitigation meeting.

The ball rheometer and void fraction instrument were first run in SY-101 between December 1994 and April 1995. Each went into two risers, 4A, 22 feet from the pump center line, and nearby 11B, 30 feet from the pump. The data solidly confirmed what was already accepted: there was not much gas there and the waste was pretty well mixed. The void meter found the gas volume fraction to be less than 1 percent down to about 100 inches, around 5 percent at 50 inches, and approaching 10 percent only near the bottom of the tank. The ball rheometer revealed the mixed waste slurry was like a creamy soup with a viscosity about ten times that of water.

> When the ball rheometer first went down Riser 4A into SY-101 on March 28, 1995, we thought the heavy tungsten ball would easily pass through the crust layer. After all, the hole that had been water lanced for the void fraction instrument only two months earlier should still be there. But the hole had closed up, and the ball just sat on the re-formed crust. So the test director wound it back up and tried lowering it faster. The crust was still impenetrable. He spun the cable drum almost fast enough for free-fall. It just hit the crust with a thud making only a small dent. Besides failing to break through the crust, the effort had also tangled the cable around the drum into a "bird's nest" and the

 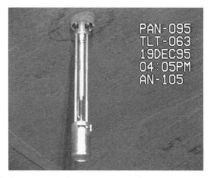

LEFT: In-tank video frame showing the ball rheometer about to enter the waste in Tank AN-105 about a year after it was run in SY-101. RIGHT: Void fraction instrument entering the dome in Tank AN-105. Closed sample chamber is on the bottom and elbow is at the riser lip.

ball had to be hauled back out hand-over-hand. The next time, applying discretion instead of valor, we punched a fresh hole through the crust with a water lance to let the ball pass through.

The ball made it all the way to the tank bottom in Riser 4A, but stopped 2 feet above the floor in Riser 11B. Likewise, the crane holding the void meter unloaded when the rotating arm was about 2 feet from the bottom in Riser 11B. This showed that the pump jet had swept the bottom clean beyond a radius of 22 feet, but not much farther. Was this where the broken air lance from Window G lay buried?

From these data, we calculated that the tank might contain 7,800 cubic feet of gas, about a third of the 24,000 cubic feet that Los Alamos had estimated before the big burps of the past. However, only 1,600 cubic feet of this was in undisturbed sediment that might be considered "releasable." The rest was floating in the crust layer or as tiny bubbles in the mixed slurry. This proved to most observers that the tank was mitigated and the pump was working fine.

I presented the void fractions and ball rheometer data on SY-101 to the SubTAP on May 24, 1995. Though they continued to fret about the outer region of the tank where no instrument could reach for lack of a riser, they could no longer seriously doubt the effectiveness of the mixer pump. The SubTAP closeout comments called long-term mitigation of SY-101 a "major accomplishment." More importantly, they recommended that the ball rheometer, void meter, and the lessons learned from SY-101 mitigation be applied to other tanks.

These recommendations marked the transition of SY-101 from a major issue to just an operation. Emphasis shifted to the overall flammable gas safety issue and the other tanks on Senator Wyden's flammable gas watch list. This new scope gave a new and broader understanding of gas generation, retention, and release mechanisms than we could get from SY-101 by itself. This knowledge would eventually enable us to plan the final remediation of SY-101 and allow the entire flammable gas safety issue to be closed.

About this time, Dr. Joe Shepherd from CalTech made a discovery that might have seriously altered SY-101's history, had it been made ten years earlier. Jerry Johnson had CalTech doing experiments on the flammability of mixtures of hydrogen, nitrous oxide, and ammonia. SY-101's burps became an emergency in 1990 based on the belief that a mixture of hydrogen and nitrous oxide was flammable without oxygen. Well, Dr. Shepherd found that this was only true *after* the gas heated to about 1,300 degrees Fahrenheit! That would only happen in a really strong, hot spark or if a burn had already been ignited. At normal tank temperatures nitrous oxide was essentially inert! The hazard just was not there!

After mixer pump operation became a normal scheduled activity at the end of 1994, the TRG dutifully met once a month to look at tank and pump data, and Westinghouse Process Engineering published a nice quarterly report on the data trends. The SY-101 USQ was closed in June 1996, and SY-101 faded into the background, except for its annual pump operating budget, which remained a serious drain. At that point, the tank was turned over to tank farm operations, and the hydrogen mitigation project ended. But troubles with SY-101 had not ended. In fact, new phenomena were already at work setting up the next emergency.

1995–1997

The Approaching Eye Wall

Those who have been through a hurricane know the eye is only a momentary lull. The rest of the storm will soon return with almost as much torment as before. But those without this experience would believe the storm was spent. Still, things would not be quite right. The high and confused surf would not match expectations for storm passage. Soon a mountain of northward-racing clouds would rise on the eastern horizon. Seeing this one would begin to think about finding shelter again. That's the way it was with SY-101. Something was not quite right.

NOT QUITE RIGHT

After all the excavation runs of 1994 had stirred up the most waste and the pump had released all the gas it was capable of, the waste level, accepted as the primary measure of the volume of gas trapped in the sediment, was slightly below 399 inches, the lowest it had ever been since the tank was filled. But about the end of October 1994 the predicted waste level, calculated from daily gas release estimates, began to creep upward almost imperceptibly. It gained only about 2/10 of an inch in the last quarter of 1994, but it still gained.

In mid-December 1994, a new level gauge came online. A new Enraf-Nonius buoyancy gauge was installed in Riser 1A. Instead of using electrical contact like the old FIC, the "Enraf" just measured the force on the wire suspending a white nylon bob from the dome. When the force dropped, the device deduced it had hit a solid waste surface or was being buoyed up by liquid. The Enraf did not have to be lifted up and dropped back onto the waste each measurement. It could just sit there while its computer gently pulled on the wire to check its position. Thus the Enraf did not tend to gather a "waste-cicle" that had to be flushed off. This would become an extremely important distinction in the years ahead.

The first readings from the Enraf put the waste level at 402.5 inches while the FIC was giving 399 inches. Why the 3-inch difference? The waste surface was known to be highly irregular because of all the ancient wastebergs that had been tossed around the tank for years. And the in-tank video had already shown that repeated flushes had dissolved a depression, some described it as a hole, around the FIC. It all seemed pretty reasonable. By June the difference between the FIC and Enraf had steadied out at 3.4 to 3.5 inches, increasing just a bit when the FIC was flushed. But the calculated gas release continued to be slightly less than the estimated gas generation rate, which gave a slight rise in the predicted waste level.

This overall trend was nicely summed up in PNNL's quarterly review of SY-101 data for June through September 1995. The following quote from this report shows how clearly the trend was understood even this early in the drama.

"The trend of rising waste level is a continuation of that observed over the entire nine-month period since the Enraf gauge was installed. From December 15, 1994, to September 13, 1995, the Enraf level has risen about 0.8 inches. During this same period, the level predicted from gas release versus the estimated generation rate has increased about 1.5 inches. On the other hand, the FIC and manual tape levels began and ended the period around 399 inches. . . . If the level rise represents gas retention, it is likely occurring in the crust layer."

Comparison of these two level measurements was to become both the main evidence for and measure of crust growth. It was also the greatest challenge to Hanford's credibility. The Enraf began registering a steady or accelerating level rise from the start. But the FIC stayed roughly constant or rose much less by virtue of the frequent flushes necessary to keep it working. At the same time, the FIC was

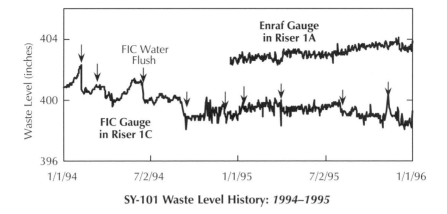

SY-101 Waste Level History: *1994–1995*

the most accurate waste level instrument when Los Alamos wrote the SA, when mitigation was being monitored during Phase B, and when the long-term operation plan was first written. In official language, the FIC was the "instrument of record" by which controls and operations were officially evaluated.

The credibility challenge came when the amount of level rise was also officially referenced to the FIC, even though the Enraf gave a much higher growth, and was universally agreed to be a better instrument. Comments and questions from the SubTAP and Defense Nuclear Facilities Safety Board staff became quite pointed, demanding an explanation as to why the FIC could be the "instrument of record" when the Enraf was exhibiting so much more growth. It almost looked like a deliberate coverup. By the fall of 1998 it had become so embarrassing that DOE Richland admonished Lockheed Martin Hanford Company, which now owned the SY-101 legacy, to not even use the term "instrument of record" when talking to the State Department of Ecology. By then the concept had lost its meaning anyway.

> In the spring of 1995, after the Enraf had begun operating, I had an opportunity to raise the crust growth trend to issue status. Los Alamos was revising the SA to allow the Enraf to serve as the primary level gauge if the old FIC failed. To do this, Bob White needed PNNL to supply an "official" offset to apply to the Enraf reading to transpose it into an "official" level measurement. That was my job.

304 | Part Three: Surface Level Rise

> In my naiveté I thought it was most accurate and reasonable to calculate the offset by averaging the difference between the FIC and Enraf over its first few months of operation. The problem was that applying this offset to the current readings pushed the "official" level above the threshold requiring "aggressive" mixer pump operation to release gas and lower the level. But this would cost tank farm operations a lot of extra time and money. They kept waiting for the "right" offset—one that would not automatically require aggressive mixing if it had to be used.
>
> Not only did I challenge organizational inertia with the politically incorrect offset, but I also told the TRG that it looked like gas was accumulating faster than the pump was releasing it, and we'd better start paying attention to it. Jack Lentsch pooh-poohed my concern, thinking the Enraf rise wasn't big enough yet to be statistically significant. Besides, it was the only one of four instruments (the radar gauge was still working, barely) showing a clear level rise. So the offset issue faded away and level rise had to wait another 2 years to become important. Jack said years later he should have listened to me.

By early 1996 it was no longer possible to ignore the fact that the Enraf was, indeed, showing a steady and consistent increase in waste level. At the same time the difference between the calculated gas release rate and the estimated gas generation rate could no longer be explained by errors and uncertainties. On top of that, the FIC level gauge had become erratic, though it sort of agreed with the manual tape. The PNNL quarterly reports began carrying a disclaimer, "The FIC became erratic and began a decreasing trend about August 1, 1995. FIC level should not be used as an indicator of stored gas volume after that date." The radar gauge data had become unusable in the spring of 1995, the instrument having succumbed to radiation damage.

DENIAL

Investigation of the discrepancy between gas release and generation and the rising Enraf level became the main items of discussion in the quarterly SY-101 reports. Though everything was pointing ever more strongly towards gas buildup, probably in the crust, nobody was ready to raise the issue to DOE Richland or Westinghouse management. Management was totally distracted anyway. DOE had become increasingly dissatisfied with Westinghouse and announced their contract would not be renewed. At the same time the Clinton Administration wanted to try "privatizing" and diversifying Hanford operations. DOE thus invited bids for teams of contractors desiring to participate in the "Project Hanford Management Contract" or PHMC.

The winning PHMC team was announced in August 1996. Fluor Daniel Hanford would provide "management integration services" and coordinate the whole team. Lockheed Martin Hanford Company (Lockheed) got the tank farms and, therefore, ownership of the SY-101 legacy, while Duke Engineering Services Hanford (Duke) took over process engineering. French companies Numatec and Cogema handled major projects, while Waste Management Federal Services and Babcock & Wilcox Hanford got the job of environmental cleanup.

Under the new DOE concept, Fluor Daniel Hanford was "management and integration" contractor instead of the traditional "management and operations" contractor. The other firms subcontracted to Fluor to perform the operations, while Fluor assigned their work, integrated their activities, and answered directly to DOE Richland. The companies that took over the principal activities of running major Hanford facilities like Lockheed Martin were considered "inside the fence." Others that provided specific services were "outside the fence." Those outside would be expected eventually to compete for their contracts with other private firms whereas the inside subcontractors would not. But those inside had to maintain the same benefits packages that existed under Westinghouse, while those outside did not.

This could be confusing. Numatec was "inside" and Cogema was "outside." But each inside contractor had an outside component. Lockheed Martin Services, for example, ran the Hanford computer networks. Each outside company brought its own policies and procedures that were different from each other and from those of the inside companies. Former Westinghouse staff were divvied up between the new companies on both "sides of the fence" or laid off altogether. The resulting stress and confusion paralyzed the site for some time.

Part Three: Surface Level Rise

> The shift to PHMC shuffled people more ruthlessly than past contract changes had. Jerry Johnson got his paycheck from Duke, Jack Lentsch became a Numatec employee, and Carl Hanson got an office at Cogema. In the spirit of privatization, many former employees were encouraged to become instant entrepreneurs and form their own companies. A few even made a go of it— Troy Stokes' HiLine Engineering was one. But many found it very stressful to be abruptly assigned to a new company with a Spartan benefits package whose name they couldn't even pronounce.

The PNNL, and the now Lockheed and Duke, engineers did not find the problem with SY-101 dire enough to trouble management about. It seems that some didn't want to believe the evidence already assembled. One theory was that mixing might have made the gas generation rate go up. The first 1996 quarterly report, issued in March, reported suspicions of an error in the gas release calculation or in the ventilation flow rate used as input. Then the Enraf data also aroused suspicion. The third quarterly report said, "At this time the values from the Enraf are being questioned, even though until recently they were believed to be the most reliable." But none of these investigations turned up any fundamental flaws that could explain the data as anything other than gas retention.

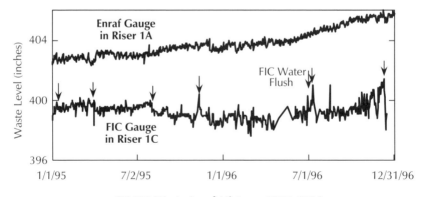

SY-101 Waste Level History: *1995–1996*

Though the evidence could not be made to go away, the level rise and the gas retention rate were not yet very large. The gas release deficit was running about 10 percent of the generation rate. If all this gas were being retained at 2 atmospheres of pressure, it would produce a level rise of 0.013 inches per day or only 4.7 inches per year. The quarterly report admitted, "Such a level increase is *somewhat* indicated by the Enraf gauge." But the party line was still, "At this time we do not see any immediate serious consequences from this gas retention because it is so minor. Enhanced use of the pump may be needed in the future to release additional gas or to demonstrate that no additional gas is actually being retained in the waste below the crust."

The level rise issue first got official attention at Jerry Johnson's November 19, 1996, weekly flammable gas meeting. His notes say, "SY-101 Enraf shows growth for the last several months. Need to verify FIC and Enraf readings!" The FIC was giving more and more trouble so, on December 18, 1996, it too was replaced by an Enraf gauge identical to the one in Riser 1A. The "Enraf 1C" data came from the Enraf that replaced the FIC in Riser 1C, while the "Enraf 1A" was the original one installed in December 1994. The final FIC reading was 398.1 inches, identical with the first Enraf 1C reading. Thus encouraged, the TRG continued to call the 1C Enraf the "instrument of record" for evaluating controls and the officially reported level. As far as confirming level rise, Jerry would just have to wait and see.

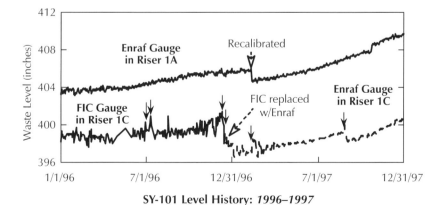

SY-101 Level History: *1996–1997*

Here is where SY-101 showed some of its personality by deciding to go back into hiding and lull the Hanford community back into complacency. Neither Enraf gauge showed a level rise for the first quarter of 1997! It really was just an instrumentation problem, wasn't it? It was slightly disappointing that, towards the end of the quarter, the brand new Enraf required a flush, just as the FIC had. Apparently the hole it was resting in needed to be reamed out occasionally to keep the bob from rubbing against the sides.

The tank's nefarious delusion worked! Even though both Enrafs showed a definite and parallel level rise all through the second quarter and the gas release deficit increased to a full 20 percent, it was all ascribed to instrument error. The SY-101 quarterly report of July 31, 1997, said, "There were indications that the waste level increased about an inch during the quarter. But, based on past experience with Enraf level gauges, the increase is thought to be due to loss of instrument calibration." The report explained the gas release deficit in the same terms. "Although the calculated gas release this quarter was only about 80 percent of the historical gas generation rate, we believe instrument problems account for most of the 20 percent difference and therefore do not support an actual increase in level. We believe gases were released at about the same rate they were generated and that additional gases were not retained in the waste." This was a complete change from what the quarterly reports had been stating for the last 2 years!

IT REALLY IS GROWING!

The July 1997 quarterly report was the last one produced by PNNL. The tank was now owned by tank farm operations and quarterly reports were in the hands of Lockheed. In the next quarterly, compiled in October 1997, Nancy Wilkins and Gerry Koreski were forced to admit the level rise was real. Both Enrafs continued to go up. Each rose about an inch in the quarter, just like the last one. It could not be pushed off on instrument error anymore. The gas release deficit stayed around 20 percent, but another trend now became visible. The calculated release rates of the more soluble nitrous oxide and very soluble ammonia were also clearly decreasing. Because these gases were released more by evaporation, this trend could mean the crust was somehow becoming more of a barrier. Also for the first time, increasing crust buoyancy due to gas buildup was mentioned as a possible mechanism for the rising level. The crust more and more appeared to be source of all of the changes people had been seeing.

The now undeniable level growth got the attention of the TRG. The official Enraf 1C level was approaching the threshold value of 399.5 inches where the TRG was supposed to consider more aggressive mixer pump operation to bring the level down. At 402 inches, aggressive operation was mandatory by the SA. Accordingly, the TRG directed that the mixer pump be run four times per week instead of three times, maintaining the same 25-minute, 1,000 revolutions per minute runs, rotating the nozzle 30 degrees each time.

The new, more aggressive schedule began on October 27, 1997, and ran until November 30. The result was a big ugly surprise. Instead of releasing more gas and driving the level lower, or at least reducing the growth rate, increasing pump operation actually *accelerated* the level growth from 0.01 inches per day to 0.03 inches per day! The gas release rate did not change. This was exactly the opposite effect that the mixer pump was supposed to have. It violated the very basis that had made mitigation by mixing so attractive. Now we were getting evidence almost strong enough to raise as a true safety issue. The TRG decided that running the pump more was not a good idea and pulled back to the nominal three runs per day.

To add to the dismay, after one run on November 10, mixer pump operator Mitsy White reported that she had the camera on and saw foam coming up around the pump! It was a gradual ooze of a gray, creamy substance through discrete vents with small bubbles occasionally blurping up. It flowed out sort of like pahoehoe lava and appeared to soak slowly back into the crust. This foaming behavior began happening often, sometimes from vents away from the pump. Though not very dramatic, this was clearly and seriously different.

The eye of the storm had passed, the new emergency was imminent, and the worry propagated to higher and higher levels of management. As SY-101 lay warm and cozy beneath the gravel on that cold December, it might have been snickering to itself, "You ain't seen nuthin' yet!"

1997–1998

Richland, We Have a Problem

This phase of SY-101's surface level rise issue is an interesting study in management dynamics. It was tank data and data analysis that drove DOE to establish both the original hydrogen mitigation and new level rise projects in SY-101. But such a change does not happen easily or quickly. At Hanford and other big, complex organizations, especially those funded by the government, there is no emergency bank account set aside that you can borrow from when things like SY-101 come up. If you need big money to fix unexpected problems, other activities have to stop to pay for them. Managers whose projects are stopped don't like it, and there are large inefficiency costs in stopping and restarting a project. Managers must make hard and serious decisions that require very powerful evidence. They can't raise an issue because they *think* there's a problem—there had better *be* a real problem.

The first step in starting up a new project of this kind is to make the key decision to stop business as usual and start spending serious money on the new problem. There must be enough pressure to force DOE and contractor managers over a threshold where the consequences of not taking action to fix the problem are worse than dropping a bunch of current projects and starting a new one.

The surface level rise issue was now ripe for its own attempt to cross the action threshold. There was clear evidence something new

312 | **Part Three: Surface Level Rise**

was happening, but it was not clear how bad it might be. The system was primed to make it an issue. In fact, tank behavior contrary to what was believed or assumed in the mixer pump SA was enough to trigger a USQ, even without well-defined consequences. But a USQ by itself was not enough. Only a serious act of will by management and DOE together could create safety issue status. Ten years of experience with SY-101 taught managers to take it seriously. But it also taught them how costly and disruptive it could be. So they had to be very careful. As a result, rather than bursting suddenly on the Hanford conscience in full force like the original flammability flap, the surface level rise issue had to endure a long period of rather tedious slogging to build a strong enough justification for an outright attack on the problem.

THE SURFACE LEVEL RISE USQ

On December 8, 1997, after finding that more mixing made the level rise faster, the TRG reviewed the requirements of the long-term operating plan and the SA in view of the unexpected result. They saw they were in a difficult position. If the level rose above 402 inches, the SA would make aggressive pump operation mandatory to knock the level back down. But recent experience indicated this would not only fail, but might even make the level rise faster. If the level kept rising, only pump bumps were allowed above 406 inches, and all pump operations were prohibited above 422 inches. It certainly looked like a no-win scenario. So the TRG called a meeting with Lockheed's Nuclear Safety & Licensing people to get some help.

The tank farm engineers and Lockheed's safety and licensing people met December 16 to consider the TRG's predicament. They agreed that the evidence showed the safety basis, as stated in the long-term operating plan and SA, might be ineffective in this case. This worry prompted them to do a formal screening to see if it was a USQ. In this process one or two senior people who are formally qualified for USQ screening sit down with a thick manual and answer a set of specific questions. The answers tell them whether the issue is a candidate USQ that should be sent up the chain of command. The screening left no doubt about SY-101 level rise. All the answers were "yes," which indicated "we have a problem," and the issue went upstairs to the Plant Review Committee.

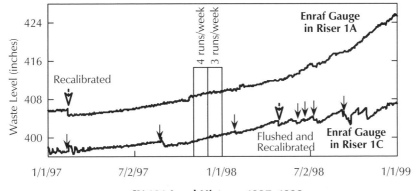

SY-101 Level History: *1997–1998*

The Plant Review Committee met right after Christmas on Monday, December 29, 1997, to consider the screening results and the evidence so far. They saw by the rising level registered by both Enrafs that the pump was pushing the level opposite the way it was supposed to. They also noted the growing deficit in gas release versus generation, and the decreasing concentrations of nitrous oxide and ammonia. Then there was the bubbly foam coming up through the crust during pump runs. All these were probably related to level growth but nobody yet knew how. The Plant Review Committee concluded, in the arcane acronymic language of Hanford safety, that there was a "discovery" regarding SY-101 because of a "PIAB" (pronounced PEE-ab) or "*p*otential *i*nadequacy in the *a*uthorization *b*asis." The PIAB was that "the observed rise in the surface of the waste in this tank is not consistent with the behavior analyzed in the authorization basis."

Meanwhile, on January 7, 1998, the TRG took a hard look at the accumulated tank data and observations and decide what do next. Between December 9 and December 31, 1997, returning to the old three-runs-per-week schedule actually *reduced* the level growth rate to 0.005 inches per day compared to 0.03 inches per day with four per week, again just opposite of what was supposed to happen. Because three pump runs per week created a slower level rise than four, maybe running it even less could slow the rise further or even reverse it. So the TRG decided to run the pump only twice per week plus a bump. The plan was OK with the SA because the 1C Enraf level was

314 | **Part Three: Surface Level Rise**

still below 402 inches. This new schedule began January 9. The level rise didn't slow down but neither did it speed up immediately.

So, what if the level kept going up? It didn't make sense to do the SA-mandated aggressive operation when it hit 402 inches. That had only made things worse when they tried it in November. But not doing it would violate the controls! The TRG was out of options. Making use of the PIAB, tank farm management untangled the dilemma with a standing order to suspend the unworkable controls. The order, approved February 12, 1998, told them to keep running the mixer pump in accordance with the SA and the long-term operating plan except that the specific operations required at 402 and 406 inches no longer applied. The order also required the TRG to review tank data at least monthly and specifically authorize operations to continue running the pump.

Finally, on February 18, with the PIAB declared, the standing order in place, and everyone in agreement, Lockheed management formally recommended that DOE Richland declare the SY-101 Surface Level Rise USQ. They did so on February 26, 1998.

DO WE REALLY HAVE A PROBLEM?

The USQ did not start the actual remediation work. It was only the beginning of the beginning, an excuse to start spending a little time and money on the problem. What happened then was strikingly similar to the scene after Mike Lawrence's famous flammable gas news conference of March 1990. Like Westinghouse did then, Lockheed set up a task force to study the issue. Also as before, Jerry Johnson and Nick Kirch were the main players. On the other hand, there was not the avalanche of public pressure and criticism that fell in 1990. The task force could actually concentrate on the work at hand without too many distractions.

Nick Kirch began by calling a series of meetings, appropriately in the old SY-101 "war room" on the second floor of 2750E. The TV monitors and data boards were gone by then, and there were folding vinyl dividers so you could have two very small conference rooms, or one not very large room obstructed with pushed back dividers. Besides Nick and Jerry, the task team included Jim Bates and me from PNNL, Bob White from Los Alamos, Roger Bauer, Carl Hanson, and Doug Larsen from Lockheed, among others. The meetings' goal was not to decide how to stop the level rise, but how to get data. The data would be used to map hazards and analyze accidents to see how bad the problem was, devise mitigation or remediation plans to fix it, and set new controls to prevent it. If all went well, these things would lead to closing the USQ pretty soon.

The task force report came out February 23, 1998. It stated the main question that needed to be answered was, "Is there a new or greater hazard . . . due to the observed waste behavior in SY-101?" Finding the source of the level growth was the first step in answering this question. Because no waste and relatively little water had been added to the tank, the only really plausible source of the level growth was gas buildup. If gas was the problem, we needed to know where it was, how much was there, and whether the two Enrafs were measuring it correctly.

The task force report set out a schedule for acquiring the required knowledge. We assumed that DOE would approve an official change request to redistribute some money from other projects to do the work. This would be the first step by DOE and Lockheed management across the action threshold. It was a relatively small step of only a few million dollars that was only a minor frustration to other DOE programs and priorities. Major damage would come a little later with the big decision to spend the tens of millions needed to stop the level growth after we figured out how to do it.

The first task, due by the end of April, was to confirm that the level rise indicated by the two Enrafs was real. To find and quantify the gas buildup, we recommended that the old void fraction instrument be dusted off and run in several risers during May. If it didn't give enough data, core samples with the retained gas sampler could be run in September and October.

The retained gas sampler was one of the triad of new instruments born in 1993 out of repeated TAP recommendations and the sweeping Defense Nuclear Facilities Safety Board Recommendation 93-5 demanding that Hanford get a better understanding of the flammable gas hazard in all their tanks. The first two, the ball rheometer and the void fraction instrument, were described in the last chapter. The retained gas sampler was just a standard core sampler modified with better seals and valves to be gas-tight. Three or four retained gas samplers, each capturing a tube of waste with its trapped gas, were interspersed with normal samplers in a standard core-drilling operation. The retained gas samples, each about 19 inches long and one inch in diameter, went to a separate hot cell where their

316 | **Part Three: Surface Level Rise**

contents were extracted under vacuum and the gases analyzed in great detail. The results not only gave the gas volume fraction, like the void fraction instrument, but also the gas composition. The retained gas sampler first went into a tank in 1996 but, because the flammable gas USQ was already closed by that time, it hadn't yet sampled SY-101.

Besides making these direct measurements, we recommended using the mixer pump itself as an instrument. We thought that the tank would give up some of its secrets by comparing what it did when the pump was run hard to when it was run gently or not at all. But, though Nick Kirch and I were both TRG members, tank farm operations never bought into this philosophy and the plan was never executed.

The most important and influential aspect of the task force report was to tie the predicted waste level, extrapolated from the current growth rate, to the work schedule. Managers, even DOE, liked it, and as the level rise rate accelerated, they all watched it closely and demanded fresh extrapolations at every meeting. The report predicted that the Enraf 1C level would go from the current 402 inches to the pump-bump-only limit of 406 inches by October 3. By virtue of a flurry of water flushes, that turned out about right. During the same period, the 1A Enraf rose from 410 to about 417 inches.

IS THE WASTE REALLY RISING?

Roger Bauer got the job of confirming the level rise. Many of us felt this task was a needless delay, and the time would be much better spent starting work on solving the problem. After all, level rise was totally obvious, and no competent person could remain unconcerned after looking at the measurements. But we didn't understand the need to carefully build an unassailable case to justify spending the kind of money we would need. The technical faction seldom understands the fact of human nature that political considerations are usually just as important, and often more important, than science!

To validate the data from the two Enraf gauges and the old FIC, Roger had to find other independent measurements of similar precision to compare them against. He quickly came upon four of them: two velocity-density-temperature trees, or VDTTs, in Risers 4B and 11B on the northwest side of the tank and the two MITs in

17B and 17C on the north and southeast. Each of the "trees" had white or black rings painted around them every 2 inches. Using several in-tank video scans and these four "dipsticks"—combined with the 1A Enraf to the south, the 1C Enraf and old FIC on the northeast, and the manual tape on the southwest, he could pretty accurately plot what the entire waste surface had done since late 1994. Incidentally, this was probably the very best and most useful data the multi-million-dollar VDTTs ever produced.

LEFT: Frame from 1994 video scan zoomed in on VDTT "dip stick" in Riser 14A. RIGHT: Frame from 1995 in-tank video scan showing VDTT in Riser 1B. Waste level had risen about an inch since 1994.

Zen Antoniak from PNNL figured out a very simple and very accurate method to calculate the level change from the video. Zen digitized the clearest video frames showing closeups of the dipsticks for each tank scan with a resolution of 640 pixels wide by 480 pixels high. He made the measurement by magnifying the image so he could count each pixel and calculating the pixel height from the number of pixels between the rings. For example, let's say a feature on the waste surface around a VDTT was 50 pixels below the lowest visible white ring, and two adjacent white marks were 10 pixels apart. Because the rings are 2 inches apart, each pixel on that frame must represent a height of 2 inches/10 pixels = 0.2 inches/pixel. This would make the waste surface 0.2 inches/pixel x 50 pixels = 10 inches below the lowest ring. Though Zen had to do some slightly more complicated calculations to correct for the tilt angle of the camera, it was a simple and reliable measurement.

Using these measurements, Roger and Zen plotted the level rise since October 10, 1994, for June 1996, May 1997, and December 1997, at the four dipstick locations, along with the level rise at the two Enrafs. The Enraf data had to be corrected for the level drops that happened each time the bob was flushed, much more so for 1C than 1A. To nobody's surprise, the level rise was comparable at all six locations. Their report made three conclusions: 1) the level rise indicated by the two Enrafs was correct, 2) the waste level really was rising nearly uniformly, and 3) the rate of level rise was increasing. The one exception was that the level growth appeared to have slowed around the MIT in Riser 17C in 1997, and they recommended a fresh full tank video scan to check it out.

But we could not get down to work just yet! The SubTAP never trusted level measurements at a single point on top of an irregular crust of unknown character. As they had in 1991, they "strongly recommended" that some method be developed to give them a "free liquid level" that would be free from the perplexing influence of the crust. This was more difficult than it might appear because the liquid waste was so loaded with dissolved solids that it quickly crusted over any device exposed to it, much like a very exaggerated hard water deposit. At the same time, altering the liquid to stop this nasty habit, diluting it with water for instance, would make the level measurement erroneous.

Nevertheless, to keep SubTAP happy, Carl Hanson set one of his young and very creative engineers to design a free liquid level system. His proposal came out in September 1998. After considering a number of alternatives, from water spray to installing a big mixer pump at the surface to physically break up the crust, he settled on a pipe extending below the crust, heated with glycol to keep solids in solution. An Enraf gauge inside the pipe would provide the level measurement. He believed a temperature of 115 degrees Fahrenheit, about the temperature of the liquid slurry beneath the crust, would make the waste stay liquid. Unless the slurry below the crust was as foamy as the stuff oozing up around the pump looked, it might even have worked. But the estimated cost was $645,000, and it would not give any data until June 1999! By then the waste level was predicted to have risen to truly alarming heights! Nobody thought too seriously about a free liquid level system after that.

There were other alternatives besides a free liquid level that some believed might be able to confirm level rise or quantify the gas buildup. The hydrostatic pressure near the tank bottom could be read directly from the mixer pump nozzle pressure gauges when the pump wasn't being run. This method, another one that had been perennially pushed by the SubTAP, was easy and cost nothing, but it had two fatal flaws. First, hydrostatic pressure is fundamentally the measure of the weight of fluid above the pressure sensor. The gas causing the level rise, however, weighs essentially nothing compared to the liquid and solids. Because the amount of solid and liquid waste was nearly constant, the hydrostatic pressure measurement would also be nearly constant regardless of waste level rise, unless the gas was accumulating in stationary sediment. However, the evidence of mixing was compelling, and the probability of a gas-trapping sediment layer was low.

The second flaw was that the nozzle pressure gauge was calibrated to measure large pressure differences and would have trouble sensing small ones that might be expected of a level rise. Besides their insensitivity, something about the gauge's installation had made them sensitive to daily temperature swings. The indicated pressure might swing as much as 3 pounds per square inch from a cool night to a hot sunny day. Jim Bates carefully examined not only the nozzle pressures, but also the pump volute pressures about midway up the waste column. He could find no evidence of a steady hydrostatic pressure rise anywhere in this clutter of fluctuating pressure. Jim suggested that a more rigorous statistical examination of the data might reveal a trend but admitted such results would probably be uncertain and "less than fully convincing" in confirming the level rise.

The pressure gauges weren't the only ones that seemed affected by the temperature. Waste temperature readings from the thermocouple trees in the old single-shell tanks had historically varied between night and daytime much more than any physically possible change in the actual waste temperature. But the evidence was mystifying. The diurnal change didn't always happen and some tanks didn't see it at all. In some tanks, the effect was a sawtooth fluctuation, beginning in late morning, instead of a single large daily swing. After a spate of some especially severe diurnal effects, Jerry Johnson sent a team out to the tank farms to try to find out what might be happening.

The temperature change could not be explained by any plausible physical mechanism in the waste, so attention turned to the measuring device itself. A thermocouple measures temperature using the phenomenon that a junction of two dissimilar metals creates a tiny voltage that increases with temperature. A thermocouple is the point where two different kinds of wire are welded together. The temperature is read by measuring the voltage between the two wires with a sensitive meter. But to get an accurate measurement requires a "reference junction" at a known temperature. In the tank farms, the reference junction is usually placed in a metal box, above ground on top of the thermocouple tree mast.

The box was the problem. Some of the boxes had a window on one side, presumably to allow a technician to inspect the innards without having to unfasten a lid. Now get this. The standard installation procedure inexplicably aligned the windows to face south. At our 45 degree latitude, this admits a lot of sunshine around noontime. Sure enough, the temperature swings were largest on clear days during the time when the sun would shine directly in the windows and heat up the reference junctions. On cloudy days, the effect was much smaller! The few tanks without a diurnal effect happened to have their windows aligned to face north in perpetual shade! What about the tanks with the sawtooth temperature fluctuations? These were the new and improved units with a little air conditioning system to keep the reference junctions at a constant temperature. The trouble was, the units were installed so the cold air from the A/C blew right at the reference junctions! So the system designed to keep the temperature constant was actually creating the fluctuation!

Some of the systems were changed, but it was considered too costly to do them all. This is another instance that supports the Hanford axiom that you will always learn more about the peculiarities and contrariness of an instrument than about the quantity it is intended to measure.

The final hope for measuring gas buildup was the old barometric pressure effect method. Although this method had never been useful in SY-101 before, some believed that, if a lot of gas were building up, the method ought to at least show some trend in that direction, even if it could not give a precise numerical value. I asked PNNL statistician Guang Chen to have a look. We decided the only hope was to look at periods of very large pressure changes when major storms came through. In theory, the large pressure swings would create a sufficient level change to swamp out whatever other effects had been obscuring the results in the past. It didn't work! Guang found several large pressure changes over a period of several years and calculated the correlation between waste level and pressure for each. The correlations were reasonably large and negative, indicating the presence of quite a bit of gas. But, if anything, the trend they showed was actually a *decrease* in gas volume from 1996 to 1998! That didn't make sense at all. We concluded that it was not possible to get a reliable retained gas estimate in SY-101 from barometric pressure and waste level changes.

So, what had been accomplished from February through May of 1998 with all this work? Roger and Zen had produced the only really successful result that confirmed that the level was rising fairly uniformly and the rise rate was accelerating as already had been shown by the two Enrafs. No other useful information had come from applying the barometric pressure method or studying the mixer pump pressure instruments. We were tempted to say "we told you so" but refrained, hoping that we could now get down to some serious work that actually attacked the problem.

Serious work began in June when Jerry Johnson brought the now confirmed level rise problem to Lockheed Director of Engineering Mike Payne. The next planned effort, per the task force's February report, was to break out the void fraction instrument to feel around in the tank to see where the gas might be hiding. Or, like most of us believed, to confirm that the gas was gathering in the crust layer. Because Jerry already had his hands full trying to finally close the flammable gas safety issue, the additional work was more than he wanted to handle. Someone else would have to carry the SY-101 ball from here.

Payne met with Director of Operations Dale Allen; they agreed something had to be done over and above what was already in the works. They began strategizing a full-blown project to deal with surface level rise, the last big step over the management action threshold. In the meantime, Dale and Mike asked John Kryzstofski (pronounced kris-TOFF-skee), mid-level manager from the Lockheed Tank Characterization Program, to carry on what Jerry had begun. In June of 1998, the highest priority was to get the void fraction instrument re-started and find out where the gas was and how much had already accumulated.

THE GREAT VOID FRACTION INSTRUMENT DATA PUZZLE

Sometimes the old Hanford axiom that one learns little from measurements except bizarre instrument behavior does not hold true. On very rare occasions, an instrument tells you a lot more than expected. This was the case with the void fraction instrument over the summer of 1998. But the machine was telling us much more than we could accept and interpret all at once. Only after the third run did we understand what the first two were trying to say. Here's what happened.

Troy Stokes and Jim Alzheimer made the first void fraction instrument measurements in SY-101 in December 1994 and January 1995 to confirm that the mixer pump was keeping gas from building up. The void fraction instrument found a gas volume fraction of less than one percent from under the crust down to about 80 inches off the bottom. This deep layer was a solid-liquid slurry with very little trapped gas. From the 80-inch elevation down to about 40 inches, the gas fraction increased to about 4 percent in what we called a loosely settled layer that was mixed every time the pump jet hit it. Below 40 inches the gas fraction abruptly increased to 8 to 15 percent in an apparently undisturbed sediment layer.

Since then, evidence from the bottom four thermocouples on the two MITs showed that the undisturbed layer was pretty much gone, leaving maybe only 6 to 12 inches of a sediment that might have some gas in it. Data from mixer pump runs implied that the gas content of the slurry layer might have increased slightly, but no measurements had been attempted in the crust itself.

In May, Troy, now president of his own company, HiLine Engineering, and Jim, still with PNNL, were assigned to put the void fraction instrument back in the tank. They installed new O-rings everywhere they could to make sure the sample chamber would still seal, and made several test runs to make sure there were no leaks in the pressure lines and all the machinery was in order. They ran the first test on June 29, 1998. A hole had been lanced through the crust for the void fraction instrument about a week earlier. There was a pump run about 12 hours before the test. The first item on the test agenda was to find the bottom of the crust by raising the void fraction instrument into the crust and then lowering it in 6-inch increments until the lower arm could be rotated manually. The crust was a lot thicker in 1998 than it had been in 1995 and the arm

Video frame showing two bubbles about two inches in diameter coming up around the void fraction instrument mast along with a semi-liquid slurry.

had to be lowered to about 54 inches below the crust surface before the crew could rotate it. There was a lot of bubbling of soupy slurry up around the void fraction instrument mast during this time.

The first actual gas fraction measurement was about 60 inches below the crust surface. It showed a whopping 18 percent gas! The next measurement, 2 feet further down, showed almost 15 percent gas. It kept decreasing from there down, but the waste still held about 5 percent gas 18 inches from the bottom. After this first traverse, the crane pulled the sample chamber back up to try measuring the gas in the crust itself. They got one good measurement of 32 percent gas about mid-way into the crust.

My job was to interpret the void fraction instrument results. Thus, my first reaction was both exhilaration and puzzlement. It was exhilarating to suddenly be shown, we believed, where all the gas was. There was a lot of it, too! But it was very puzzling that there could be *that* much gas, especially in a material we believed to be a liquid that normally can't hold gas. On top of that, the average slurry gas fraction of 9 percent should have caused a waste level rise of 36 inches, even neglecting the extra gas that must be held in the crust. But the 1A Enraf showed only a 12-inch rise. It just couldn't be! Yet all the other void fraction instrument measurements to date appeared to be consistent between tanks and with the other data.

I could hardly wait for the next void fraction instrument run on July 22. It was in the same riser as the prior one, but we did not feel it was necessary to disturb the crust layer this time. Because there was already a hole there, the crust was not lanced. A pump run

324 | **Part Three: Surface Level Rise**

aimed within 30 degrees of the riser was made three days before. The three-day "resting period" after the pump run should have allowed extra gas stirred up by the pump to dissipate.

What a surprise! The average gas fraction for this run was only 1.2 percent, consistent with the 1994/95 data for the slurry layer. In fact, the gas fraction was fairly constant all the way down, with no evidence of any undisturbed waste at all. Had this run been the first, we would have been gratified that nothing much had changed from four years earlier, though we would probably been a little frustrated at not finding where the gas was hiding. As it was, this run came only a month after one showing a *huge* gas volume! How could the gas fraction in the tank *decrease* from an average 9 percent to an average of just over 1 percent while the waste level *rose* about an inch? There should have been a drop of at least 30 inches and a catastrophic gas release twice as big as the most fearsome pre-pump burp!

Because neither the high average gas fraction of the first run nor the drastic decrease between the first and second runs was physically possible, the only reasonable conclusion was that the June results were evidence of a *local* and also *temporary* event. A liquid simply cannot hold a high local gas fraction. The lighter gassy region would have risen quickly to the top, just like a cloud of bubbles exhaled by a SCUBA diver. But, as yet, nobody could point to a cause for a temporary local surge in the gas fraction.

The bad part was that I had to present these results to the Sub-TAP on August 12. They were more confused with the results than I was. They didn't understand or accept the theory that the first test had measured some kind of a temporary local disturbance. They wondered if the slurry was really a liquid. Maybe it was a solid in which gas could be accumulating nonuniformly, allowing pockets of much higher gas fraction. Others disagreed, voicing suspicion that the void fraction instrument itself was not measuring the gas correctly. They could not agree on the specifics, so they threw up their collective hands and concluded that "further VFI [void fraction instrument] tests would be of little value."

I'm indebted to John Kryzstofski for going ahead with the third test in spite of the SubTAP exasperation. But what would the result show? How could we run the test to discover what caused the difference between the first and second tests? Driving home to Kennewick one evening, I tried to think of everything that was different between the two void fraction instrument runs that could change the results. Was it the time after the pump run? No, they weren't that different. Was it lancing before one and not the other? No, the lancing was a week before the test.

Deep in thought while waiting for the light at Jadwin Avenue, I remembered one big difference! In the first run in June, we messed around in the lower part of the crust pretty violently right before other measurements started. There was no crust disruption at all in the July tests. That had to be the explanation! Disturbing the crust may have released enough gas to cause a gob of it to lose its buoyancy and sink. It would still contain quite a bit of gas, and, if it sank slowly, the void fraction instrument would capture some of it to show a high gas fraction. Now, how would you prove that?

By the time I passed Swift Avenue on the bypass, I had a plan. What if we made two top-to-bottom traverses in the third test? If there were no crust disturbance on the first one, the results ought to look like the second test in July. Then we would go up into the crust, deliberately muck around, and make a few measurements. A second full traverse after this crust disturbance should show high gas fractions like the June test. All this ought to be done in a brand new riser so, if it all went as predicted, we could claim it was universal behavior, not just an artifact of the waste around one riser.

Kryzstofski's task team and the TRG accepted the plan for the third void fraction instrument test to make one pass with no crust disturbance (to simulate the July test), followed by crust measurements, and then a final pass (to re-enact the June test). It happened on September 11, 1998, in Riser 1C, the home of the famous waste level "instrument of record." The Enraf was temporarily removed so the void instrument could go in. To minimize the disturbance, the crust was not lanced. Because the area around 1C had been flushed with water many times over the years, Troy and Jim thought the void instrument would go through the crust without hanging up.

The void fraction instrument went in smoothly as expected. On the first traverse, four measurements were made in the mixed slurry shortly after a pump run, but without any prior disturbance of the crust. The measured void fraction for these tests ranged from 1.2 to 3.2 percent, with an average of 1.4 percent. These results were

326 | Part Three: Surface Level Rise

consistent with the 1994/95 and July measurements. More importantly, they confirmed that the void fraction below the crust was generally quite low! So far, so good.

Next, several void fraction instrument measurements were made up in the crust. The gas fractions ranged from 2.1 percent to 43 percent with an average of 30 percent, consistent with the 32 percent measured in June. Then three measurements were made in the slurry layer to see if the crust disturbance dislodged gas-bearing material that could be measured with the void fraction instrument. Eureka! The gas fractions were 22.6 percent, 12.2 percent, and 8.4 percent, following about the same profile as the June test! This confirmed, without a doubt, that the high gas fractions were caused by the crust disturbance!

Looking at the results of all three 1998 void fraction instrument tests together, it was clear that the crust was accumulating gas while the rest of the waste from near the tank bottom to just under the crust still had quite low gas fractions. There was no indication of gas buildup in the loosely settled layer, nor any evidence of remaining undisturbed sediment. While the void fraction instrument had already provided very valuable data in five other double-shell tanks that eventually enabled closing the flammable gas safety issue, this new data in SY-101 may be the most valuable.

The results answered the enabling question posed in Nick Kirch's task team report. We knew where the gas was, we knew where it wasn't, and we knew about how much was there. Subsequent sampling and measurements filled in the details to complete the description of the crust and build a model to predict its behavior. This would eventually answer Nick's primary question as to whether the gas in the crust was "a new or greater hazard." The managers had enough now to go forward another step.

THE TECHNICAL REVIEW COMMITTEE

Immediately following the puzzling results of the June void fraction instrument test, we decided some collective head scratching was in order. So a group of the wisest heads from PNNL, Lockheed, and the subcontractors that knew something about waste in general and SY-101 specifically gathered into a group modeled after the old data management committee that strove to understand and improve the data flowing from the data acquisition and control system, or DACS, during mixer pump days. We invited scientists, engineers, managers, and folks from operations so that we could get all sides of a question and get everyone to buy into the answer. In most meetings, we had Jerry Johnson, Roger Bauer, Nick Kirch, Carl Hanson, Blaine Barton, Dan Reynolds, JR Biggs, and Doug Larsen from Lockheed and Perry Meyer, Lenna Mahoney, Phil Gauglitz, Jim Bates, and me from PNNL. Initially we called ourselves the Committee for Insight, Understanding and Alleviation of Level Growth in SY-101 or CIUALG. It makes an awkward acronym, but you can pronounce it by loosely following the Irish Gaelic as K^{y}OO-l^{i}g.

In the process of trying to understand the first sets of void fraction instrument data, the committee logically progressed to planning subsequent void fraction instrument tests and later deciding whether and how the retained gas sampler should be used. Gaining credibility from this work, managers began to refer DOE and TAP questions and concerns directly to the CIUALG and generally took its recommendations seriously. They even asked our opinion on some big decisions involving a lot of money. Though it never got any official status, the CIUALG soon became the primary technical basis generating organ of the SY-101 level rise remediation project.

After the project was officially authorized about January 1999, Jerry Johnson thought CIUALG was too flippant a title for the serious business the committee was doing. Accordingly, we changed to the Technical Review Committee or TRC, not to be confused with the Test Review Group, or TRG, that had to officially approve what the TRC wanted to do with the tank. It helped that several of the TRC were also members of the TRG.

The CIUALG was fearless, even if it wasn't feared. Because most of its members were involved with the early level rise confirmation studies and tank data collection exercises, besides figuring out the void fraction instrument results, we thought we were best qualified to boil down the evidence into a formal conclusion on what was causing the level rise. On September 2, 1998, even before the third void fraction instrument test, the CIUALG somewhat overconfidently crafted this consensus statement in its informal minutes: "There is no substantial challenge to the hypothesis that the level growth is due to gas retention mainly in the crust layer. While this cannot be formally proved, alternate explanations can be disproved, and it is most consistent with the body of knowledge. The committee recommends that further actions regarding SY-101 level growth be based on this finding." Now all we had to do was figure out what the "further actions regarding SY-101 level growth" would be. That's where the big money would have to go. It was time to start up a formal project.

THE PROJECT IS BORN

In September of 1998, with the cause of surface level rise now officially stated, Mike Payne and Dale Allen asked Rick Raymond to take over management of the SY-101 level growth activities. Rick started October 15, 1998. His first job was to negotiate a "performance incentive" with DOE Richland. A performance incentive is attractive to both parties. The contractor agrees to assume some risk and perform a big job quickly. That's the performance part. At the same time, DOE agrees to compensate the contractor very well if the job gets done on time. That's the incentive part.

The first SY-101 surface level rise performance incentive was simply to form the project and develop the budget, work plan, and schedule. This performance incentive required Lockheed to: 1) by December 31, 1998, set up a focused project team and evaluate options to close the surface level rise USQ; 2) develop a workable technical plan, schedule, and budget by March 1, 1999; and 3) perform the work according to the plan, with specific expectations to be set later. More than a million dollars was available to Lockheed if they could accomplish all this to DOE's satisfaction. A whopping 60 percent of the money was reserved for the last goal.

Smooth, urbane Rick Raymond was not exactly the rough-tough Type-A project manager you'd expect for an SY-101 remediation project. He wasn't entirely the antithesis of Jack Lentsch, who could be smooth and urbane when he had to, but Rick had quite a different style and personality. Rick's motto was, "Trust your scientists and engineers. If you can't trust them, get new ones!" That trust allowed him to delegate the technical decisions which pulled people, including potential critics, into the project. It's hard to complain about what you're responsible for. Rick didn't scare people like Jack did either.

Rick called himself a "one meeting kind of guy," who would rather talk to everyone together at one big meeting than be distracted by a bunch of little meetings. While the project was forming in the winter of 1998/99, Rick commandeered a big, triangular conference room on the second floor of 2430 Stevens Center as his command center and ran it like a sort of continuous meeting. When the work actually started, he moved the entire team down to a suite of offices above the Hanford Central Stores building and had a big weekly meeting that everyone working on the project was expected to attend.

Rick worked to make teamwork and shared purpose contagious. At the very beginning of the project, he put out ten "Team Commandments" for the project. They included:
- Goals are understood, all are committed to completing them
- Trust the team members
- A sense of belonging and pride in accomplishments
- Diversity of opinions and ideas is encouraged
- Creativity and risk-taking are encouraged
- Decisions are supported and made together

In most projects, such mottos are just pieces of paper yellowing at the corners of cork boards. But Rick really lived them. This made the project one of the most uniquely effective and enjoyable in the Hanford experience.

Because Rick had been in on the early part of the original SY-101 flammable gas emergency back in 1990, he naturally chose people he had seen do that work, along with others he had come to trust along the way. For the core project team, Rick gathered Ken Jordan as deputy project manager and Roger Bauer to manage the administrative stuff. Roger eventually replaced Ken as Rick's deputy in mid-1999. On the engineering side he called up some of the old hands that had done the mixer pump job seven years before. Carl Hanson was the chief design engineer, assisted by Tony Benegas. JR Biggs pushed operations in the farm assisted by Dan Niebuhr. Fred Schmorde, who had installed the radar gauge in SY-101 back in 1992, helped Rick with interface operations and later became the SY farm facility coordinator and TRG chairman. Mike Grigsby handled the nuclear safety and licensing issues, and Nick Kirch led process engineering. Joe Brothers brought in science and technology from the CIUALG-cum-TRC team already led by PNNL

330 | Part Three: Surface Level Rise

plus others as needed. Rick's direct co-conspirator at DOE Richland was Craig Groendyke, who had arrived on the scene as I did about the time of Phase B mixer pump testing.

On the PNNL side, Perry Meyer took the technical lead with assistance from Beric Wells, who had joined us from Washington State University a year before, Lenna Mahoney, me, and several others. Joe Brothers handled the overall project management chores for PNNL and became a primary technical resource for Rick during the intense project planning exercise in December 1998. This small team grew to ten to twenty PNNL folks by the summer and fall of 1999. Perry thought it was a "project managers dream come true." He had all the money he needed, and Joe Brothers handled every conceivable project management hassle. High-powered science and engineering help was available from PNNL to answer any question he couldn't handle. He couldn't help looking good! Perry's own technical horsepower, work ethic, and good sense were a big part of it too.

> Rick began a unique team-building, morale boosting event at his weekly project meetings. The first item of business was always the "Most Important Person" award. The MIP was the one whose task was on the "critical path" that week. Everyone was supposed to help this person out, or at least get out of the way, so the most important person could get enough work done to get off the critical path the next week.
>
> The award was a white T-shirt with a humorous cartoon and a growing list of names and dates of past MIPs. The MIP had to walk up front to receive the shirt from Rick to the cheers and jeers of the rest of the team. Nick Kirch, JR Biggs, and Carl Hanson were frequent awardees, though Rick himself and Craig Groendyke also shared the honor at least once. When names had filled all available space on the first T-shirt, Rick supplied another. There were eventually three, each now hung proudly in the last MIP's office.

While all this organization was going on, DOE Richland decided that the SY-101 level rise issue needed a higher level of review than the SubTAP, which by then was entrenched as an adversary and largely ineffective. Craig wisely brought back Dr. Mujid Kazimi to lead a resurrected full TAP made up of the combined Chemical

Ken Jordan, Nick Kirch, and JR Biggs holding the three "Most Important Person" T-shirts that had accumulated at the end of the project.
Courtesy Rick Raymond.

Reactions and Worker Health and Safety SubTAPs. Kazimi's credibility and leadership steered the new TAP into general agreement with the project's plans and conclusions. TAP backing turned out to be absolutely vital in fending off hostile challenges from inside DOE Headquarters and in keeping the Defense Nuclear Facilities Safety Board friendly. Without it, the project would not have been nearly the success it was.

Courtesy Rick Raymond

We were very fortunate to have Craig Groendyke as our primary DOE overseer and customer. He came to DOE after 24 years as a civilian project manager in the Navy where he worked on weapons systems in nuclear submarines during overhauls. Seeing a sub come in, get fixed, and go out again gave Craig great satisfaction. SY-101 level rise remediation was a similar accomplishment to him. The Navy training made Craig smart and reasonable and gave him the political savvy to survive in a sometimes murky environment where appearances and opinions often override physical facts.

Tall and wiry, Craig was a competitive cyclist. Waiting for meetings to start, we'd listen in jaw-dropping incredulity while he described how really serious bikers would shave their legs for a deep massage or answer calls of nature in the saddle to save a little time during races. Craig carried this focus and intensity to his DOE work, spending many late nights working the technical details of the performance incentives, sometimes hours working on one word!

DESPAIR AT HANFORD SQUARE

The top priority for the new project team was to work out the technical strategy that would actually fix the surface level rise problem and satisfy the second expectation of the performance incentive. To do this, they organized a three-day "value engineering session" at the Hanford Square conference center in North Richland, November 10-12. The "Despair at Hanford Square" was a sequel to the famous "Fiasco in Pasco" meeting of February 1992, which had crafted the technical plan for mitigating the flammable gas releases in SY-101.

Just as the Fiasco in Pasco was anything but a fiasco, the Despair at Hanford Square didn't end in despair. In both, people came together and worked out a plan to do what had to be done. As one could expect, some of the same people were involved in both meetings. The Hanford Square participants included Nick Kirch, Carl Hanson, and Dan Herting, old hands from what used to be Westinghouse; Bill Kubic from Los Alamos; and Perry Meyer, Phil Gauglitz, and me from PNNL. Additional "ad hoc" team members included Rick Raymond and the project staff, Craig Groendyke, Jerry Johnson, JR Biggs, and Doug Larsen. The TAP was also represented, of course, as well as the Defense Nuclear Facilities Safety Board and Washington Ecology. Certified Value Engineering Specialist Rich Harrington facilitated the whole show.

> Value engineering had become a big deal at Hanford in 1998–2000. It was the accepted management technique for working out knotty problems and putting big plans together. In a value engineering workshop, you lock a group of people together in a conference room for three days or a week with a no-nonsense facilitator and slide pizzas under the door until the job is done.
>
> It's quite an experience. All through the first day or even two, the facilitator drags the team through a cumbersome and frustrating process called the function analysis system technique, or FAST. The objective is to make a FAST diagram to describe the problem and its solution. The conflicting push and pull of differing views make this a very discouraging time that feels extremely SLOW. But it's also a necessary team-building catharsis to enable the group to create a useful conclusion. After the FAST phase comes brainstorming. This is a great relief when everyone gets to let loose ideas, silly or sound. Then the team sorts the better ones into a plausible solution using various clever rating schemes. The final step is a presentation to management, sometimes right in the meeting room, requesting their approval to start work.

Here's how the SY-101 level rise value engineering workshop went. Most of the concepts proposed for mitigating gas releases at the Fiasco in Pasco in 1992 were dusted off and trotted around the brainstorming session. Heating the waste, changing its chemistry, liquefying it with sonic vibration, and disrupting it with air bubbles or various kinds of mechanical and hydraulic machines were on the list. We seriously considered shutting the pump off for a while because mixing was what apparently caused the crust to grow. Some even thought we should let the tank go back to burping again for a few times to break up the crust. This concept lost credibility after Craig Groendyke reminded us of the in-tank video of bending air lances and thermocouple trees during the last big burps in 1991 and 1992. A new concept that looked almost reasonable was to just let the crust grow until it became self-limiting. Phil Gauglitz was so convinced that this would happen that he vowed to shave his head if the waste level went above 438 inches. As it turned out, Phil kept his hair, but not by much. It was a truly creative session that produced a stock of good ideas to work with.

Before evaluating and organizing the brainstorming ideas, the team derived a "must-do" list from their FAST diagram of things that had to be accomplished at all costs. The most important function was to remediate level rise by November 1, 1999, when the waste level was projected to reach 458 inches. At this level the primary and secondary tank come together and there is no double-shell containment, which Washington State Department of Ecology requires. The level was already above the 422-inch maximum operating level, so some prompt temporary action was also needed to mitigate level growth while design of the actual remediation hardware was in progress. A concurrent must-do function was to prevent the tank from reverting to its historic large gas releases by either permanently changing the waste so it couldn't burp or by continuing to run the mixer pump to prevent them.

The stated double goal of remediating both level growth and gas releases created a political quandary. The immediate problem exciting DOE and Lockheed management and generating the funding for the solution was surface level rise. But it made no sense to do a costly project on SY-101 to fix the level rise and still leave it prone to burps. Yet the mixer pump had been publicly proclaimed as the cure for burps and Hanford enemies at DOE Headquarters could raise eyebrows and complain, "You told us that tank was mitigated in 1993! Why do you want to spend another $30 million doing it again?"

Accordingly, Craig and Rick made a joint decision to state the official project goal as curing the level rise problem. But they also held onto the unofficial, but "real," goal - to fix flammable gas releases permanently, and an even more closely guarded objective of putting the tank back into useful service! Accordingly, if a concept would both take care of level rise and let Rick shut off the mixer pump, it got a much higher rating than one aimed only at level rise. Also, if the same piece of equipment could be used for normal tank farm operations, that "slightly better" hardware would be chosen.

The team split up into groups to recommend specific plans for the must-do items. The group considering temporary crust growth mitigation had several ideas to work on. One was the "big foot" to break up the crust by pushing down on it. But, recalling that the recent void fraction instrument runs disturbing the bottom of the crust apparently made a lot of it rain back down to the bottom, the big foot concept evolved to a "big lift." An umbrella-like device would push through the crust, unfold, and lift up, rotating as it went to scrape the most gassy part away. Because JR Biggs was its most enthusiastic proponent, it morphed from big lift to "Biggs' Lift." Another temporary concept arose from the many times we had seen a little flush water dissolve a hole in the waste. So all we had to do was use a little more water, 500 to 1,000 gallons, and make a big hole to let some of the accumulating gas out. For the time being, however, both short-term mitigation concepts were put on hold in case of runaway crust growth that couldn't wait. The wait was not very long.

The team assigned to figure out permanent level growth remediation went to work already focused on waste transfer and dilution. The full workshop had already concluded that dilution was the solution. If you add water, the soluble salts making up the crust and most other undesirable features in the tank would simply dissolve. Because removing water in the evaporators caused all the troubles in SY-101 in the first place, reversing this process ought to remove the trouble. Dilute the waste enough and the crust would go away, the waste would lose its ability to trap gas, and the tank would cease to burp. But the tank was already over full, and there was no room to add anything, even water. Some of the pumpable slurry under the crust would have to be removed first to make room for dilution water. At the same time, transfers had to be kept to a minimum because there was very little space in other tanks to put the waste in. Tank farm operations would accept only just enough transfers and dilutions, no more.

These constraints focused the discussion. Should they do a small transfer and dilution or a large transfer and small dilution? Maybe just a large transfer would be enough to fix the crust growth problem, even if it didn't cure gas retention. That would take a larger dilution, but how large? They could not agree on a plan by the end of the day. But Perry Meyer was on this team and wasn't ready to concede. He got together with Nick Kirch and kept working on into the evening. Somehow Perry and Nick managed to tie all the transfer and dilution ideas together into a logical sequence that made good technical sense. First, do a small transfer and small dilution and wait to see what the tank did. This would answer a lot of questions and validate crust behavior models or define where they needed changed. Then, with this experience backing it up, do a large transfer and dilution sized to be the final solution. Additional

336 | Part Three: Surface Level Rise

large or small transfers and dilutions could be done if needed to guarantee success.

They sketched out a simple logic diagram and showed it to Rick Raymond, who was also working late. Rick loved it, and it became the official statement of strategy for the project! This may also have been the foundation of Rick's successful working philosophy of letting the tank tell us what to do and when. Talk about masterful project management! It was almost as if the tank itself was brought into the project and given specific responsibilities! The final event of the workshop was a formal presentation by the team leaders to top management to get them to approve the plan, at least in concept, on the spot. They did, and the SY-101 Surface Level Rise Remediation Project was off and running—almost.

THE TECHNICAL BASIS FOR REMEDIATION

Having a solid consensus on a method for remediating surface level rise was all well and good. Convincing potential critics that it was a *viable* method required a solid technical basis showing it would really work. Fortunately, we had accumulated a good store of knowledge from work on the overall flammable gas issue over the 6 years since mixer pump installation.

The basis for dissolving the crust by adding water was easy. Data from the evaporator runs showed that about 30 percent of the water had been removed to make the waste in SY-101. Tanks still containing the kind of waste that fed the evaporators also had a crust, but a thin, apparently stable one. So, adding back the 30 percent water the evaporators had taken away should reverse the process that was now creating the thick, gassy crust. Because the tank carried just over 1 million gallons of waste, about 300,000 gallons of water would be necessary. We also knew from laboratory analysis of the old SY-101 core samples taken in 1991 that the solids in crust and sediment were generally very soluble. In fact, dilution tests had already shown that one gallon of water would dissolve the all soluble solids in about one gallon of waste. If the crust were 10 feet thick and contained 50 percent gas by September 1999, it would take 165,000 gallons of volume of water to dissolve it completely. Thus it appeared that something like 200,000 gallons would be a good estimate for the required dilution, maybe 300,000 gallons to be safe. Rick estimates he must have explained to critics at least fifty times that "water tanks don't burp! If we dilute the waste enough, we MUST succeed." Many still refused to believe it.

Back in 1996 we had also figured out how to predict whether a tank would burp or not. The prediction used a model that estimated the maximum gas volume fraction in the sediment based on the tank's gas generation rate, sediment depth, and sediment and liquid densities. Simply comparing this maximum gas fraction to the gas fraction needed for neutral buoyancy told whether a tank would burp. Though the SubTAP never really understood and pretty much rejected this "buoyancy ratio" concept, it was the final nail in the coffin for the flammable gas issue. It has since become the core of the current Hanford tank farm safety authorization basis.

We gratefully acknowledge Kemal Pasamehmetoglu of Los Alamos for discovering the principles behind the buoyancy ratio model. Kemal pondered the fact that all radioactive wastes generate gas, but only a few tanks burp. The generated gas had to be escaping steadily by some unknown means, or all tanks had to eventually accumulate enough gas to burp. In the burping tanks, apparently a higher gas generation rate overwhelmed the escape process, allowing gas to build up until the sediment became buoyant. From this conclusion, and from Phil Gauglitz' experiments that showed gas was held in a sediment layer as bubbles, he surmised that individual bubbles must be somehow slowly migrating through the sediment, eventually escaping out through the liquid into the domespace. Kemal then postulated that the speed of a bubble should depend on the density and stiffness of the sediment through which it was migrating. By balancing generation against the rate at which bubbles could escape by this mechanism, he could calculate the gas fraction.

The great breakthrough came when Perry Meyer looked back into the data from the ball rheometer that by now had been run in the five other burping tanks. We frankly had not found much use for this information, but now it was what turned on the lights. The ball rheometer showed that the strength of the sediment was zero at the top and increased linearly with depth. Plugging this linear variation of waste strength into the Kemal's bubble migration model, Perry predicted a gas volume fraction that followed a curved parabolic profile through the sediment layer, similar to the temperature profile. This was very close to the gas distribution the old void fraction instrument had found in

338 | **Part Three: Surface Level Rise**

many tanks! So, even though the bubble migration mechanism was only a theory—in fact, Phil Gauglitz never did find it in the lab—the theory appeared to pull all the available data together in a way that made good sense.

It was a simple matter to extend the bubble migration model to predict which tanks would burp. If the predicted gas fraction was greater than the value required to make the sediment layer buoyant, it had to burp. The measure of burp tendency was cast in terms of the "buoyancy ratio," the predicted average gas fraction divided by the neutral buoyancy gas fraction. A buoyancy ratio greater than 1 indicated a burp was possible. Tanks with high gas generation rates, deep sediment layers, and small differences between sediment and liquid densities had the highest buoyancy ratios. We calibrated this simple model so that AN-103, the tank that just barely burped, had a buoyancy ratio of 1. With this adjustment, the buoyancy ratios of the other burping tanks increased in the same way as their historic average burp volume. Old, prepump SY-101 had by far the highest buoyancy ratio and the highest gas release volumes of all.

So what did the buoyancy ratio model tell us about how much dilution might be necessary to halt the burps forever? Using some guesswork with the available dilution data to predict the changing densities and sediment depth as a function of dilution, the buoyancy ratio for SY-101 would go to 1 when roughly 300,000 gallons of water were added after an equal waste transfer out. We were in luck! The amount of dilution needed to fix the primary level rise problem also appeared to be enough to accomplish the unstated secondary goal of preventing burps and turning off the mixer pump!

The technical plan that Perry defined at Hanford Square now had numbers. The small transfer would be in the neighborhood of 100,000 gallons with about the same volume of water added back to dissolve most of the existing crust. The large transfer would be up to 500,000 gallons with similar back dilution, the exact amount depending on what the small transfer and dilution did. Later, more refined calculations and excruciatingly tedious negotiations with the TAP confirmed that the total transfer and dilution volumes needed to be around 500,000 gallons.

PLANNING THE PROJECT

Rick did not need the kind of clever political maneuvers that Jack Lentsch and Steve Marchetti crafted at the Fiasco in Pasco to make the project work. Nevertheless, before actual work could begin, the plan, schedule, and budget details needed to be worked out and documented to DOE's satisfaction in accordance with the performance incentive agreement. The planning process was an intense flurry of meetings called a technical basis review through cold and foggy days of December and January.

A technical basis review, or TBR, is another team effort where everyone planning a task negotiates with each other so all of them are tied sensibly into the project as a whole. The technical basis being reviewed—or rather created—in this review is not of the physics and chemistry kind. It is the detailed breakdown of man hours, procurement costs, and subcontracts that go into the overall budget request and schedule. At the same time, the people who sit at the technical basis review meetings, like Nick Kirch, Carl Hanson, JR Biggs, and Joe Brothers, not only know how much time and money it takes to do a job, but they understand the science and engineering of it as well. They are also uniquely qualified to describe and work technical risks and uncertainties into the planning so that statisticians can apply the magic of Monte Carlo simulation to create a schedule and budget with a probability attached. The dreary weather and windowless meeting rooms in 2430 Stevens Center offered no distractions to this work.

Planning the project was very difficult. The waste level was rising too fast to do things in a nice, efficient sequential order. All tasks possible had to run in parallel. Engineers had to calculate flows through the transfer piping while designers sized them, safety folks set controls for the operation, and procedure writers described how to run it. Often some change far up the line forced most of the work to be re-done. But when it worked, the "parallelized" schedule made for awesome leaps of progress. The project management term for this method of schedule acceleration is "fast tracking." Fast tracking sounds simple, but it is really very hard. Neither is it efficient, but it can save the project a lot of time and money in the end.

With minute attention to detail and a lot of nights and weekends, the *Tank 241-SY-101 Surface Level Rise Remediation Project Plan* was completed, signed off, and published February 19, 1999. DOE Richland approved the budget change request, and the money started flowing in March. The schedule showed that the first transfer would be completed by late November at an 80 percent probability, allowing for known risks. The surface level rise USQ would close in late January 2002, again at an 80 percent probability allowing for risks. The total cost was projected to be just under $25 million.

Though the first transfer was not done until mid-December, the level rise USQ was closed in November 2000, more than a year ahead of the 80 percent probability schedule. The actual project cost was about $1 million under the estimate—a very good performance. In fact, it was so good that the project was submitted to the Project Management Institute in nomination for 2001 Project of the Year where it finished in second place. But this is way ahead of the story. First there was a desperate race between tank and technology to halt the rising level before the tank overflowed.

1998–1999

A Crusty Christmas

The SY-101 surface level rise remediation project began because the tank's waste level was rising and not responding to the mixer pump as the SA assumed it should, triggering the surface level rise USQ. As project planning started in the late summer and fall of 1998, the level rise had begun accelerating in a sinister way and, by early November, passed the maximum "authorized" operating level of 422 inches. The tank seemed delighted at its unauthorized status and began showing signs of more mischievous behavior. We began to understand the data trends and derived simple models that described in more detail how the crust was growing, how big it could get, and how soon. The predictions were pretty exciting.

The start of the project also brought with it the need to ferret out all the "what if" safety problems that might happen when waste was removed and water added to dissolve the crust. There was a frenzy of brainstorming, analysis, and experiments to find out which problems were real and how bad they could be. Also, as already mentioned, we had to quickly get an estimate of just how much waste needed to be transferred and how much water should be put back in to remediate the tank. All of this work happened around Christmastime in the winter of 1998/99. For Rick's project team it was a truly crusty Christmas!

342 | Part Three: Surface Level Rise

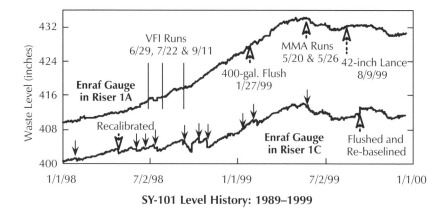

SY-101 Level History: 1989–1999

MOUNTING URGENCY

For all the high-powered technical skill PNNL believed they could bring to bear, one very mundane item turned out to be most visible and important. Shortly after SY-101's level rise got USQ status in early 1998, the obvious questions were "how fast?" and "when will the level reach X inches?" Fitting a simple polynomial function to the last 6 months or so of level data was all it took to answer these questions. Anybody could have done it, but Perry Meyer started it and became known as *the* SY-101 level forecaster. As level rise accelerated, he continually fielded urgent calls for the latest update on projected date the waste level would rise beyond the 458-inch double-containment threshold. Plots went out to Dick French, manager of the DOE Office of River Protection, Lockheed top management, and almost everyone. There was even a board in the project office in 2430 Stevens Center, posted with the last week of SY-101 level data and the latest projection plot. Rick wanted everyone to know they were in a race and who was currently ahead.

> Nick Kirch thought up a very insightful way to instruct people just how massive the projected, and later the actual, crust would be. At meetings he would point to the ceiling, in most rooms about 8 to 9 feet from the floor, and announce in a quiet tone that "The crust in SY-101 is thicker than this room!" At that point, all would usually cease talking, look up where Nick was pointing, then turn and shake their heads in amazement. It never failed to set the correct perspective for discussions of whatever someone was proposing to do to the crust.

In November 1998, it looked like the waste level might reach 458 inches by about November 1999, before the remediation system could be turned on, and the date kept creeping back week by week. This sense of urgency, more than anything else, kept most of the hostile challenges at bay and kept the team, DOE, TAP, and everyone else focused on getting on with the cure.

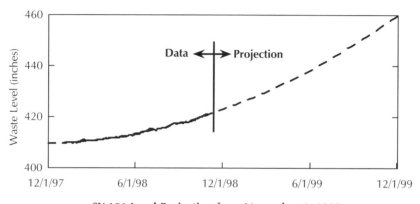

SY-101 Level Projection from November 1, 1998

Besides the level itself, other disturbing signs appeared, indicating worse things ahead. The level rise rate was up to 0.08 inches per day by November 1998. This was a milestone because that was the average growth rate between burps before the mixer pump went in! Mixer pump runs were releasing less gas too. It became noticeable back in March 1998. Projecting this trend, Perry predicted that the tank would stop releasing gas completely by the summer of 2000!

After the intensive excavation period in 1994, each pump run aimed at the thermocouples on the MIT in Riser 17B had caused a sudden jump in the temperature, registered by the bottom thermocouple 4 inches from the bottom of the tank. This meant that the pump jet was able to clean out the settled solids to a radius of about 28 feet on every run. In June 1998, the time it took for the bottom thermocouple on the 17B MIT to jump began increasing. By December 1998, it ceased jumping and showed only a steady rise. The jet could no longer clean the tank bottom as far as it had. Other subtle changes in the pump pressure and pressure drops also began showing up in mid-1998 that indicated either the pump, or the fluid it was pumping, was changing.

The most serious change was in the motor oil temperature behavior. Normally, the motor oil temperature rose from the nominal waste temperature of about 115 degrees Fahrenheit to 170 to 180 degrees by the end of each 25-minute pump run at 1,000 revolutions per minute. In June 1998, the oil temperature at the end of a run started rising faster than the seasonal rise in waste temperature. By November 1998, the oil temperature began hitting the limit of 190 degrees F in less than 25 minutes, forcing operators to shut the pump down early!

Then the average waste temperature itself departed from its historic seasonal trend. Previously, it had followed a cycle of increasing in response to summer heat to a peak in October then decreasing to a minimum in April. In December of 1998, it just kept going up instead of beginning its seasonal decrease. Apparently the crust layer had become sufficiently thick and gassy to act as an insulating blanket, keeping the radioactive decay heat in.

All these changes seemed to be showing that the tank was doing something more than just growing crust. Or were the changes just the result of a growing crust? There were lots more questions. Was the trend inexorable, or would whatever process was causing it eventually slow down? Would the remediation system be ready in time to stop it? Would the pump keep running or become useless before then? More importantly, just what kind of danger did all these changes imply? How much gas was being stored in the crust, anyway?

THE CRUST FLOATS!

Our understanding of the crust and its gas retention potential got a big boost in November and December 1998. The void fraction instrument runs of June and September and initial retained gas sampler data gave us an idea of how much gas was stored in the crust. There was also plenty of information on the amount of level rise and growth in the crust thickness. But how were gas and crust dimensions related?

There were clues. The two level gauges had always given different readings, with the new Enraf in Riser 1A always higher, now by 18 inches, than the one in Riser 1C. But the 1C Enraf and the FIC before it needed to be flushed with water every few months for many years. The salty crust was known to be very soluble, and the water had to be dissolving whatever it hit. But, because water was much lighter than brine, it could only dissolve crust above the free-brine level, the level of the liquid that would rise in a hole drilled through the crust. So, as long as regular flushing continued, we could assume the 1C Enraf was approximately measuring the liquid level while the dry, unflushed 1A Enraf showed the level of the very top surface of the crust. At the same time, we could get a reasonably accurate measure of the total crust thickness by looking at the temperature profiles from the two MITs.

We now knew the top surface, the liquid level, and the total crust thickness. Suddenly it became obvious that this information made it possible to describe the crust as a simple floating object. Actually, it was a sort of porous floating object like a bag of ping-pong balls. We had always known that the crust was floating, but now we had a measure of its draft, the part under the liquid, and freeboard, the part held above the surface. Archimedes discovered 2,250 years ago that a body immersed in a fluid feels a buoyant force equal to the weight of fluid it displaces. Appealing to Archimedes and applying reasonable assumptions about the density of crust material, we could calculate exactly how much gas it would need to float the way the Enrafs said it did. In simple terms, anything lying above the liquid level had to be pushed up there by a buoyant force below the liquid level. The recent crust core samples showed that the non-gas part of the crust had about the same properties as the non-gas part of the sediment at the tank bottom. All the buoyancy had to come from gas. An increasing difference between the 1A Enraf, measuring the crust top, and the 1C Enraf, measuring the liquid level, had to mean an increased gas volume in the crust.

The first crust buoyancy calculations were very encouraging because the crust gas volume matched what the void fraction instrument and retained gas sampler data said had to be there. The next step was to apply the same kind of curve-fit extrapolation that Perry had become so adept at using and forecast future gas volumes. The numbers immediately grabbed our attention. By about September 1999, the earliest date for the small transfer, the crust could be 10 feet thick and contain almost 20,000 cubic feet of gas! This was as much gas as the tank had ever contained on its worst burping days before the mixer pump! It was getting pretty obvious that the answer to the original question posed at Nick's task force the year before, "Is there a new or greater hazard...due to the observed waste behavior in SY-101?" was resoundingly "Yes!"

I sent out some plots describing the crust projection in an e-mail to the project team December 30 under the subject "A Christmas Chrust." The message recommended that the temporary crust mitigation plans be activated. It sure looked like we needed to do something to slow crust growth down before it got completely out of hand. Though Rick was a little irritated that I didn't consult the team before sending the message, it was clearly correct and he had to agree with the recommendation. Accordingly, JR Biggs got the go ahead to start work on the "Biggs' Lift" and water spray crust mitigation concepts. More on these a little later.

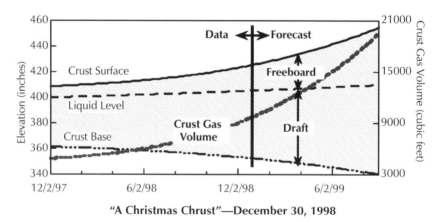

"A Christmas Chrust"—December 30, 1998

NEW INSIGHTS

Besides its gas content and dimensions, we were learning more about the properties and personality of the different layers of the crust. The void fraction instrument gave us a first physical feel for the clay-like material with the 30-inch arm that sticks out from its long vertical mast like the foot of an "L." In the June 1998 test, the operators found the base of the crust by pulling the arm up and rotating it with a short strap wrench until it met resistance. This happened about where the temperature profile showed the crust base to be. After this, the instrument was drawn up into the crust about 15 inches when pressure started showing on the hydraulic cylinder that kept the arm horizontal. Both these tests showed that the bottom part of the crust had some strength all the way down to the base.

In the September 1998 void fraction instrument run, the operators used the crane load readout to decide how hard to pull upward on the arm. They stopped pulling when the crane load hit 2,000 pounds, about 400 pounds more than the weight of the void fraction instrument system. With this higher force, the sample chamber penetrated about halfway through the now 65-inch crust, only 16 inches below the liquid level. From this test, it was clear that the upper part of the crust was quite stiff. Data from both the June and September void fraction instrument runs showed that the gas fraction in the crust was highest near the bottom, up to 43 percent. Up towards the middle it was somewhat below 30 percent. A little later the retained gas sampler would show this trend more clearly!

John Kryzstofski suggested that we get core samples to answer critics who still believed that a change in waste composition might be the source of the level rise. He also wanted to use the retained gas sampler on several core segments, and the CIUALG quickly jumped into discussions on where and how the retained gas sampler segments should be used. We were excited because the retained gas sampler was better suited to measuring the gas volume fraction up in the stiffer part of the crust that the void fraction instrument could not reach.

Core samples with several retained gas sampler segments were taken in December 1998 and January 1999 with special emphasis on the crust. The core sample analysis showed that the waste chemical composition was essentially the same as it was in 1991 and also about the same from top to bottom. Retained gas samples came from three levels in the crust, now more than a foot thicker than during the last void fraction instrument run in September 1998. The gas composition in the retained gas sampler segments showed the crust gas to be somewhat richer in hydrogen than lower in the tank but still about the same composition as estimated back in 1992. All this proved that crust growth was not from some mysterious chemical reaction. It was a mechanical, not a chemical, phenomenon.

The gas volume fractions in the retained gas sampler segments ranged from 20 percent up near the liquid level, 30 percent close to the mid-point, just like the void fraction instrument, and over 50 percent at the crust base. A 50 percent gas fraction was almost a foam! Recalling the bubbly, foamy-looking stuff that had started seeping up around the pump column during pump runs and around the void fraction instrument mast during the summer of 1998, we speculated that gas was collecting near the base of the crust. We called it the "bubble slurry" layer. We believed that the bubble slurry layer, by its extremely high gas content, was a definite safety hazard. There might be a large gas release if something caused bubble slurry to come rushing up through a hole in the crust or if a big piece of crust were to roll over.

We got a free experiment in crust dissolution in late January 1999 when tank farm operators dumped 400 gallons of water on the crust under Riser 5B. The rising crust was getting too close to the hardware at the bottom of the main in-tank video camera. The water flush was intended to dissolve a hole to make room to keep the camera operating awhile longer. A temporary, portable video camera showed us the effect of the water dump. It was pretty much as expected. The water splashed on the hard chunks of salt, but it sunk below the surface quickly. On the way down, the water dissolved a round hole with a roughly flat bottom. The depth of the hole appeared to be about the same as the 18-inch difference between the 1A and 1C Enrafs, so its bottom was probably at the liquid level. The appearance of the waste exposed around the sides of the hole and the action of dissolution we saw on the video revealed the "dry" crust above the liquid level as a collection of salt chunks, individually hard and strong, but collectively weak, like crushed rock.

In early February 1999, Scott Cannon and Dave Barnes gave us a new measurement capability. They designed small-bore neutron and gamma probes to go down the 1-inch pipe inside the MITs. Bigger neutron and gamma probes had been used for many years inside the 4-inch fiberglass liquid observation wells in the single-shell tanks to measure how much of the waste was saturated with liquid. The neutron probe has a capsule of radioactive americium and beryllium to flood a small volume of waste with fast neutrons. Hydrogen atoms are very good at slowing down these neutrons, so they can be detected. Because each water molecule has two hydrogen atoms, the count of "slow" thermal neutrons is a measure of the local water content in the waste. This meant that parts of the crust with a lot of gas bubbles and the dried-out upper layer should have a low neutron count because less water was there. The gamma detector simply counts the gamma ray hits coming from the waste. Because gamma radiation in Hanford tanks comes mostly from cesium-137, and because cesium-137 is very soluble, the gamma count is also a measure of the liquid fraction in the vicinity of the detector. Both probes were run in both MITs in SY-101 every few weeks, but the neutron probe gave the most detailed picture of the crust.

> Before he came to Hanford, Dave Barnes' job had been interpreting logs from oil exploration wells. He applied this experience to find the "interstitial liquid level" from neutron and gamma logs in single-shell tanks that were being saltwell pumped. This is where you'd find liquid if you sunk a well into the waste. Dave was a real expert at this, but he wasn't used to seeing a lot of gas below the liquid level. He had always believed that, if there was any gas at all, the "soil" was not saturated. He couldn't stomach the idea that there might be gas bubbles trapped by liquid-saturated waste, and always wanted to put the interstitial liquid level at the base of the crust! We had long arguments about this. Once I almost convinced him the liquid level was near the top of the crust at the level measured by the 1C Enraf gauge, but I don't think he believed me.
>
> Dave was also one of the few among Hanford contractor staff who had a Macintosh personal computer on his desk. CH2M HILL, Westinghouse, and Lockheed before it, and even DOE Richland, all enforced a uniform, company-wide Microsoft® Windows environment. On the other hand, PNNL actively supported Macintoshes and UNIX machines, as well as Windows boxes. As a dedicated Mac enthusiast, I supplied Dave with our excess Macintoshes through something called "inter-contractor transfer." The last machine I gave him was my old non-USB G3 when I upgraded to a G4 in 2001. Sadly, CH2M HILL finally took Dave's unauthorized Mac away in mid-2003.

The first neutron logs in the two MITs came out in February 1999. It was like a window opening to show us what was inside the crust. As we expected, the upper 20-inch layer above the liquid level was really dry. The first 20 to 30 inches below the liquid level appeared to be almost saturated with liquid, but it looked like the bottom 40 inches held a lot of gas. In fact, there was a thin layer of what we thought might be bubble slurry. Oddly, this little spike in the neutron log appeared only in the 17B MIT on the north side of the tank, not 17C in the southeast. Later, the bubble slurry layer would grow much more obvious and appear in both 17B and 17C neutron logs.

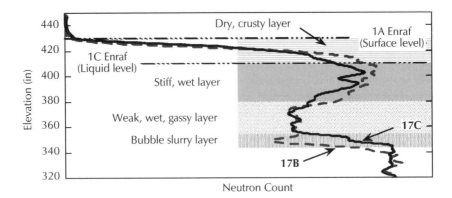

NEW HAZARDS

After confirming that the crust held a lot of gas and was rapidly accumulating more gas, we set out to discover, some would say "imagine," all plausible ways this gas might come out. If a gas release mechanism looked real, we then had to figure out how fast and how big the release might be. Those few that might, at the far end of the probability scale, be able to make the tank domespace flammable, would need controls or mitigation schemes. The list of potential problems seemed endless. New ones appeared every week. It was hard work but also some of the most rewarding. For once we seemed to be able to keep ahead of all those shadowy critics that come out from under rocks in the middle of big projects shouting "but what IF . . ."?

The first intense period of safety analysis ran from January through about May 1999. Most of the crust gas release issues came out of a series of brainstorming workshops at the end of December 1998. By May most of the initial crop had been dismissed or understood well enough to work around.

The bubble slurry layer was the source of flammable gas in the crust that we believed was the primary hazard. It had a measured gas fraction exceeding 50 percent, it was very likely weak enough to flow readily, and it appeared to be getting thicker. So brainstorming meetings concentrated on mechanisms by which the bubble slurry could flow out from under the crust onto the surface and release its gas.

We posed many potential bubble slurry release mechanisms that could be discounted immediately by prior observations, measurements, or just plain common sense. One of the first was a potential "geyser" through a small hole. But there had been a lot of small holes punched in the crust recently and, though there was sometimes a little bubbling and frothing, none of them created a geyser. Small holes or cracks seemed to seal themselves very effectively. Another was the suggestion that the whole crust might collapse like a failed soufflé, suddenly releasing all its gas in a great sigh. But all the experience so far showed the crust was nothing like a soufflé. It had a hard, dry chunky top over stiff, wet clay-like material with little gas in it. The only region with a culinary character might be the bubble slurry itself. If it collapsed, it would only create flat bubbles that would probably "blurp" up through "chimneys" in the crust a few at a time without any danger of flammability. Besides, nobody could conceive what would make the bubble slurry collapse suddenly.

One possible scenario we could not dismiss was rooted in our ancient observation from the 1991 in-tank video that the crust was a collection of "wastebergs." Marine architects know that a stable boat has to be heavier on the bottom than the top. That's why sailing ships had to carry rocks in the hold as ballast to keep from capsizing. Now, if SY-101's wastebergs were losing ballast by building up a thicker and thicker layer of buoyant bubble slurry, they might capsize and dump a load of flammable foam in the dome. If one wasteberg capsized, would it tip its neighbors over and start a chain reaction where all the bubble slurry would end up on top? This could be downright dangerous, though probably quite exciting to watch.

Rick supplied us with plenty of money to research the most plausible hazards, so we brought Rudy Allemann back from retirement and turned him and Scot Rassat loose in a well-equipped lab with miracle-working support from Donny Mendoza to check it out. To represent wastebergs, they bought wooden blocks from the toy store and glued steel washers on top to make them unstable. They carefully added the weighted blocks in a flat-bottomed round bowl until no more would fit. Adding water to the bowl to float the blocks created a simple but intuitive and accurate simulation of the supposed collection of tippy wastebergs. Rudy immediately found out that, while individual blocks flipped over instantly, the collection of blocks in the bowl gently held each other up so the whole stayed

352 | Part Three: Surface Level Rise

stable. Even poking individual blocks under or violently wiggling the whole collection on a shaker table could not create a capsize. In fact, a good number of blocks had to be removed to make space enough to allow a capsize. While nobody could claim that the floating blocks were a match for the actual properties of the crust, neither could they argue anymore that capsizing wastebergs were a serious threat.

> The floating block simulation was very convincing, but it was also a little comical. What was the public to think? PNNL was a U.S. government national laboratory bringing the best possible science to bear on serious problems! Why were they playing with children's toys in a dishpan, for goodness' sake?
>
> Nevertheless, the block capsize experiments were shown off to more people than anything except the actual dilution campaigns. Besides Lockheed and DOE Richland upper management, the TAP, Defense Nuclear Facilities Safety Board, and important DOE Headquarters people got to watch the bowl of wooden blocks shaking. After Rudy was done, Scot passed out the blocks with the rusty washers still attached as souvenirs. All the project folks had one or two on their desks. Rick still has a red one with an "R" on it and some extra ones he shows when he recounts the story.
>
> Donny Mendoza tried a more realistic crust simulant of blocks cut out of a bowl of Jello with a bottom layer suffused with small Styrofoam beads that looked like cottage cheese. On a leaf of lettuce, the experiment might have made a good looking salad. The problem was that the blocks glued themselves back together after a short time, so individual wastebergs ceased to exist. This might actually have been more like the real crust.

The other possibility that we had to take seriously was bubble slurry flow through a large hole. It was obvious from experience and theory that small holes were self-sealing. But, though reasonable arguments could be made that a foam flow through large holes would also be self-limiting, it could not be proven. This scenario was more difficult to evaluate than the crust capsize and not quite as much fun. Rudy, Scot, and Donny went to work in the lab building an experiment to simulate foam flowing up through a hole.

Because the slurry had some small strength, not all of it would flow. Think of a rectangular box with sand flowing out through a hole in the bottom. When sand stopped flowing, a cone shaped depression would remain with the sides of the cone defining the angle of repose of the sand. Turn that picture upside down and you have the idea for upward foam flow. The thicker the crust and the thicker the bubble slurry layer, the more foam was available to flow and the faster it would flow. The experiment used light mineral oil with hollow glass micro-spheres to simulate a thick bubble slurry. They trapped the slurry under an inverted flat-bottomed cylindrical tray submerged in a larger tank of water. Foam flow started when they pulled a rubber stopper from a hole in the middle of the bowl.

Rudy worked out a simple theory that predicted the overall behavior of the experiment. Perry refined it to include the strength of the slurry. The model predictions and experimental results matched better than either Perry or Rudy had hoped. Even better, when scaled up to full tank size with a 6-foot diameter hole and an 18-inch-thick bubble slurry layer, the model predicted that far less gas would be released than needed for flammability. Perry was very pleased with the work and the way it turned out. In his words, the whole foam flow investigation "was very nicely done. It had the highest technical quality, both experiment and modeling, [of any hazard we looked at. Foam flow was] probably the greatest real risk."

A related safety problem concerned the mixer pump. It had to keep running to protect the tank from burping. But the crust might grow thick enough for its base to feel the circulation induced by the mixing jet. The hydraulic force could sweep the gassy bubble slurry inward and down into the pump inlet. We feared that the pump would not work if it was sucking bubble slurry containing 50 percent gas. To see what the circulation pattern under a thick crust might be, Perry and Kurt Recknagle dusted off the TEMPEST code that Don Trent and Tom Michener had used to predict pump-induced burps before the mixer pump was installed. The results of a series of TEMPEST runs in March 1999 showed that, as expected, the circulation currents up at the base of a thick crust at 250 to 300 inches from the bottom were very gentle. In fact, there was not even enough force to stir material with the consistency of a thin milkshake. Apparently it didn't matter if the crust base came right down to the pump inlet. But, Perry recommended that the pump not be run if the crust base came down to less than 12 inches above the

354 | Part Three: Surface Level Rise

inlet, an elevation of 270 inches. That was plenty of margin if the crust didn't change drastically.

LICENSE TO KILL

At the end of April, the DOE Office of River Protection bestowed on its contractor an amazing gift, the likes of which had not been seen in all of Hanford history and likely would never be seen again. A letter from DOE on April 27, 1999, authorized the contractor to "conduct activities which exceed previously accepted risk, so long as [flammable gas concentrations do not exceed 25% of the lower flammability limit] and the contractor imposes additional prudent controls necessary to safely conduct such operations." In other words, we were pre-authorized to do anything we thought necessary with no external controls except to be "prudent" and stay below 25 percent of flammability! James Bond should be so fortunate!

We all called it the "Wagoner Letter," even though John Wagoner had retired four months earlier and never saw it. The letter was actually signed by both Dick French, Manager of the Office of River Protection, and James Hall, manager for DOE Richland. Craig says that Steve Wiegman actually put the words on paper and Jackson Kinzer took it up the line to French. Ralph Arcaro, local Defense Nuclear Facilities Safety Board representative, said, "DOE is finally giving the contractor some room to operate." On the other hand, Bob Marusich was amazed and aghast that DOE would dare to do such a thing. Bob thought that the project team would have worked the same way without the letter, it just would have taken longer.

The risk of taking longer was what drove the Office of River Protection to issue the letter. The blanket authorization was issued because of "the urgent need to mitigate the growing hazards within this tank." They recognized that "the hazards in Tank 241-SY-101 are increasing and the frequencies and consequences of potential accidents are increased over those previously accepted by DOE. These increased risks cannot be reasonably quantified at this time." The Office of River Protection wisely looked at the accelerating level growth and concluded that the risk of not doing something, or doing something too late, was much higher than giving us freedom to act. The formal controls that had been proposed for remediation were extremely complex and prone to confusion. If something happened that was outside the analysis, reanalysis and reapproval would have stopped us dead while the surface level kept rising!

This new freedom did not lessen the need to anticipate hazards, quantify their consequences, and control them if necessary. Lockheed was *very* concerned about meeting the letter and intent of those "prudent controls" necessary to ensure safety. However, we could now concentrate on *real* hazards and *reasonable* controls, bypassing the groaningly tedious formal approval process usually required of safety documentation.

> Craig believes the Wagoner Letter was responsible in a major way for the success of the project. It worked so well, in fact, that the Office of River Protection applied it to other projects. They had recognized and the surface level rise project proved that you don't need a lot of controls all the time, if you just keep it below 25 percent of the lower flammability limit. Saltwell pumping in tanks on the flammable gas watch list was an example. Imagine how much easier it would have been to build, install, and run the mixer pump with a Wagoner Letter in 1992/93!

THE GREAT CRUST-BASE DROP

Maybe all this activity convinced the tank it had made a strategic error by accelerating level growth in the fall of 1998. Could SY-101 have realized, like Japanese Admiral Nagumo did after Pearl Harbor, that a rash attack can inspire a powerful adversary to heroic effort that guarantees the attacker's eventual defeat? If so, the tank knew it ought to back off a little. Maybe without the driving urgency of a rapid-level rise, DOE would decide not to spend so much money and delay remediation for year or two, even cancel it altogether! Maybe this scenario pushes anthropopathism too far. But the timing was uncanny. Here's what happened.

The Great Crust-base Drop was a period of about a month from mid-March to mid-April 1999 when the base of the crust layer lost its clear-cut definition and apparently fell to a lower elevation. The neutron logs of March 11 and April 16 showed us what had happened. The crust base on March 11 was a relatively sharply defined 335 inches. On April 16 it was a vague 320 inches. The crust as a whole grew about 15 inches thicker than it had been the month before.

The crust-base drop was apparently a symptom of a general change in the character of the crust. The first clear indications that something was different showed up, appropriately, around April Fools' Day. The background hydrogen concentration between pump runs jumped back up to 20 to 30 parts per million about March 24 after reaching an all-time low in mid-February. On April 1, pump-induced gas releases also began a steady rise. Peak hydrogen concentrations following pump runs rose from 100 to 150 parts per million to 250 to 350 parts per million. Similarly, the nitrous oxide concentration reversed a 5-year decreasing trend and began to rise again. More dramatically, the level growth rate on the 1A Enraf gauge took a big drop about April 8 and kept going down, eventually going negative in late May.

We never could explain the causes and mechanisms of the crust-base drop to anyone's satisfaction. Like the great temperature drop following Event E in December 1991, the hydrogen suppression phenomenon following pump bumps in 1994, and the great nitrous oxide reduction from 1993 through 1999, this aspect of SY-101's personality will also remain a mystery. Actually, it may have come from the tank's tendency towards practical jokes. The crust-base drop that reversed the acceleration in level rise happened only a little over a month before the project tried to do the same thing. It was as if the tank saw what we were preparing and deliberately beat us to it.

BIGGS' LIFT

After the dire projections of crust growth and gas volume in the "Christmas Chrust" message of December 30, 1999, showed how bad things might get, the project decided to initiate temporary mitigation action, and JR Biggs got to work on the Big Lift and the cruciform lance. The Big Lift was to have a lifting and scraping effect on the base of the crust like the void fraction instrument did in the summer of 1998. The cruciform lance was the big, high-pressure spray system initially used to bore a hole for the mixer pump in 1993. It was supposed to bore big holes in the crust in the same way it did then.

JR really believed in the Big Lift. In his mind it was simple—the crust was obviously not letting enough gas out, so it needed to be broken up. The idea had died three times at the November workshop, but JR kept reminding the group that the crust was growing faster than a permanent remediation system could be brought into action. Something needed to be done quickly. With this argument, he managed to drag it back on to the table each time, even though actual design and construction were put on hold until now.

JR (James R) Biggs is a uniquely effective individual who should share the credit for making the SY-101 level rise remediation projects happen. He works as comfortably and expertly out in the tank farm in whites, describing a new piece of hardware to union journeymen, as he does in a conference room full of overbearing scientists, explaining how that hardware has to work in the farm.

Courtesy Rick Raymond

From hard experience JR strives to keep equipment stone simple so less can go wrong, and the operators have fewer distractions. He believed a device should be manually operated as much as possible. If it had to be electric, there should be only an "off-on" switch, and maybe a light to indicate it's "on."

JR has been around Hanford quite a while. He began training as a tank farms shift supervisor August 12, 1992, went on shift and immediately got involved with window work. He remembers not being very impressed with the in-tank video of Event E until he realized the scale of the action—a 3-foot wave of wet concrete was bending thermocouple trees. In July 1998, after SY-101 mitigation was proven and the excitement was over, JR left Hanford to work at the first chemical weapons destruction plant in the world. The site was at Johnston Atoll, 2 miles long by a half mile wide, and roughly 900 nautical miles west of Hawaii. Like Hanford in the Manhattan Project days, it was chosen for it's remoteness from population. The best part of the assignment was snorkeling and fishing in the sizable lagoon with giant manta rays, sea turtles, tropical fish of all kinds, and sharks! After a 4-month tour on the atoll, JR returned as operations lead for the SY-101 surface level rise remediation team where the only salt water was in the tank and no swimming was allowed!

Biggs' Lift was ready to go in May 1999. The formal name for this device was the Mechanical Mitigation Arm. It was an L-shaped arm similar to the void fraction instrument but a lot bigger. The arm was a pipe 6 feet long, compared to the void fraction instrument's 30 inches. Al Kostelnick of Cogema Engineering designed and built it in only 12 weeks! By JR's demand, it was entirely man-powered. The arm was extended and braced by a cable pulled by a hand-pumped hydraulic cylinder. There was a water spray nozzle on the mast where the cable entered it and a water flush to clean the pivot bearing where the arm attached to the mast. About 900 gallons of water squirted into the crust at the end of the arm during each deployment.

It took four sturdy operators to rotate the arm, each pushing on a 3-foot lever, like sailors hoisting an anchor with a capstan. JR called them "the ox team" because they were like a yoke of oxen walking in a circle threshing grain. The handles could have, and maybe should have, been made longer to increase the torque, but Al designed them to break before anything else did. He didn't want to be embarrassed by breaking the mast and dropping the arm irretrievably into the tank.

We planned to run the mechanical mitigation arm on April 20 in a riser that was close to the camera so we could see what was happening clearly. We were concerned that the lowest elevation the mechanical arm could reach, 332 inches, might still be within the crust. The ox team might have trouble getting rotation started. The plan was to do a series of short lifts, rotating the arm several times after each one, if possible. If the crust was too stiff for rotation, the crew was supposed to try rocking the arm back and forth in hopes it would break up the crust enough for a rotation. If that didn't work, the crane would simply pull the mechanical mitigation arm straight up until reaching the limiting load, then drop the arm to vertical and pull it out of the tank. We hoped that the quarter-inch cable angling up from halfway out the arm to the mast would disturb quite a bit of waste, too.

The first mechanical mitigation arm run was on graveyard shift on Thursday, May 20, 1999. It took a crane and a crew of 50 people to do the operation. The ox team started their first walk around the mast at 4:00 AM. At this lowest elevation, rotation was easy and not much happened. In an hour the team had made six lifts with rotations and oscillations and the hydrogen was up to 600 parts per million. The arm had to be lowered back into softer waste several times so it could be rotated. It was such strenuous work that the ox team got blisters on their feet pushing the mast around and a new "yoke of oxen" had to be called in.

With great exertion, the fresh team was able to rock and rotate the arm through several more small lifts before they had to admit the crust was just too stiff. Besides, it was getting late and they had to leave time to take the thing out of the tank and lay it down before shift change. At 5:55 AM the crane began lifting the arm up through the crust very slowly. It finally hit the load limit of 2,200 pounds and had to be let drop, exiting the tank at 6:45 AM.

This first Mechanical Mitigation Arm run released a total of 116 cubic feet of gas and disturbed about 216 cubic feet of crust. They had been able to do rotations 18 to 20 inches up into the crust, about as far as the void fraction instrument had made it. The TRC considered it a success and asked for another run in the same riser the following week, hoping that the added water would soften up more crust. Maybe they could get all the way through this time!

The second Mechanical Mitigation Arm trial was done on swing shift on Wednesday, May 26, 1999. The extended arm made its first rotation 26 inches below the crust surface at 8:00 PM. Half an hour and two lifts later, they were running around the track doing six rotations per minute, one every 10 seconds! But, when the next lift brought the arm up to about where they had to stop last time, the crust was still too stiff to rotate the arm all the way around. The ox crew kept working, up a little and down a little, for an hour but progressed only a few inches higher than they had penetrated May 20. It was time to call it quits. The arm was out of the tank at 10:09 PM.

The hydrogen concentration peaked at 1,300 parts per million, over twice the concentration of the first run! Perry calculated a total of 173 cubic feet of gas release with only a little additional crust disturbance. The crust around the riser subsided a little and a large area on the left side of the riser, probably beyond the 6-foot radius of the arm, also subsided during the run. Several large bubbles blooped up around the mast during the early part of the operation. The mast carved out a "clean" hole more than a foot in diameter in which it rattled around with little resistance. The liquid visible around the mast was very watery with no appearance of bubble slurry flow at any time. They had done a good job. Rick said, "The

360 | Part Three: Surface Level Rise

safety consciousness, dedication, hard work, and attention to detail of this crew can not be over stated. Without this great crew, led by Butch Hall, this work could not have been done!"

> The project brought all kinds of people together. Perry Meyer and Jim Bates were stationed in the DACS trailer with a laptop during the Mechanical Mitigation Arm runs, keeping track of gas releases, making predictions, and taking notes. Everyone involved was at the pre-job briefing: PhD scientists, engineers, managers, and union workers. Perry found some of the slogans and graphics on the workers' t-shirts "verrry interesting." At the other extreme, Lockheed Vice President Dale Allen took Perry and Jim to his office for coffee about 3 AM while they waited for the job to get started.
>
> Sometimes bringing different kinds of people together causes a culture shock. For the second Mechanical Mitigation Arm run, a bunch of engineers were in the DACS trailer with shift supervisor Dan Niebuhr. Field work supervisor Butch Hall was in the farm running the operation. Butch was said to be a hard workin', hard cussin' miner who thought engineers were unnecessary at best. The engineers were like kids with a new toy and they wanted to play, asking for different lifts, rotations, oscillations, and water sprays. Butch tried to accommodate them, but it was hard to communicate. Engineers were talking typical technical lingo while Butch yelled back on the radio through his mask things that sounded like "++squawk++ Wagh chha rrr dchhou wahht ++squeel++" meaning "What the #@!% did you want?"
>
> The two groups had butted heads on the first mechanical arm run, and JR Biggs saw disaster in the making. He asked Dan Niebuhr to come in and take charge. Dan backed the engineers off and told them to "tell me what you want to do and I'll communicate it to the farm." It worked well. The engineers didn't get their feelings hurt, and Butch didn't throw things at them. Perry said they had a lot of fun anyway.

The two Mechanical Mitigation Arm runs affected the behavior of the tank for a long time afterward. Mixer pump runs released a lot more gas for a while. More important, the background hydrogen concentration stayed high for two weeks. This three-week gas release dropped the level measured by the 1A Enraf almost 2 inches from its all-time high of 434.3 inches. Not only that, but the surface level kept falling slowly, on the average, all the way up to the first transfer and dilution in mid-December 1999. It wasn't rising anymore!

Did Biggs' Lift really stop level growth, or did the April crust-base drop do it? We'll never know for sure. The crust-base drop certainly started increasing gas releases and reducing the level rise rate, but the actual level drop came only after the Mechanical Mitigation Arm did its work. The TRC was not in full agreement with what the Biggs' Lift experience meant. There was no evidence of bubble slurry being stirred by rotating the arm or coming up in the "clean" hole around the mast. Maybe the all-pervasive bubble slurry didn't even exist. On the other hand, a lot of gas was released from a larger region than the mechanical arm actually disturbed.

Because crust growth had stopped for the present, the TRC decided not to run the Mechanical Mitigation Arm again. But they now had what they believed was an effective method to release gas and halt crust growth if it should start again. It turned out there was no further need of it.

BIG LANCE

Besides putting Biggs' Lift back into mothballs, the TRC decided that there was no rush to use the water lance either. It would eventually have to be run anyway to make a hole in the crust for the new generation transfer pump. That would be soon enough. The 42-inch cruciform lance was originally built to make a big hole through the waste for the mixer pump. It sprayed water at high pressure from many tiny nozzles screwed into the bottom surface of two 40-inch lengths of 1-inch pipe. The two spray pipes were fastened across each other horizontally and bolted to a 60-foot mast of 2-inch pipe that required two cranes to raise to vertical. If the water spray wasn't enough, the full weight of the mast and lance, about 2,500 pounds, could be applied to the crust.

The lance went into the tank through Riser 5B, about 20 feet east of the tank center, early Monday morning, August 9, 1999, almost 3 months after Biggs' Lift. By that time the crust was a full 10 feet thick with what appeared to be a nicely restored bubble slurry layer on the bottom. The water came on with the lance just above the waste surface at 5:15 AM. The pressure ramped up to 250 pounds per square inch, and the fine mist made some fog. The lance pushed into the crust with sprays on at 5:18 AM with the crane supporting about half its weight.

The strategy was to slowly lift the lance up and down without trying to rotate it to let the sprays do their work. It took about 20 minutes to penetrate through the hard part of the crust into the pasty layer. The lance bored through this much faster, passing all the way through the crust 5 minutes later. Boring through the crust did not release any gas. Where was that bubble slurry layer? After pulling it back out for a quick look, the crane operator lowered the lance all the way down, reaching the tank bottom at 5:48 AM. The lance rested there about 15 minutes and squirted 200 gallons of water before being lifted back up above the crust again. The hydrogen concentration only now began to rise slightly, reaching 212 parts per million when the lance came out at 6:06 AM.

The plan was now to twist the lance back and forth while plunging it down through the crust again to make sure the hole was good and round. Though the lance didn't feel any resistance on the way down, the spray nozzles apparently plugged by the time it reached 8 feet and it took about 35 minutes to clear them. Then the water flow stopped again when the water supply to the pump failed. After taking a look at the hole, the operators made the decision to quit and pulled the lance out of the tank at 8:05 AM.

The lance had successfully prepared a hole for inserting the transfer pump, disturbing at least 100 cubic feet of crust. It used 1,200 gallons of water and released 80 cubic feet of gas altogether. The hydrogen concentration stayed quite low, hitting just 400 parts per million the last time the lance was pulled out.

This didn't match our expectations at all! Both Mechanical Mitigation Arm runs had released quite a bit more gas even though they didn't bore through the crust completely. Also, there was no sign of the bubble slurry that the neutron logs so clearly showed. Either there was no bubble slurry in the vicinity of the riser that the lance went into, or it was not the foamy fluid we believed it to be. Maybe it was a lot stronger and simply could not flow. In any case, it was clear that "foam flow through a large hole" was simply not the dangerous accident we had postulated back in March.

THE CRUST VS. THE PUMP

By the summer of 1999, whether the Mechanical Mitigation Arm or the crust base drop caused it, there was no doubt the tank had changed a lot. It was good that the surface level stopped growing and even went down a few inches. But the crust was also at least 15 inches thicker. That was bad. Even worse, the accelerating level rise that supplied the moral force for the project was gone and never returned! But enough momentum had apparently built to keep the wheels of the Hanford and national bureaucracies rolling along at good speed on their own. The timing was just right. The project was authorized, money allocated, and the schedule set by the end of March while the pressure was still on. By June, when it became clear that level rise was not going to return soon, it was too late to reverse the process. Besides, even if it had stopped growing, SY-101 was still a problem tank. It had a thick, ugly crust with a lot of gas, and it still needed regular mixer pump runs to keep it from burping. DOE still wanted the crust removed and hoped to be able to shut the pump off in the bargain.

With the realization that the crust was now quite a bit thicker, the summer of 1999 brought renewed concerns about whether and how the crust might interfere with the mixer pump. The post-drop crust base seemed to be stationary at about 300 to 310 inches, 4 to 5 feet above the pump inlet at 258 inches. TEMPEST runs had already shown that this was plenty of room to keep the pump from sucking crust. But what if there was another crust drop event after a 100,000-gallon transfer that lowered the crust base by 3 feet? This might drop the gassy bubble slurry low enough that the pump might suck foam. Or, it might plug up after ingesting gobs of clay-like solids above the bubble slurry. In either case, the pump would be at least temporarily ineffective.

Nobody wanted to risk breaking, plugging, or even degrading the mixer pump until it was clear we didn't need it anymore. Somehow we had to tighten our assumptions and opinions about how or whether these calamities might happen into hard, technically defensible conclusions. To attack the issue of crust ingestion, Perry organized a workshop at PNNL for August 5, 1999. Besides the Hanford graybeards, he brought nationally renowned pump expert Chris Brennan up from the California Institute of Technology to pass judgment. Our old friend Walter Haentjens from the company that made the pump came out from Pennsylvania. Both experts quickly dismissed any conceived problem from ingesting solids. Haentjens reminded us that the pump itself had been designed to "suck rocks" and shouldn't have any problem with pieces of soft clay we believed the crust was made of. The structure of the pump column also formed a sort of cage around the pump inlet blocking any chunk large enough to plug the inlet.

Gas ingestion was another story. The group concluded that the pump could operate with up to, maybe, 10 percent gas, but its performance would degrade quite a bit. The growing bubble slurry layer could potentially supply a lot more gas than that. But this kind of gas ingestion was easily fixed by simply doing a little back dilution to raise the base of the crust back up away from the pump inlet. In the end, the recommended strategy was not to worry—just run the pump, but keep monitoring pump pressures and flows for any pangs of degradation.

So far, so good. But another worry surfaced in July that the initial transfer might disturb the crust in such a way that it would release enough gas to lose buoyancy and sink to the tank bottom! Adding water might do the same thing. However it happened, the sinking crust might release a large amount of its gas on the way down. Would this be a flammability hazard? Once on the bottom, the mixer pump jet might not be able to mobilize a 10-foot-thick sunken crust that was much stronger than a naturally settled sediment layer. If the waste didn't get mobilized, how big a burp could be expected from the remainder?

We realized that the process of gas release from a sinking crust was just the reverse of gas release from the rising gobs that had been analyzed back in the tank's burping days. Back then we found that the work done by the expanding bubbles in the rising gob was what broke it apart and let the gas out. So, if the crust were sinking, the work done to compress the bubbles would tend to liquefy the solid material and release some gas. But the calculation showed that a lot less energy was expended in the sinking crust than in a rising gob. Not much gas would be released in the sinking process. In fact, a sizeable fraction of the original gas volume in the crust had to be released before it could even begin to sink. So we concluded that gas release from a sinking crust was not a hazard.

Assuming the crust kept at least a shade of its original structure and character like the two large pieces of the Titanic resting two miles under the Atlantic, how it reacted to the mixer pump jet depended on how it sank. If it had dissolved from the top down from water sprayed on top, the stronger top layers would be gone, leaving a little of the much weaker lower part to sink. The pump could mobilize this portion of the crust out to almost 30 feet. The thin ring of crust remaining around the tank wall could not accumulate enough gas to make a future burp hazardous.

On the other hand, if the crust had dissolved from the bottom up, the weaker gassy layers would be gone and the strong hard crusty material would sink to the bottom. The pump jet would reach out less than 15 feet in this stronger, deeper layer. If the ring of the crust that the jet couldn't excavate were to burp eventually, it could be a big one! If it all went at once, the tank domespace could theoretically reach 12 percent hydrogen, three times the concentration needed for flammability! But it was not likely that more than a quarter of the sunken crust would participate in a single burp. Also, the hard stuff would soften quickly when saturated with warm liquid, so the pump could eventually dig out a lot more of it. Finally, the 300,000 gallons of dilution water planned at this time was more than sufficient to dissolve all the crust, plus the rest of the soluble solids. The bulk of the sunken crust would disappear soon after dilution was done, and there was essentially no danger of future burps from that quarter.

WILL IT SINK OR SWIM?

But would removing waste and adding water back in cause the crust to sink in the first place? There was a widely and strongly held belief that simply dumping water on top of the crust would sink it just like filling a boat with water. If dissolution released enough gas, the remaining part of the crust could lose its buoyancy and sink. But how much had to dissolve before this could happen? And, if it didn't sink, how would dissolution change the remaining crust's character? We needed a good model to answer these questions and Archimedes' principle seemed like the right place to start.

First we had to refute the assertion that added water would sink the crust by its mere weight. Archimedes reminded us that displacing any fluid creates a buoyant force whether the fluid is water or heavy salt brine. Picture a solid object floating in a pool of brine. The buoyant force of the displaced brine pushes part of the object above the liquid surface. Now pour water gently over the brine to minimize mixing. The less-dense water stays on top of the brine and surrounds the portion of the object protruding above the brine. The buoyant force of the displaced water lifts the object out of the brine a little farther than when it was surrounded by air. It cannot sink! In fact, if the object is less dense than water it can leave the brine completely and float in the water layer!

Because hardened skeptics seldom believe calculations alone, Scot Rassat did a little experiment to demonstrate the fundamental physics of buoyancy. He added lead shot to a small sealed bottle so it would float in heavy oil, but not in water. As Scott added water on top of the oil, the little bottle rose farther and farther out of the oil until it was completely covered with water. He repeated the test with a lighter bottle that would float on water. As before, the bottle rose higher and higher until it pulled out of the oil completely when there was enough water to float it. The same thing happened when Scott modeled a porous crust by gluing together Styrofoam beads in a cylindrical mold with metal pellets mixed in to get the right bulk density.

In every case the simple Archimedes calculation predicted what the test showed almost exactly! This proved that, ignoring dissolution, pouring water on the crust could not make it sink. Not only that, but we could predict the waste level as a function of how much water was added. Including the effects of dissolution was a little harder. First we had to know how much crust dissolved for a given volume of water added. Then we had to figure out how this dissolution changed the crust buoyancy depending on whether the top or bottom of the crust dissolved. All this came together in October and November 1999 into a general model we used to predict the results of the first transfer and back dilution.

Dan Herting's dilution studies on the core samples taken in December 1998 showed how the chemical composition of the liquid and solid changed as the waste dissolved. From Dan's data, Beric Wells and Bill Kuhn deduced that 1 gallon of water would cause around 3 gallons of crust to disintegrate. For "top dilution," where water was put on top of the crust, the insoluble solids that did not dissolve would stay on top, adding weight that would have to be supported by buoyancy from below. All of the gas would be released from the portion of crust in which soluble solids dissolved. Water added below the crust, called "low dilution," would come in through the transfer pump inlet dilution system about 100 inches from the bottom. The rising stream of water would mix pretty thoroughly with the much heavier slurry on the way up to the crust. However, the water would first have to dissolve the solids suspended in the slurry before it could begin dissolving the crust. The insoluble solids would fall into the slurry below. Again, all gas in the portion of crust affected by dissolution would be released.

Top dilution looked like the best method to dissolve the crust. It would remove weight while leaving most of the crust's buoyancy intact, allowing it to float higher. This would keep the top exposed to the most dilute liquid while the heavier brine created by dissolution would drain under the crust and get out of the way. There was little danger of the remaining crust sinking until most of it had dissolved.

Because all the measurements showed that the crust held most of its gas near its base, low dilution would tend to remove more buoyancy than weight. After the bubble slurry layer was gone, the remaining crust would not have much excess buoyancy. A fairly thick layer of crust could potentially be made to sink with relatively little low dilution. However, much of the lighter water would more likely find its way up through the crust and end up on top anyway. Thus, we began to believe that either top or low dilution would end up as top dissolution. The first top and low dilutions would be very closely watched to see if this were true.

Given the best information we had about the crust structure and properties, the final Archimedes buoyancy model said that about 95,000 gallons of top dilution would dissolve all of the crust. Low dilution would need 160,000 gallons because it also had to dissolve the solids in the slurry. Dissolving all the soluble solids in the tank from bottom to top would need a total of just under 200,000 gallons of water assuming an initial transfer of 200,000 gallons of waste. This was just about what we had estimated at the beginning!

By Thanksgiving 1999 we had a model that could predict how the crust would behave when waste was removed, water was added on top or beneath it, and what would be left afterwards. Most of the major safety issues were put to rest, and we were ready to accomplish the remediation plan that came out of the Despair at Hanford Square almost exactly a year earlier. Now let's step back and look at the physical system that Carl Hanson and his crew had been working on while all the science was going on. Though we now had an estimate of how much dilution would be needed to permanently prevent the SY-101 from burping, we did not yet have agreement from the TAP and DOE Headquarters. In fact, we didn't get a consensus on the final transfer and dilution volumes until the transfer and dilution campaigns were almost done!

1999

Ready to Remediate

While the crust was being probed, pictured, and predicted, the rest of the project was building the innovative hardware to dissolve it, training the gung ho work crew to run the process, and gathering the political will to start and sustain the actual remediation campaigns. There were grand accomplishments in all these areas that fundamentally changed the way Hanford operates. Like the Hydrogen Mitigation Project in 1992 and 1993, the Surface Level Rise Remediation Project in the year 1999 was another one of those times and places that defines careers, makes reputations, and creates memories against which all other experiences will be measured.

CREATIVITY REIGNS

Since early spring in 1999, when the work was officially authorized, Carl Hanson, Tony Benegas, and many others had begun designing and building the components of the hardware that would perform the transfer and dilution. The need for speed drove the team to find creative shortcuts and make things simpler wherever possible. Rick encouraged non-traditional thinking while giving originators the responsibility to make sure their ideas would work. Carl thought it was the high point of his career. Freedom to be innovative led to some outstanding new concepts, some of which have become the norm for new projects.

> The atmosphere of creative freedom led to some interesting acronyms for the system and its components. The Surface Level Rise Remediation Project itself was SLRRP, vividly describing how the waste would be transferred out of the tank. It had to be done quickly, and the pumping and piping system was accordingly termed RAPID because it enabled the project to *Respond And Pump In Days*. The system was installed in and controlled from the PPP, or *Prefabricated Pump Pit*. Way out on the back end of the system, the ASSD, or *Anti-Siphon Slurry Distributor*, dumped transferred waste into SY-102. The project team was obviously having too much fun.

To make room to add dilution water, a large volume of the thick, heavy slurry first had to be removed from the tank. This would require a very capable pump. Betting on his reputation for acquiring and adapting the original mixer pump, Carl tapped Tony Benegas to find the right one. Tony chose the Westinghouse New Generation Transfer Pump out of a list of several options. It was based on earlier designs for naval nuclear reactor coolant pumps and was supposed to have bullet-proof reliability to justify its high cost. It was one of those flashy items of new technology that was supposed to solve everyone's problems. Marshall Hauck, senior engineer at Kaiser, had been promoting it at Hanford for a long time. He had a lot of slick slides, a transparent scale model, and a good sales pitch, but the pump was way ahead of its time. Nobody wanted to take the risk of being the first to use it, so the pump had been lying dormant in a warehouse, waiting for SY-101 to come along.

The 60-horsepower submersible pump could move 250 gallons of slurry per minute at a pressure of 60 pounds per square inch. Its two canned rotors had air-tight seals to protect all vulnerable internals from the waste. Tony fastened a high-pressure water spray ring around the inlet to dilute the thick, muddy slurry one-to-one to make sure the pump could move it. The dilution water was the first thing turned on and the last turned off, so the pump would be automatically flushed with water before and after every run.

The new pump looked good, but Perry Meyer saw a subtle weakness. The density of the slurry-water mixture increased if the transfer flow rate increased while the inlet dilution flow was held fixed. The higher the transfer flow, the higher the density of the fluid in the pipeline and the higher the power required to pump it. But the pump would not feel the higher power demand until the long transfer line filled with the higher density fluid. Perry was concerned that this might cause the pump to slow down and speed up in uncontrollable cycles. But the operators learned that, by increasing or decreasing the dilution flow in advance of changes in flow rate, the pump stayed very stable and well-behaved.

One oversight created some embarrassing moments when the new transfer pump was installed. The crane operator lowered it through the riser and into the waste smoothly and expertly. The big 42-inch spray lance had been run all the way to the tank bottom a few weeks before to make sure there was nothing in the way. But the pump would not go all the way down! It didn't seem to be binding in the riser or hitting any hard obstruction. It just gradually unloaded the crane and stopped! Fred Schmorde, drawing on his submarine experience, finally figured out that it was floating! The end caps were left on the discharge pipes and they were still full of air. They had to take the caps off and let waste in. This change of plan was duly approved as being within the original work scope, and the pump finally submerged to its intended place.

Standard practice would have been to mount the pump column in one of the existing pump pits that had been built into the tank more than 30 years ago. But these old concrete bunkers were notoriously contaminated with waste from past leaks and cluttered with debris. Carl and Tony had encountered high radiation and other troubles when they put the mixer pump into the old central pit in 1993. Given the tight schedule, they were very reluctant to risk trying to put the new pump in another old pit.

Someone on Tony's team devised a way out of the difficulty. They would build a new, clean pump "pit" system that would sit above ground to avoid contamination and radiation shine that would occur in a real pit. It could be tailored to the specific needs of the project so they didn't have to worry about what fittings might be lacking or in the way. Then, when the work was done, it could be simply removed with the rest of the equipment, maybe even re-used on another tank! Thus, the Prefabricated Pump Pit, PPP or "Triple-P," concept was born.

> The PPP required four hefty concrete footings to support its weight, plus the weight of the transfer pump hanging from it. But the ground above SY-101 was so heavily populated with buried pipes, conduits, ventilation ducts, and junction boxes it was hard to find places to dig. They ended up having to put the footings wherever there was space and build support legs to them. It was like the old game of "Twister" where you spin a little pointer that tells you where to put your hands and feet on colored circles and try to keep from falling over.

Using small temporary steel pits above ground instead of big con-crete pits below ground was such a good idea that the PPP has become a normal tank farm design feature. The single-shell tank retrieval projects in S farm are using small PPPs. The original PPP in SY-101 recently had its piping re-routed so operators could transfer waste all the way to 200 East Area with the New Generation Trans-fer Pump.

The PPP needed a goodly flow of hot water to supply inlet dilution to the transfer pump and to flush the transfer lines, and it had to be available without onsite electrical power to run pumps. Following the same philosophy of flexibility and simplic-ity as used in the PPP, Tony designed the "water skid" on an 8- by 20-foot metal sled. It had a 2,000 gallon holding tank, two 20-horsepower pumps with associated valves and gauges, and a 2,000 pounds per square inch nitrogen bottle to provide pressure for emergency flushes. The water skid connected to the water supply and the PPP with heavy-duty heated and insulated flexi-ble hose. During transfer and dilution operations, the "Water Skid Operator" was every bit as important a position as the "Triple-P Operator."

Just like the old pump pits, the normal practice for transferring waste out of SY-101 would have been to plug into the old under-ground steel piping system. But the old hard piping might be

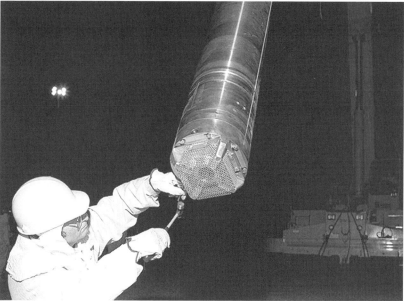

TOP: Prefabricated Pump Pit being positioned for installation. The shiny cylinder in the center fits into the riser. BOTTOM: Inlet of the New Generation Transfer Pump with dilution sprays and screen.
Courtesy Rick Raymond.

374 | Part Three: Surface Level Rise

plugged, more likely it would leak, and there was always doubt as to whether the end connections were useable. You could theoretically dig a trench and put in new pipe, but something already in the ground was sure to be in the way. Even if the path was clear, safety controls demanded that any trenches over a waste tank be dug by hand!

> A horror story about old waste transfer pipes happened during saltwell pumping in S farm in January 2000. Liquid waste started squirting out of an electrical junction box! But how could an electrical box leak waste? After some careful investigation and excavation, the box was found to be connected to conduit carrying heat trace wire to the transfer line. Transfer lines were designed with heaters to prevent the waste from cooling enough to precipitate dissolved solids and plug the line. In this case, water had gotten into the box, flowed down the conduit and rusted a hole into the transfer line. So when liquid was pumped through the line, it naturally flowed through the hole, up the conduit and out the box at the surface.

If it was too risky to use old hard piping, and either impossible or impractical to put in new underground pipe, why not just lay new pipe on top of the ground? It would have to be encased inside a larger pipe for leak protection, but it would be possible. Better yet, why not use an over ground flexible hose-in-hose? You could run it anywhere it needed to go and even adjust it after everything was hooked up. It was temporary, so it would not add to the junk yard already cluttering the top of the tank. It was definitely the right idea at the right time!

Tony used 125 feet of 2-inch black reinforced plastic fiber hose (made of polyethylene propylene diene, if anyone is interested!) for the transfer line and encased it in a 4-inch secondary containment hose of the same material. It looked just like a big automotive radiator hose. He heat traced the outside hose to keep the waste warm and less likely to clog the line between tanks. Tony put the hose-in-hose system through extensive torture tests and it passed them all. He then cut the test hoses into short lengths and passed them out to the project team as souvenirs. I still have mine somewhere. One real hazard of over-ground transfer was that it ran really radioactive waste right out in the open where people might be walking around. After calculating the potential radiation dose from the hose, the safety people required lead blankets for shielding with little plywood bridges so people could cross the hoses without scuffing the lead blankets.

The hose-in-hose over-ground transfer line worked extremely well on SY-101. It was another revolutionary concept that found a lot of uses later. It probably saved the saltwell pumping program and some primary milestones in the Tri-Party Agreement. Most likely the hose-in-hose will make retrieval of single-shell tank waste practical. It was a breakthrough that Rick and his crew were rightfully proud of.

One last problem challenged Tony's design crew. The waste transferred out of SY-101 was supposed to go into SY-102 next door. But they didn't want the high ammonia releases that might happen if the SY-101 waste just splashed in from the top of the dome. So the same engineer that recommended the hose-in-hose transfer line also designed a long vertical pipe called a "drop leg" to carry the waste down below the waste surface in SY-102. There was only one little problem. Though SY-101 and SY-102 are at almost the same elevation, the drop leg would become a siphon when carrying the heavy slurry from SY-101. There was no shutoff valve on the transfer line and, once siphoning started, up to half the contents of SY-101 could flow into SY-102 before it stopped! This problem had to be corrected in a hurry. A new set of designers tried to make a siphon break, but their first version would still have allowed siphoning if the transfer pump were not shut down just right. Perry Meyer and JR Biggs teamed up to write a report on the requirements for a siphon break to clear up the confusion. This combination of PNNL scientist and no-nonsense field engineer was unusual, but it is an example of the kind of team effort common in the project.

The final design of the siphon break, called the Anti-Siphoning Slurry Distributor, or ASSD, did prevent siphoning. But the air gap cut into the line above the SY-102 waste level to break vacuum and prevent siphoning also sucked prodigious volumes of air into the line when waste was flowing. The air bubbles flowing down the drop leg saturated themselves with ammonia vapor. They were excellent ammonia scavengers and rapidly pumped ammonia vapor into the SY-102 domespace. It was almost as if the siphon break had been designed specifically for that purpose. Transfer actually had to

376 | Part Three: Surface Level Rise

be slowed down on one or two occasions to keep ammonia releases within limits.

Because the transfer and dilution processes were to be controlled partly by direct visual observation, there had to be several cameras mounted and ready to record the scene. The existing color video camera, installed 20 feet west of the tank center, could be panned and tilted to see essentially the entire tank, except for a blind spot behind the massive old mixer pump. Unfortunately, the transfer pump and dilution inlet were right in the blind spot. To fill in the blind spot, a portable color video camera was hung on the southeast side of center at a 20-foot radius. A fixed, 180-degree "fish-eye" video camera was added to the existing camera to give a constant view of the central part of the tank, mainly the mixer pump. This allowed operators to see what was happening to the mixer pump while the other two cameras were looking elsewhere. A union video operator controlled the cameras and recorded video in a tiny "video shack" trailer that had been in use in the SY farm since 1991.

One innovation that might have had a profound effect on tank farm operations had a very different fate than the hose-in-hose transfer line, for example. Judith Bamberger and her team at PNNL had been working on measuring the density of a flowing slurry using ultrasonic waves. In concept their "densimeter" measured how much of an ultrasonic wave was absorbed or reflected off tiny particles in passing through a pipe in which slurry was flowing. The more absorption and reflection, the more particles, and the more particles in the flow, the higher the density. At the same time it also measured the flow rate, temperatures, pressures, and maybe even the viscosity.

It looked like a wonderful instrument that would give scientists and engineers all the data they could wish for. But JR Biggs, who was in charge of hooking up all the new stuff in the farm to the DACS, was not impressed. He thought the operators needed only an "on-off" switch—not a fancy, probably unreliable, densimeter! He was also quite skeptical of anything ultrasonic, remembering the trouble Tony Benegas had with the ultrasonic flow meters on the mixer pump nozzles, and the fiasco of the ultrasonic Ames "fish finder."

Nevertheless, under pressure from the TRC, Rick, and others, JR reluctantly agreed that the densimeter should be installed in the PPP and wires run to its own little data logger in the DACS trailer. The densimeter faithfully measured and sent data on dilution water and transfer flow rates and their temperatures during the remediation campaigns, but there was no time to qualify the density feature, and it was never turned on. A couple of years later, when SY-101 was about to make its first cross-site transfer, there was a concern about whether the transfer pump could pump a slurry through the 5-mile pipeline. So Judith got another chance to try the densimeter. She turned it on during an initial transfer line flush, and it correctly recorded the density of water. But, for some reason, it was not used during the actual transfer, so we still don't know if it can measure slurry density. It's still part of the transfer line on SY-101.

THE FAILED READINESS REVIEW

It's not enough to be innovative in building the parts of a system. The whole thing has to be hooked together, wired up properly, and people trained to run it all. Finally, to make sure every item of hardware, software, and documentation is in place and useable, and to ensure that each worker knows how and when to operate each valve, switch, or lever involved with his job, there has to be some kind of overall demonstration exercise of the whole system. This "dress rehearsal" is part of an internal process called a "readiness review." Because the Surface Level Rise Remediation Project had to do the design, fabrication, installation, procedure writing, training, and testing in parallel, a readiness review was pretty important. But it turned into much more than a quick dress rehearsal. It was an "agony and ecstasy" process that was the birth of one of the most powerful legacies of SY-101.

Imperfect human nature brings every big organization or project an occasional embarrassment to keep it humble. Hanford and SY-101 are certainly not exceptions. The old SY-101 mitigation project was embarrassed at the quick failure of the VDTTs after their installation in Window G. Finding the mixer pump too big to fit the riser in Window H was another. Then, after trimming the pump and installing it in Window I, the unplanned pump start and Rock-on-a-Rope incidents embarrassed Westinghouse Vice President Bill Alumkal into shutting down the tank farms.

378 | Part Three: Surface Level Rise

Though it was really unpleasant, the Rock-on-a-Rope embarrassment was a catharsis that, after years of anger and frustration, tightened up procedure and attitude across the entire Hanford culture. The Surface Level Rise Remediation Project was similarly refined in the fire of humiliation by failing a DOE readiness review in the fall of 1999. Though painful, the experience made a profound change in operational discipline that still affects tank farm work to this day. One might say that it was the final step of the cultural change that began with the Rock-on-a-Rope, 6 years before. Here's how it happened.

The original project plan included detailed preparations for a plant readiness review and independent readiness assessment by DOE. But, given the urgency of the growing level, at least up through April 1999, and the Wagoner Letter relieving the project from much of the formal safety assessment and control effort, Rick Raymond hoped and maybe tacitly assumed that the readiness review might be waved or curtailed. But when the urgency subsided after the surface level stopped rising in the summer of 1999, DOE decided that there was time for a readiness review. It happened in September.

There are four major stages to a readiness review. The first is actually getting ready. This means building, installing, and testing all the hardware, writing procedures to operate the hardware, and training crews to use them. The second stage is to collect a stack of "affidavits" formally stating that all requirements have been met. In the third stage the crews actually run the procedures to demonstrate their ability to operate the hardware. Finally, assuming the demonstration is successful, minor deficiencies are collected in a "punch list" to be corrected before operations begin for real. The project failed the demonstration part.

The planned operations to run the SY-101 transfers and dilutions were sufficiently complex that Rick decided he needed a dedicated crew and targeted the group that had just completed the C-106 sluicing project successfully. He brought the operators in to work with the project team, writing procedures before the design was complete, and helping the designers improve the equipment while it was being built. A cold test facility was dedicated to the project, and Rick decreed that all spare parts not needed in the field be assembled there in a mockup system. The crew worked on the mockup for weeks, exercising the system and practicing different operations. They got to know the equipment very well indeed.

Because everyone was so familiar with it, Rick decided to hold the readiness review demonstration at the cold test facility. In retrospect this was a strategic error. The cold test facility and mockup weren't a very accurate representation of the actual tank system. Worse yet, because it was pumping water instead of radioactive waste, reviewers felt free to invent accident scenarios at will during the review instead of evaluating a well-prepared, scripted demonstration on the actual system. Nobody had taken the time to define exactly what the crew was going to demonstrate; thus, the reviewers took control by default.

The operators were giddy with all the shiny new equipment, but they knew it well. They thought the operation was simple, and the readiness review would be easy. But they were not ready. Supervisors at the cold test facility were not exercising leadership to prepare the crews for the review. Some of the project staff were concerned that the team was not ready and tried to convince Rick not to start the review, but the pressure of schedule overruled them.

So it started. The engineer assigned to brief the crews for the review was new and didn't have time to come up to speed. He was not coached on presentation techniques and didn't even have an overhead projector! None of the managers showed up for the briefing to emphasize the seriousness of the operation. At that point Fred Schmorde knew "we were dead." When the managers did show up, taking pictures of the occasion seemed more important than the review itself. The casual attitude started the downhill slide to failure. Rick admitted that he "failed to explain his expectations, so they were not met."

Conduct of operations was getting to be a big thing, and the reviewers were sensitized to it. Procedures were supposed to be executed crisply with good, clear communication so everybody knew exactly what they should be doing and what the equipment was doing. But things got out of hand. The informality of the reviewers concocting simulated accidents off the cuff encouraged the crew to be equally informal in responding to them. As a result, communications became too lax. People called each other by their first names instead of by position. Instrument readings from the tank farm were reported as numbers without including their units of measure.

380 | **Part Three: Surface Level Rise**

DOE soon halted the demonstration as unsatisfactory. The crew obviously knew the equipment extremely well, but their conduct of the operation did not pass muster. If Rick had only been there and said, "This isn't good enough. Please come back later," the reviewers would have applauded and probably passed them on the next try. But Rick didn't do that; in fact, he left halfway through the demonstration to take care of another fire drill. He had assumed the C-106 crew came with the communications and paperwork discipline built in. But if they had been trained that way, it did not come over to SY-101.

The project had to change the crew's attitude and quickly! Lockheed Vice President Bill Ross was assigned to form an internal "drill team" to shape up SY-101 operations. He promised that his drill team would be tougher than DOE, and he put the crews through sternly monitored training and repetitive drills to tighten up their operations discipline. In fact, the SY-101 re-training regimen became a standing joke in the 200 Area: "They make you do WHAT?" JR expected strong resistance from the operators at the new levels of discipline. But the crew surprised JR and everyone else with their response. They took responsibility for the failed review and went above and beyond expectations to set a new standard altogether! They got to the point where they were comfortable correcting each other and accomplished a complete turnaround in only a few weeks.

Here's an illustration of how the crew might have been communicating and how they did it after re-training.

How they did it the first time:	How they did it after re-training:
DACS: Hey, Bill!	DACS: Triple P Operator, DACS Operator
PPP: Yo!	PPP: Triple P Operator, go ahead.
DACS: What's the flow rate?	DACS: Please read transfer flow rate F0035
PPP: Ahh, 135.	PPP: Transfer flow rate F0035 is 135 gallons per minute.
DACS: OK.	DACS: Understand transfer flow rate F0035 is 135 gallons per minute.
	PPP: That is correct.

When they repeated the demonstration in November, the DOE readiness review team was impressed and passed the team with high praise! Rick considers Bill Ross the unsung hero of the Surface Level Rise Remediation Project. Bill's team went on to drill the crews for other projects in the same way. SY-101 became THE standard for operations discipline at Hanford. That standard is still there, especially for single-shell tank retrieval where many of the SY-101 folks still work.

HOW MUCH IS ENOUGH?

The fundamental question in planning the transfer and dilution process was how much? How much waste needed to be removed and how much water had to be put back in to dissolve the crust and enough of the other solids to remove the thick crust and forever keep the tank from burping? We did not get full technical and political consensus on the total transfer and dilution volumes until just before the last campaign in March 2000. Discussion began on this topic almost as soon as transfer and dilution became the preferred remediation option and continued until June 2000, two months *after* the last transfer and dilution!

The first rough estimate for how much transfer and dilution would fix the tank was 250,000 to 300,000 gallons as stated back in February 1999 in the technical basis document. This was about what would restore the water removed in the evaporators and about twice what the Archimedes model predicted would be necessary to dissolve the crust. It was also the volume of water that would have to be added to halt burps as predicted by the buoyancy ratio model derived several years earlier by Perry Meyer from Kemal Pasamehmetoglu's breakthrough bubble migration concept.

But now, faced with the prospect of actually turning off the mixer pump in a few months, we had to be *very* sure the dilution would be enough. We also had to make the case so convincing that the reviewers would believe it too. This meant we had to show the details of how much transfer, top dilution, and low dilution would be done in each campaign, how we would run the mixer pump between them, and what criteria would eventually be used to show success.

The TRC had a marathon series of meetings about twice a week from mid-October through mid-November to work this out. We already had a bare outline in the project plan of a small transfer and dilution to show us how the tank would behave, followed by a large transfer and dilution to actually do the job. We initially asked for a small transfer of 150,000 gallons to lower the waste level about 6 feet. But, to keep enough space in SY-102 to maintain saltwell pumping, operations would allow only an 88,000-gallon transfer. No matter, that was plenty of room for the small dilution. We decided to dribble about 25,000 gallons of water on top and inject another 35,000 gallons through the transfer pump inlet. According to the crust buoyancy model, the top dilution would be enough to dissolve essentially all the current hard crust freeboard, and the low dilution would start dissolving solids in the slurry. After all this was done, we'd run the mixer pump and watch things to see if all behaved as expected. There was still enough room to add another 30,000 gallons to raise the crust a foot, if the pump gave us trouble.

The TRC also did a little brainstorming on what could possibly go wrong during the first campaign and what could or should be done about it. If the crust or a piece of it hung up on the tank walls during the first transfer, we'd stop the transfer and estimate the added dome loading to see if the tank could stand it. If not, we'd wash off local "lollipops" with a spray lance or add 25,000 gallons of water on top to loosen bigger sections. Other potential anomalies included another crust-base drop event occurring just before the first transfer, unexpectedly high or low gas releases, changes in released gas composition, a surface level change inconsistent with transfer, dilution, and gas release, or the crust sinking early.

After much discussion, we realized that these or almost any other conceivable problem during transfer and dilution was most surely cured by continuing transfer and dilution! So the best "reasonable and prudent" control demanded by the Wagoner Letter was simply to go ahead. This had the double benefit of silencing the ever-present bystanders who always yell "stop the operation" because of some last-minute concern. Now we could quietly counter with "continuing the operation is the safest thing to do!"

The second campaign would be the large transfer and dilution to bring the total to 300,000 gallons. We wanted to transfer at least 212,000 gallons out then add 45,000 gallons of water on top and 200,000 gallons through the transfer pump inlet. The crust buoyancy model said that this campaign would dissolve the entire crust and all the readily soluble solids in the slurry. Dissolving the crust should release all its gas and similarly for the solids in the slurry. There was little doubt that this would cure the surface level rise problem.

> The TRC worried over what to do about large floating dumplings or wastebergs left over after the bulk of the crust was gone. We talked awhile about how to attack them through other risers with high-pressure sprays or mechanical devices. Nick Kirch brought us back to reality by reminding us that, because the most dilute liquid will always be on top, a big floating berg will eventually dissolve if we just add water, no matter where we add it. Besides, they presented no hazard at all! Thoroughly chastised, we promised not to worry about leftover dumplings anymore!

But would a third campaign be needed to stop burps so we could shut off the mixer pump? This was the main question we had to answer completely, leaving no nagging doubts. The 300,000-gallon goal had been set with the buoyancy ratio model a year earlier using predictions about what dilution would do to the waste. We still believed that the buoyancy ratio was the best measure, but we brought out all the other models and rules of thumb that had been used in the past to see how the totality of evidence lined up. Besides the buoyancy ratio, there were three other methods for determining a tank's burp potential. Like the buoyancy ratio, each of these methods calculated a number or set of numbers that, when plotted with actual tank data, divided the tanks into two distinct groups—those known to burp and the rest. We called this "discrimination" or "separation" between burping and non-burping tanks. The wider the space between burping and non-burping tanks on the plot, the better the discrimination. What we wanted, then, was for the expected waste conditions in SY-101 after the last transfer and dilution campaign to plot well down into the non-burping group, or at least well out of the burping group.

Back in the mid-1990s, Steve Agnew at Los Alamos looked at the overall chemical compositions of the tanks and found that the concentration of four species—sodium, aluminum, nitrite ion, and organic carbon—seemed to be good indicators of burp behavior. The criteria were only indirectly related to physical phenomena happening in the tank. The concentrations of sodium and aluminum might have been an indirect measure of the liquid density and the depth of the sediment, and the organic carbon and nitrite concentrations were known to influence gas generation.

In 1996 Scott Estey and Mike Guthrie of Nick's Process Engineering group searched for combinations of physical measurements that might separate burping from non-burping tanks in the same way as Agnew investigated the chemistry. They found that the product of the liquid density (expressed as the "specific gravity,", the ratio of actual density to the density of water) and the sediment depth were the best discriminators. The "Estey criterion" was that tanks with a product of specific gravity and sediment depth of less than 150 inches would not burp. While not intended to be a physical model, the two pieces of the Estey criterion were also important in the more detailed buoyancy ratio. The bubble migration theory said that the sediment depth set the maximum gas volume fraction while the liquid specific gravity controlled how much gas was needed to make a burp.

The most historical measure of burp potential was invented by Dan Reynolds based on his long experience with evaporators and observing their long-term effects on tank behavior. Dan found that tanks with an average waste density of less than 1.41 times that of water would not burp. Because it was both very simple and a pretty good discriminator, Nick Kirch wrote it into the formal waste compatibility criteria for inter-tank transfers.

When the best estimates of SY-101 data and the effects of dilution were plugged into the four sets of criteria, the buoyancy ratio and Estey criterion were satisfied after a 300,000-gallon transfer and dilution, confirming the original calculation in February 1999. But Dan needed 430,000 gallons to reach his 1.41 average specific gravity, and Agnew's chemistry criteria demanded 500,000 gallons. Apparently, the 300,000-gallon transfer and dilution would cure big burps and melt the crust, but it might take 500,000 gallons to remove the tank from the flammable gas watch list and return it to normal service. But there was tremendous pressure to keep transfers and dilutions to the absolute minimum to conserve precious tank space.

We came up with a two-step plan to keep everyone happy. After the first two campaigns brought the tank to the 300,000-gallon mark, we would re-analyze the waste conditions and re-compute the burp/no-burp predictors. If they showed that SY-101 had not yet moved well out of the burping group of tanks, a third campaign would raise the total transfer and dilution to 500,000 gallons and the calculations would be repeated to confirm that it was enough. Finally, after one last look around, and a deep breath, we would shut the mixer pump off for a test period of around 90 days. If there was no sign of returning crust or dangerous gas accumulation, or any other bad thing happening, we could assume the tank was safe for use.

That was the plan we had on the books for the first and second campaigns. But the details of the second dilution, the third campaign, final mixer pump runs, and how we'd justify shutting the mixer pump off, would depend on the results of the first campaign.

LAST DOUBTS AT DOE

In addition to the questions we subjected ourselves to by our own brainstorming, DOE Headquarters had a few questions of their own. Back in December 1998, Jim Owendoff, acting Assistant Secretary for Environmental Management, was concerned that the level rise USQ might turn into a huge disaster like the flammable gas issue had back in 1990. People in his own office were warning him that it might. Craig Groendyke, Rick, and Billy Hudson from the TAP went back to Washington, D.C., to give him with a thorough briefing.

The briefing seemed to help alleviate concerns, and design and planning for the transfer and dilution got up to speed. But a little later, on April 1, 1999, Craig was surprised to receive a list of 30 questions from Bob Alvarez at DOE Headquarters, dealing with a broad collection of concerns. It was clear that whoever composed the questions had a very confused notion of what the problem was and the physics of gas retention and release. The Hanford reply to many of the questions had to be "please clarify." The clarified questions were eventually answered, but it was clear that Alvarez still had doubts.

One of the most persistent criticisms in this period concerned colloids, tiny particles that can form a solid gel in a liquid at very low concentrations. Some of the less soluble tank wastes formed colloids when washed with various solvents. Assuming that SY-101's surface level was rising because the whole tank had become a solidified colloid gel that was not letting gas escape,

> critics at Headquarters concluded that Hanford had blown it and their remediation plan would be a disaster!
>
> Fortunately, the manager of our little group at PNNL, Bill Kuhn, is a really smart guy who knows a lot about colloids. In April 1999, Craig and Rick orchestrated a showdown conference call where Bill explained politely but authoritatively to doubters at DOE Headquarters that the soluble saltcake waste in SY-101 was very unlikely to form colloids, and that the surface level rise had been essentially proven to be by crust growth. Therefore, the project's dilution method was sound and was sure to eventually effect remediation.

The majority view at DOE Headquarters was that Hanford had misdiagnosed the problem and their proposed solution would be ineffective. Leo Duffy, the former Assistant Secretary for Environmental Management who oversaw mixer pump installation, offered his opinion that "Hanford is in a reaction mode rather than a planning mode. I gather they're baffled by what's going on." TAP member Don Oakley chimed in, unconvinced that "anyone understands the chemistry and physics involved in this crust." Steve Agnew from Los Alamos, who had been a part of the initial furor in 1990, was amazed that "the level has reached 435 inches and no one is freaked out yet!" *New York Times* reporter Matt Wald also voiced his view of the problem, "A giant radioactive soufflé is rising toward the top of a million-gallon tank of nuclear waste buried near Richland. Whipped up unexpectedly by a pump that was supposed to dissipate pockets of hydrogen gas, the waste . . . could eventually overflow. . . . They are rushing to pump some of the waste into another tank."

Though they were loud, the critics couldn't back up their assertions with tank data. A few at DOE Headquarters even favored our plan. But the battle lines were drawn, and Owendoff couldn't resolve the differing opinions of his staff. He became more and more nervous as the time for the first transfer and dilution got closer. Maybe he had a right to be. After all, we were actually planning to shut down the mixer pump! What if our plans really were wrong? What if the mixer pump couldn't be restarted and the tank started burping again? What if...?

We had been trying to combat DOE Headquarters with technical power. We confidently stuck out our collective chins and challenged, "We can prove your objections are wrong!" We should have known that would not work. Doing the same thing to the SubTAP had only hardened their opinions and driven them to enmity. Technical folks are too often blind to the more powerful political influences that need to be considered when the stakes are high. As Bismark recognized, "Politics is the art of the possible." The project had to somehow make its plan not only correct, but possible.

Political issues are sometimes too easy to overlook. By late summer of 1999, with the surface level no longer rising, the major safety issues put to rest, and Carl Hanson making good progress on the hardware, Craig Groendyke was confident that the project was going well and arranged a meeting September 14, 1999, to bring the TAP up to speed on tank conditions and start building their support for the upcoming remediation campaign. But, during the introductory presentation at the Worker Safety & Health session, Lockheed president Fran Delozier stated that "standard procedures are in place [to insure worker safety] and a TAP review is not needed."

Dr. Morton Corn, who had been on the TAP since its inception in 1990, objected. He contended that Fran's statement showed insensitivity to perceptions from the outside. Besides, why have a Worker Safety & Health SubTAP if they didn't have anything to review? Mike Payne tried to explain that the Tank Waste Rememdiation System had made great improvements in worker health and safety and he did not feel it needed to be reviewed for the Surface Level Rise Remediation Project. He advised the TAP to focus on technical issues. Mike's explanation did no good. Dr. Corn resigned from the SubTAP in protest after the meeting, causing a huge flap within DOE Richland and Headquarters.

Craig and Rick did some quick political damage control to convince Lockheed management that they had no choice but to submit their "standard procedures" for a review by what was left of the Worker Safety & Health SubTAP. Then Craig called them to a special meeting where Rick gave a personal apology and presentations on the safety procedures and controls. Though painful and a little humiliating, this therapy restored full TAP backing for the project.

388 | Part Three: Surface Level Rise

Fortunately, Rick and Craig had garnered strong political backing from other powers that DOE Headquarters could not ignore. By November 1999 Rick had the full TAP, including the mollified Worker Health and Safety faction, behind him. Even the most hostile member of the old Chemical Reactions SubTAP lauded the project as "the best planned than anything (sic) in the past!" Rick convinced the Defense Nuclear Facilities Safety Board that by taking small steps, doing a small transfer and small dilution to see how the tank responded before doing the big ones, we would not be getting in over our heads. Board member Herb Kouts had also conferred with Billy Hudson of the SubTAP, who endorsed Rick's plan. After formal discussions, Defense Nuclear Facilities Safety Board recorded a resounding endorsement. Their chief Hanford staffer, Steve Stokes, told Rick that the "Board agreed that preparations for SY-101 were proper and we should do the transfer. The contractor is sensitive to unknowns and has properly anticipated them." Adding weight to this recommendation, the full board had officially closed Recommendation 93-5 on sampling and understanding the waste on November 15, 1999.

Also in mid-November DOE Richland and Lockheed had agreed on a performance incentive entitled "SY-101 Safety Mitigation" to perform the waste transfers and water back dilutions that all hoped would tame the tank for good. Lockheed's potential reward was a huge $2,500,000, second only to the C-106 sluicing incentive from the year before. For $875,000, the project had to finish the first transfer and add 60,000 gallons of water back in by December 15, 1999, and do additional transfers and dilutions to bring the total dilution to 300,000 gallons by April 28, 2000. A full million dollars rode on shutting off the mixer pump and beginning a test to see whether the transfers and dilutions really remediated the big gas releases. Lockheed would get a final $625,000 for adjusting the safety authorization basis documents so the mixer pump would no longer be required to keep SY-101 safe. But big rewards carry big risks, and each carrot has its stick. DOE's stick was a loss of $625,000 if we didn't get 300,000 gallons of back dilution done by July 14, 2000.

With all his political support in place and DOE Richland formally committed to the plan, Rick went back to the Forrestal Building in Washington, D.C., and laid it all on the line with Owendoff. This was in December 1999, shortly before the first transfer was supposed to happen, so Rick also had schedule pressure on his side.

The gist of Rick's message was, "This is the path we are on. We own the safety problem. We believe the plan will be effective and safe. We understand the process that caused the problem [concentrating the waste], we believe the process is reversible, and we are reversing it [by dilution]. If you think the plan is faulty, stop us." That simple explanation, along with the political challenge, convinced DOE Headquarters.

There was another major change at the end of December that should have complicated the project, but it was not allowed to. In 1999 Lockheed's corporate management in California had come to realize that, though they were experts at building airplanes and space flight equipment, the Hanford contract had become an unproductive distraction. They told DOE in early 1999 that they wanted out and DOE put their contract out for bid. The big civil engineering corporation CH2M HILL of Denver, Colorado, won the bid. CH2M HILL formed in Corvallis, Oregon, in 1946 as CH2M, named from the initials of its founders. Cornell, Howland and Hayes were former students of Merryfield, an Oregon State University (then College) Civil Engineering professor. In 1971 CH2M merged with Hill and Associates to become CH2M HILL. Because the big project that made them famous was building the advanced sewage treatment system for Lake Tahoe, it was probably appropriate for them to tackle cleanup of Hanford's radioactive waste.

CH2M HILL officially took the reins from Lockheed January 1, 2000. Though there were some all-employee meetings and some orientation sessions for managers at Rick's level, the project went ahead as if nothing much had happened. It had to. We were about to change the world.

December 1999–April 2000
Remediation Happens

The series of transfers and dilutions that finally halted gas retention in SY-101 had a character similar to the series of mixer pump runs that mitigated gas retention and stopped the historic burps 6 years earlier. Mitigation took almost 2 months, from the end of October through mid-December 1993, while remediation took about 3½ months, from mid-December 1999 through early April 2000 when the pump was shut off. Both were of great interest to DOE, though the latter was not nearly as popular with the news media. Both had some initial setbacks, both operations turned up some surprises, and both were very successful. You can be the judge of how closely the fine details of the two efforts compare.

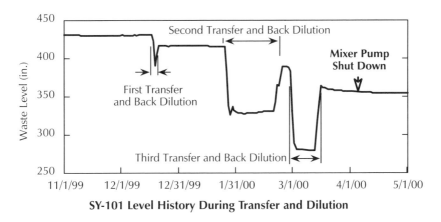

SY-101 Level History During Transfer and Dilution

392 | Part Three: Surface Level Rise

So it was time to begin. The first milestone of the DOE performance incentive was approaching, and the delays were over. The SY-101 Surface Level Rise Remediation Project would finally get to do what its name implied. The stage was carefully set. Craig Groendyke arranged a public video viewing room in the 2440 Stevens building in North Richland. There were real-time video and radio communications relays, plots of important quantities, some of the nice models of the new generation transfer pump, a piece of the hose-in-hose transfer pipeline, and slick posters to explain what was happening. Craig considered this public communication outlet a vital piece of the operation. The press and TV reporters were there, along with out-of-town dignitaries, and families of the operators doing the work out in the tank farm. Harry Harmon even came out from back East to see it. Craig held a formal press conference there.

Rick made sure there was a senior management presence in the DACS trailer at all times during operations to demonstrate Lockheed's corporate commitment. Rick also wanted senior project management, tank farms management, and science and engineering presence as well. People filled in a master roster of who was supposed to stand watches in the DACS, video shack, or visitor center on every shift every day that a transfer or dilution was going on. Some of those typically on the roster were Nick Kirch, Roger Bauer, Blaine Barton, John Conner, Beric Wells, Joe Brothers, and I.

> In keeping with the hard-nosed discipline that had been reinforced among the operations crew after the failed readiness review, there were strict rules for those of us standing watch. First, the SY Tank Farm was barricaded and a guard would let in only those whose names were on the roster. We found the DACS trailer segregated into two sections. The official DACS operators, operations engineer, and senior supervisor manned the control room. It was strictly off limits to the rest of us who crowded into the narrow corridors in the other half of the trailer. We were not even allowed to stand and watch what the operators were doing. There was even a chain across the control room entry to emphasize its "sacredness." We were instructed to be quiet, communicating only with the Senior Management Coordinator, who was usually a vice president like Rick Raymond. It was a scary, military atmosphere that emphasized the solemnity and seriousness of the work.

SY-101 Transfer and Back-Dilution Campaigns

Campaign	Start Date and Time		End Date and Time		Transfer to SY-102 (gal)	Top Back Dilution (gal)	Low Back Dilution (gal)
1	12/18	0645	12/19	0545	89,500		
	12/19	1834	12/20	0430		26,000	
	12/20	0445	12/20	1458			36,000
2	1/25	1622	1/27	2207	240,500		
	1/28	0143	1/28	2219		78,500	
	2/21	0240	2/21	2232			99,500
	2/23	1037	2/23	2053			50,500
	2/23	2057	2/23	2330		11,000	
Totals #2					*240,500*	*89,500*	*150,000*
3	2/29	0635	3/1	0523	160,500		
	3/1	1418	3/2	1326	125,500		
	3/13	1126	3/14	1300			127,000
	3/14	1300	3/14	2030		22,000	
	3/14	2030	3/15	1030			60,000
	3/15	1030	3/15	1500		14,500	
Totals #3					*286,000*	*36,500*	*187,000*
Grand Total					*616,000*	*152,000*	*373,000*

FIRST TRANSFER AND DILUTION—TRIAL RUN

The first campaign was supposed to start on the swing shift of December 17, 1999. But one last little embarrassment delayed it until the next morning. A few hundred gallons of dilution water had been pumped into the transfer line by 11:30 PM when it was time to start the transfer pump and move some waste. But it would not start! Frantic troubleshooting uncovered a minor glitch in the start-up procedure. Two switches had to be turned on in the right order to correctly set up the variable frequency drive that controlled the pump motor. The written startup procedure had the order of the switches reversed, preventing power from reaching the pump. The problem was so simple that they could have just flipped the switches in reverse order. Instead, to preserve operational discipline and set a good example, a formal change to officially correct the procedure was rammed through the system.

Re-writing the startup procedure and rousting managers out of bed for approval signatures took a few hours but everyone was back on duty at 6:05 AM on December 18, 1999. At 6:45 AM, the transfer pump began moving waste through the transfer line, officially starting the first remediation campaign. The transfer was completed 23 hours later, after moving 89,500 gallons of SY-101 slurry mixed with about 80,000 gallons of dilution water into SY-102.

During the transfer the waste level dropped steadily at a little over an inch per hour. It was too slow to see, but you couldn't help staring at the video monitor, waiting in rapt anticipation for some hint of the movement you knew was there. After 15 minutes a crack appeared in the crust around the pump column. After 2 hours you could begin to see clear evidence of the crust's descent. It went down smoothly, with no large cracks, "tilts," waste "hang ups" on the walls, or "lollipops" on suspended hardware. The appearance of the crust surface did not change during transfer except near the tank wall. There a ring of waste remained attached after transfer stopped, eventually building an irregular bank about 3 feet high.

I was on the DACS watch during that graveyard shift when the first transfer began. One of my jobs was to monitor the progress of the crust by watching it descend around the VDTT and MIT "dipsticks" and the mixer pump to make sure it wasn't hanging up, tilting, or doing anything else undesirable. Reading the camera tilt angle off the screen and knowing the distance from the camera to the object of interest, I used simple trigonometry to see how far the crust had dropped.

But my calculations soon showed that the crust was staying higher on the west side than elsewhere and the difference kept growing. Could the TAP's persistent worry about the crust hanging up have been correct? Should I tell Rick, who had the Senior Management Coordinator watch, to stop the transfer? But wait! The closer to the camera the less the crust drop. Was that a coincidence? After quickly sketching some triangles, I discovered that the tilt angle should have been included in the calculation, because the camera had to look down more steeply at the closer objects. The corrected numbers showed the crust drop to be quite uniform, within an inch or two. Only the area around the mixer pump appeared to drop a little more. What a relief!

The transfer dropped the surface level about 40 inches, but an 89,500-gallon transfer should have lowered the surface level only 32.5 inches. The 6- to 7-inch difference implied that a "bathtub ring" of waste about 3 feet thick remained stuck on the wall. We didn't expect this, but the thickness was about what we had predicted earlier from the estimated crust strength. This ring would show up very clearly after the big transfers in the second campaign.

At about 2 PM on December 19, a mixer pump run was attempted but aborted after 5 minutes because the DACS software sensed that the volute pressure was too low and set off an alarm. The software didn't account for the reduced hydrostatic pressure around the pump after the transfer. As the waste level rose during dilution, the hydrostatic pressure recovered and brought the volute pressure back within range to shut off the alarm. But the DACS had to be reprogrammed to allow the mixer pump to run after the larger transfers to come.

The first top dilution began at 6:34 PM on December 19, 1999, about 13 hours after the transfer. Now we'd get to see if all our calculations and predictions were correct. In 10 hours, 26,000 gallons of dilution water just dribbled onto the crust out of the riser holding the transfer pump column. It was non-dramatic and unceremonious, but we didn't think we needed a complicated spray system just to lay water on.

Within the first half-hour the depression beneath the transfer pump filled with moving water. We looked on with excitement; this

LEFT: In-tank video frame showing a "bathtub ring" of waste attached to the tank wall after the first transfer. RIGHT: In-tank video frame of MIT 17B (north side of tank) showing "islands" of crust exposed after the first top dilution.

was the first time humans were able to view a waste tank's crust layer being dissolved in real time. It was more like watching grass grow. The pool around the transfer pump enlarged gradually, but most of the water was apparently moving beneath the top surface. After 3 hours pools started appearing on the crust at several places. Eventually water covered most of the surface, leaving islands sticking out. Within a few hours after water flow ceased, the liquid found its way beneath the crust, and surface pools disappeared. The dissolution and erosion around the islands left cliffs and mesas on the crust surface with a kind of Monument Valley look.

A 10-hour, 36,000-gallon low dilution began at 4:45 AM on December 20, 1999, almost immediately after top dilution ended. The low dilution water was injected 100 inches from the tank bottom by simply turning on the in-line dilution flow while leaving the transfer pump off. Unlike the top dilution there was no evidence of water being added. The waste just crept back up to about the level it had started from two days earlier.

COMPARING "SPECTATIONS" TO EXPECTATIONS

So how did our predictions compare with the observations and measurements? It was almost too good to believe. The post-dilution surface level was supposed to be 408 inches, a perfect match with the neutron logs that showed a top surface level of around 408 inches at both MITs. At the same time the two Enrafs measured surface levels of 404 and 416 inches. The new crust thickness was predicted to be 94 inches. The neutron logs showed 95 to 102 inches. Likewise, the crust base was predicted to lie at 314 inches and the neutron logs gave 305 to 313 inches.

We also got a more satisfying and dramatic, though less exact, confirmation of our understanding of the dilution process from the temperature measurements on the two MITs. We had claimed that after the top dilution water had become saturated with dissolved solids, it would sink through the crust into the slurry below. Lo and behold, towards the end of top dilution, a sharp temperature decrease showed up on the MIT thermocouples at 268 inches, right under the crust base. Cool brine, at under 100 degrees Fahrenheit, had come from the surface and suddenly showed up at a level where the temperature had been 120 degrees. The subsequent low dilution raised the pool of cold fluid back up above the 268-inch thermocouple and it disappeared from view. Perfect! The crust and dilution water behaved just as we predicted!

A striking and unexpected observation during low dilution was the sudden rise in the slurry temperature above the water injection point. It began as soon as water started flowing from the transfer pump inlet and raised the slurry temperature from its initial 122 degrees to almost 126 degrees Fahrenheit. This temperature rise affected the entire tank, because both MITs measured it. The slurry stayed at its higher temperature until a 25-minute pump run on December 21 mixed it with the cooler waste under the 96-inch injection point.

Some quick research revealed that dissolving aluminum hydroxide solids give off a lot of heat. Most other solids absorb heat when dissolved, like those sports ice packs that get cold when crushed, mixing water and a nitrate salt. While SY-101 surely contained a lot of aluminum we thought most of it was already dissolved as sodium aluminate. But that some kind of reaction was heating the slurry and aluminum hydroxide dissolution was as good an explanation as any.

We didn't expect a large gas release during the first campaign, and none happened. Unexpectedly, though, more gas was released during transfer than during dilution. The hydrogen concentration started rising as soon as the transfer started. It leveled off at just under 400 parts per million about 9 hours later and dropped abruptly when the transfer ended. Ammonia reached 300 parts per million by the end of transfer and began dropping slowly afterwards. We guessed that the disturbance of the falling center section of crust rubbing on the 3-foot ring attached to the wall may have caused both the hydrogen and ammonia releases. During top dilution, the hydrogen concentration rose to only 150 parts per million while the ammonia concentration fell sharply, scrubbed out by the water falling through the dome.

Things were different in SY-102, the tank that was receiving all this waste. Its ammonia concentration started up immediately when SY-101 waste began coming in, even though it was diluted about one-to-one with water. That ammonia scavenging drop leg was doing its job! The first transfer was slow enough and brief enough that the peak concentration reached only 500 parts per million, but it was a sign of what would happen in the big, fast, second transfer next month.

398 | Part Three: Surface Level Rise

Summing up, the crust thickness, waste level, gas releases, and dilution water behaved as we expected, sometimes remarkably so. The few unexpected events, like the attached ring of crust, higher gas release during transfer than dilution, and heating the slurry during low dilution were all benign and explainable. All things considered, we could boldly claim that the model and our accumulated understanding of SY-101's waste had correctly predicted the outcome of the first transfer and dilution. We were ready for the big one!

SECOND TRANSFER AND DILUTION—CRUST GONE

The big one was the second transfer and dilution, originally slated to be the last one. It was a two-stage affair. The first event was a 240,000-gallon transfer that would lower the waste level by 7 or 8 feet. This was to be followed immediately by a large top dilution designed to remove the entire crust as an identifiable entity. After a 3-week wait to let this water do its complete work, an even larger low dilution would dissolve all the most soluble solids below the crust. There were going to be very big changes in SY-101. We would soon have the tank at our mercy, something Hanford had yearned for since it was filled in 1980!

While we waited, the tank kept our attention with several events where a muddy, foamy fluid flowed out over fairly large areas of the crust. If the top dilution had softened the crust and opened passages through it, and if the low dilution had "melted" the base of the crust, this might have been the long-expected bubble slurry flow we had warned about 9 months ago. The largest flow was on January 7, 2000, as inferred from a headspace hydrogen spike to 300 parts per million at that time. By comparing before and after in-tank video images, the flow covered between 50 and 100 square feet, 1 to 2 percent of the surface area. A full tank video scan on January 10 showed signs of five such flows since the first transfer, though all were smaller than the January 7 flow. At least 14 spontaneous releases showed on the gas monitoring records, probably indicating these and other slurry flows.

There were also thirteen 25-minute mixer pump runs between the first and second campaigns. The first two pushed the hydrogen concentration above 500 parts per million, but the rest were unremarkable. The mixer pump didn't really run any differently with

36,000 gallons of dilution under the crust than before. There were no noticeable difficulties nor any enhancements of performance.

> The time of the second campaign was cold, and there were a few inches of snow on the ground. For those working around the tank farm the most dreaded duty was walking 50 yards in the snow to use the dark, cold portable toilet perched in the snow across the dirt parking lot. As we emptied the radioactive waste tank, the waste tank on the potty filled up. The female workers almost mutinied towards the end of the initial 3-day transfer and top dilution.

The second waste transfer started at 4:22 PM on Tuesday, January 25, 2000. The transfer pump ran steadily for almost 54 hours at full speed, removing 240,500 gallons of slurry from SY-101. Removing this waste dropped the surface level about 100 inches. The crust went down smoothly and uniformly. There was no evidence of a big "hang up" or "lollipops" on tank hardware except the ring of waste found stuck to the tank walls in the first transfer, which was still there. After the transfer you could see how thick it was in some places where the crust did not stick or had fallen off. It looked like a dark gray wall 2 or 3 feet thick. This time the whole waste surface descended at about the same rate. It was exciting to see the massive mixer pump column emerge, all packed and piled with a dirty gray clay-like crust material. The base of the still 8-foot-thick crust was now down to between 210 and 220 inches, well below the main mixer pump inlet at 232 inches. But no attempt was made to run it until after dilution raised the level back up.

Frame from the fisheye camera showing the mixer pump column exposed after the second transfer, several hours into the second top dilution. Water is entering directly behind the mixer pump.

Gas releases during the second transfer were similar to the first one, though the hydrogen concentration in SY-101 was a lot lower. It averaged about 275 parts per million during the transfer, dropping immediately at the end. In SY-102, hydrogen averaged about 80 parts per million through the transfer. Apparently some tiny gas bubbles from the SY-101 slurry were still coming through the pipeline.

Ammonia concentrations stayed relatively low in SY-101, only reaching about 500 parts per million at the end of transfer. But high concentrations in SY-102 and the SY farm ventilation exhaust caused some concern. SY-102 ammonia kept climbing and hit 5,300 parts per million when transfer ceased. The SY stack ammonia concentration rose to almost 1,800 parts per million at the end of the transfer.

> The ammonia from SY-102 eventually goes out the SY farm ventilation exhaust duct that sucks from all three SY tanks. The exhaust flows from a 12-inch pipe, 20 feet off the ground. The pipe works like a two-story fireplace chimney. With a fireplace chimney, you seldom smell smoke outside, though you'd probably suffocate if you climbed a ladder and held your face in it. Just like the chimney, the hot gas plume from the SY stack rises rapidly, moving away from the workers and mixing with fresh air as it goes. The team of industrial health folks walking around the farm never measured any worrisome ammonia concentrations, and few reported even smelling it.

LEFT: In-tank video frame showing a slab of waste adhered to the wall after the second transfer. Thickness of the slab may be up to 3 feet. RIGHT: In-tank video frame of the waste surface behind the mixer pump (foreground) after a large foam flow event late in the second top dilution.

The big top dilution was much more exciting than the transfer. This is where remediation actually began in earnest, reducing the crust to a shadow of its former fearsomeness. After the second phase of dilution at the end of February, it ceased to exist at all. It was an privilege to be there and watch it happen. There was just enough time for the shift change between completing the big transfer and starting the dilution. Warm water started flowing down the transfer pump column onto the crust at 1:43 AM on Friday, January 28, 2000. The planned 78,500 gallons was added by 10:19 PM, so the crew got the weekend off.

The excitement started quickly. By about 2:00 AM the entire crust surface was covered with liquid. Under the liquid, the crust was dissolving fast, and the ring of waste on the tank walls began to fall off. On two occasions, the video showed exceptionally big sections of it falling off the wall, throwing a massive splash all the way across the tank. At the same time the temperatures near the tank bottom dropped suddenly as a "landslide" of cooler waste slid across the floor. Several other splashes and waves appearing on the video were probably caused by similar landslides. Craig played these video clips again and again at the visitor center in 2440 Stevens to show how powerfully we were disturbing the crust.

There was a lot of action going on under the liquid surface too, that we only discovered after studying all the data. About the time the top dilution ended, a big chunk of crust on the north side of the tank sank and turned over. It stopped at about 200 inches, where it apparently found slurry still heavy enough to buoy it up. It stayed submerged there for about a day when it resurfaced, turning over once again. The whole scene played out on the temperature profiles and the neutron logs measured at the 17B MIT. It was a real wasteberg capsize event that Rudy had been trying to model with the little wooden blocks in the bowl. A similar, but much less dramatic, crust submergence happened in the southeast near the 17C MIT.

When these events happened, a white, foamy bubble slurry suddenly appeared on the waste surface. Sometimes there were small surface waves, and there was usually a sharp rise in hydrogen concentration. In fact, these events pushed hydrogen to the highest concentrations seen during the entire remediation process. It reached 2,000 parts per million several times during these events. The white, foamy material appeared in minutes, but it dissolved again in a few hours. Several more foam flow gas releases happened

402 | **Part Three: Surface Level Rise**

after the top dilution ended, but only one of them pushed the hydrogen concentration close to 2,000 parts per million.

For the big transfer and top dilution I was assigned to swing shift from 3 PM to midnight in the video shack. All I could do was sit and watch the three video screens. Non-union folks were not allowed to wiggle any video control knobs, though I could politely ask the video operator to zoom in or out or scan a specific area. I was supposed to report anything exciting to the engineer on watch in the DACS via telephone. The video shack was probably the most exciting place to be for this operation. We got to see all the action, at the expense of doing a lot of waiting. To keep busy during the transfer, I repeated the crust drop trigonometry calculations I had done during the first transfer.

The dilution was more exciting with occasional slurry flows, crust capsizes, and splashes of big slabs of waste falling off the walls. It was like watching a shipwreck break up on a beach or the death throes of some huge prehistoric sea creature. The wasteberg capsize took me by surprise in the video shack. Once you get used to staring at an area of crust when things are barely happening, it sort of takes your breath away when something really moves. During the first event, I happened to be zoomed in tight, looking intently at one of the Enraf bobs as a ridge of waste around it slowly dissolved. All of a sudden there were ripples, then some real waves on the liquid surface. The camera operator zoomed back out, and we saw a large area of white foam flowing down from the north side, eventually covering the whole tank surface like a downy white blanket. The foam dissolved in a couple of hours, leaving a calm, mirror-like lake. Awhile later it happened again. I didn't notice waves this time, just the moving blanket of foam covering the surface. I was glad I was there to see it.

When we were shutting down at the end of the big transfer and top dilution, I saw one last dramatic, even tragic event on the video. Part of the shutdown procedure was pulling the portable video camera out of the tank. It hung down with its mast wrapped in a plastic bag to prevent contamination. To pull it out, a sturdy operator just squatted down and forcibly pulled it up the riser out of the tank. As we watched on the main video camera, we saw the portable camera lurch up, dangle and then stick. It was clear that the plastic bag had bunched up in the riser somehow. The camera operator got on the radio and politely tried to suggest that he should try it again more gently. But it was time to go home. After one last tug jammed the camera tightly in the riser, everyone left. Nobody came back to get the poor little camera and the radiation fried it in a few days.

The big top dilution raised the waste level by 32 inches. For the first time both the 1A and 1C Enrafs were reading the same level without different crust features holding them apart. The neutron logs showed that the upper 30 to 40 inches of waste was a very dilute layer that thinned gradually as the remaining crust dissolved and mixed with the saltier liquid below. At the same time, the lake-like liquid surface was replaced by a new "crust." It may have been a "slushy" crust debris and foamy stuff left over from the many slurry flow gas releases that went on for several days. The new crust looked much softer and more mobile than the thick, stiff pre-dilution crust, but it still covered the entire surface and still held some amount of solids and gas. That would soon change.

Three weeks later, the big low dilution was accomplished. Rick did it in two stages because some of the crews were needed for other pressing work. The first stage of low dilution began at 2:40 AM, Monday, February 21 and added 99,500 gallons of water by close to midnight the same day. After a mid-dilution mixer pump run on February 22, the second low dilution started at 10:37 AM on Wednesday, February 23. It finished at 8:53 PM that evening after adding 50,500 gallons. A small 11,000-gallon top dilution immediately followed.

During these operations, tank behavior was uneventful. There were no big chunks of thick crust to capsize, and all the foamy stuff was already on the surface. There were no big gas releases to note, and the floating layer of foam and crust debris did not submerge during dilution. But the 161,000 gallons of water dissolved whatever was left of the crust that was readily dissolvable. The gamma logs showed a 40-inch layer of mostly water left on top to continue slowly dissolving the more stubborn solids, while the neutron logs showed nothing that could be clearly identified as a crust layer. It was gone, leaving only a surface scum broken by a few tiny islands for a memorial. But even this bit of evidence would soon disappear with the third and last campaign in March.

THE FINAL PLAN

In February 2000, the remediation plan finally solidified to what we had been expecting for some time. The second campaign was not going to be the last one. Rick had asked me to write a "white paper" for the TAP, Defense Nuclear Facilities Safety Board, and DOE Headquarters describing the TRC consensus on transfers and dilutions, the plan to shut off the mixer pump, and what the numbers were based on. The predictions of the required dilution had been made with the buoyancy ratio model along with some earlier criteria. But, in the several years before surface level rise took off, the old Chemical Reactions SubTAP faction had developed a strong dislike for the buoyancy ratio model and, in fact, all models for predicting burp potential. TAP Chairman Kazimi could not completely quell this residual skepticism and so began several months of argument and controversy about how we should claim the final dilution was enough.

On their first review of the white paper, the TAP asked for a formal treatment of uncertainties in the data used in the models. Accordingly, Beric Wells went back to the drawing board and repeated the predictions as a Monte Carlo statistical simulation that placed uncertainty distributions on the answers. The results were pretty surprising. With broad uncertainty bands spread around what we had seen before as a single point value, Agnew's chemistry model didn't look very good at all. It no longer discriminated between burping and non-burping tanks enough to be useable. Neither did the Estey criterion nor average specific gravity completely separate all the burping and non-burping tanks with uncertainty bounds added.

The buoyancy ratio model was the only one that discriminated burpers from non-burpers adequately with the uncertainty bounds included. But it also showed very clearly that, if we wanted to be 95 percent sure there would be no more burps, we had to go all the way to 500,000 gallons! We ran the numbers several more times but got the same result. We had no choice but to run that third campaign. Rick was reluctant because the DOE performance incentive only required 300,000 gallons. He thought we were done! We finally convinced him, and the last big transfer got under way on February 29. It seemed appropriate that something so potentially momentous should start on a leap year!

THIRD TRANSFER AND DILUTION—REMEDIATION!

The third campaign was pretty much an anticlimax. All the violence and excitement of the dying crust was done with the big top dilution in the second campaign, and there was little solid waste left to turn over, sink, or release gas. But there were still a few small surprises.

The third transfer began at 6:35 AM on Tuesday, Leap Year Day, February 29, 2000. The transfer pump started pumping at 120 gallons per minute (gpm) and ramped up to 200 gpm 3 hours later. The high transfer flow into SY-102 pushed its domespace ammonia concentration to a record high value of over 7,000 parts per million (ppm) with a full 2,000 ppm belching out the SY farm exhaust stack. While there was no hard control on the ammonia concentrations, the project's license with Washington State Ecology allowed only 100 pounds of ammonia to be released in 24 hours. This would happen at a stack concentration of 1,750 ppm. On the morning of March 1, it was clear that the SY farm exhaust would exceed the 100-pound limit before the day was over if the transfer continued with the stack ammonia concentration at its 2,000 ppm level. To avoid this violation the transfer was shut down at 5:23 AM after pumping 160,500 gallons from SY-101.

The second part of the third transfer began at 2:18 PM on the afternoon of Wednesday, March 1 at the same 200-gpm total transfer flow, one-third water and two-thirds waste, that had ended the previous session. By 10:30 PM that night the ammonia concentration in SY-102 went above 6,000 ppm, and the SY farm stack had hit 1,750 ppm, the value calculated to send the limit of 100 pounds of ammonia out the stack in 1 day. The transfer rate had to be reduced to about 100 gallons per minute to bring the stack ammonia concentration back down. When the transfer ended at 1:26 PM on Tuesday, the ammonia concentration in SY-102 again hit 7,000 ppm, and the SY stack ammonia was 2,000 ppm. As before, however, the plume from the stack rose away from the farm, and nobody complained of smelling ammonia.

The second portion of the third transfer had removed 125,000 gallons of SY-101 waste, for a total of 286,000 gallons, the biggest transfer yet. Because it began with a lower waste level than either of the prior two, it exposed more of the tank to view than ever before. The waste surface went down to 280 inches, 40 inches lower than at the end of the second transfer. The top of the waste was now actually below where the base of the crust had been at its thickest just before remediation began in December!

This time the waste level drop of 58.5 inches in the first session matched the drop expected from the transfer volume of 160,500 almost exactly, indicating all of the waste attached to the tank wall was gone. But the 48.5-inch drop of the second session was greater than the 45.5 inches that a 125,000-gallon transfer should have produced. This implies a rim of waste at least a foot thick was still attached to the tank wall down low. Sure enough, some of it was visible on the video when the transfer ended. That layer of waste must have been caked on the wall all the way to the bottom! The ancient concerns of the TAP and others in 1994, that the mixer pump was not moving waste all the way to the wall may have been true, contrary to my claims! But never mind. It would be gone soon.

The crew tried a mixer pump run at 7:52 PM on Monday, March 6, with the waste level down to 280 inches. The top of the pump motor was actually exposed at this level. On top of the motor was the flush manifold, a section of pipe on top of the motor with lines running down to the motor cooling jacket and pump volute that had been installed to run flush water to these areas in case of a bad clog. There was a 3/8-inch hole drilled in this flush manifold to allow air in the motor and pump volute to escape so that liquid could enter and prime the pump when it was installed. Ever since then, a tiny jet of waste had flowed out this hole during pump runs, but it had been invisible under 10 feet of waste and crust.

But now, with the hole exposed, a powerful little jet of liquid began spraying out when the pump started. At first people thought the pump had sprung a leak. It was not dangerous, but they thought it was not a good idea to contaminate the exhaust filters with radioactive mist from the spray. Accordingly, they aborted the pump run after five minutes and did not try to run it until after the third dilution was completed March 15. Resting the pump for nine days was not a big deal anymore. Mixing was no longer needed to mitigate burps, though the requirement was still on the books. Its main use now was as a tool for mixing in the last bit of dilution water for the final step of remediation.

The back dilution of the third campaign was done in four steps, alternating large low dilutions and smaller top dilutions. As in the second campaign, cold water at 50 degrees Fahrenheit was used to speed the natural cooling of the waste. The first large low dilution on March 13 and 14 added 127,000 gallons. A small 22,000-gallon top dilution began immediately and a second large low dilution started as soon as valves could be switched over to add 60,000 gallons by 10:30 AM on March 15. The last 14,500-gallon top dilution ended at 3:00 PM the same day, making the total back dilution 223,500 gallons, 187,000 gallons of which went in at the transfer pump inlet and 36,500 gallons on the surface.

There were no dramatic events or gas releases of any note while dilution was going on. The hydrogen concentration rose in small steps eventually went up to 400 parts per million in the last top dilution. Afterward the background hydrogen dropped to around 100 parts per million, where it stayed for about four days. There was gas-bearing material being dissolved somewhere, maybe that last layer of waste still on the wall.

The top layer of floating foam and crust debris stayed on the surface during both top dilutions but began dissolving. By March 17 the neutron and gamma profiles showed a dilute layer about 30 inches deep at the surface. By March 29, further dissolution and aggressive mixer pump operation had removed essentially all traces of a crust or dilute layer on top of the waste. Even so, the surface remained mostly covered by a thin scum of white foamy material from then on.

YOU REALLY MEAN "SHUT IT DOWN"?

After the third campaign, the mixer pump was run as aggressively as possible to mix the tank thoroughly so all the dilution water would have maximum effect. The cool, watery waste was easier to pump and more full-speed runs could be done per day without approaching the pump motor's oil temperature limit. So we could do what might be termed "power sweeps" of up to four 25-minute runs per day at 1,000 revolutions per minute, each one with the nozzles rotated 30 degrees. Thus, four runs would sweep two thirds of the tank.

The old mixer pump did a total of eight power sweeps from March 17 to April 1, plus a single run on March 21. Gas releases were progressively less, but even the last run pushed the hydrogen concentration up to 100 parts per million for half a day. By then there just wasn't much more gas for the pump to free. But there was one last little surprise. The waste had become so thin that the pump jets actually imparted motion at the surface. Though not apparent on the video, the currents were actually strong enough to move the foamy "crust" and pull the Enraf bobs out of vertical alignment, especially at 1C. This made the level reading artificially low. Operations had to adjust the Enraf procedures to lift the bobs off the waste surface and let them swing back to vertical. Also, that little 3/8-inch hole on the volute flush manifold was still squirting during pump runs, even though it had resubmerged. It also caused a surface current that moved the foam away from the pump column during runs.

By April 1, 2000, the data suggested that mixing had reached the point of diminishing returns, and it was finally time to do what had been unthinkable just months ago. At 2:48 PM on Saturday, April 1, 2000, the mixer pump was shut down for the last time. John Conner's and Gerry Koreski's quarterly review of SY-101 data says, "The mixer pump was run four times on April 1 and then shut off for the rest of the period. . . . The final run of the period was the 1194th time the pump was operated since its installation in the tank."

That's all. There were no boasting newspaper articles, no TV interviews with a smiling Rick Raymond pointing to a quiet pool of liquid in the background. A minute of silence in 200 West would have been appropriate, maybe a glass raised in toast to the famous mixer pump that had served so well and so long. There were no toasts or speeches. I don't believe anyone even raised a Styrofoam cup of stale coffee in the DACS trailer. Then again, it was Saturday afternoon and the crew probably hoped to get home early. Having the last mixer pump run on April Fools' Day seems appropriate somehow. After all, SY-101, that ornery old tank, had made fools of us many times in the past. Maybe this time the joke was at the tank's expense.

2000–2001

Are We There Yet?

The three remediation campaigns completed March 17, 2000, transferred a total of 616,000 gallons of waste out of SY-101, excluding transfer line dilution water, and added back in 525,000 gallons of water. But this is not exactly what was left in the tank because some of the fluid transferred out includes water added in earlier dilutions. Accounting for this fact, 525,000 gallons of *original* undiluted SY-101 waste was removed, leaving 539,000 gallons in the tank. At the same time, 434,000 gallons of dilution water were left in the tank, making the final dilution ratio 0.8 gallons of water per gallon of original waste.

As a result, the waste level had been brought from 430.5 inches on December 15, 1999, to 357 inches after the last mixer pump run on April 1, 2000. The highest waste level was 434.3 inches on May 18, 1999. Dilution had cooled the tank from an average of 122 degrees Fahrenheit in December 1999 down to 95 degrees in April 2000 and reduced the total gas generation rate by almost an order of magnitude, from 100 cubic feet per day to about 14. Dissolution had fundamentally altered the waste's chemical composition. Before dilution it contained about 45 percent sodium salts and 36 percent water by weight. Afterwards it was 32 percent sodium salts and 60 percent water by weight.

On the business side, as of April 1, CH2M HILL had been compensated well for the project's accomplishments in accordance with the DOE performance incentive agreement. DOE paid $500,000 for doing more than the required 300,000 gallons of dilution before April 28, 2000, and a tidy $1,000,000 for shutting the pump down and starting the evaluation period before May 31. Had the first dilution not been 5 days later than the December 15, 1999, deadline, they would have earned another $375,000.

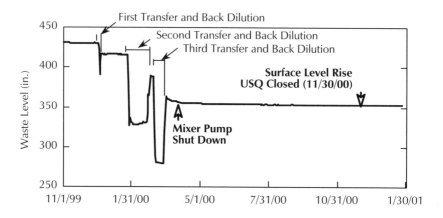

CH2M HILL still had $625,000 left to win for submitting a recommendation to remove the mixer pump requirement from the tank farms safety authorization documents. Would we be able to supply Rick and Craig with enough evidence to be able to convince everyone that SY-101 was really conquered by then? Had we done enough? How could we prove it? This was the main question we wrestled with over the summer of 2000. Fortunately, Jerry Johnson, who had been on temporary assignment full time helping set up a safety organization at the vitrification plant, came back to help guide us through it. His "view from 50,000 feet" was invaluable in getting the technical, administrative, and political ducks lined up to quack out a positive conclusion convincing to DOE Headquarters and the Defense Nuclear Facilities Safety Board.

THE EVALUATION PERIOD

Few of us remembered what it was like not to have a mixer pump run every few days. We were all gratified and a little proud of ourselves to have it shut down, but now what, exactly, were we supposed to do? It was one thing to observe the apparent absence of any hazard, but how could we prove the tank would stay non-hazardous indefinitely? That was now our job.

The best course of action we could think of was to just watch the tank for a while. Nick Kirch originally conceived this time as the mixer pump off period, or MPOP (pronounced "EM-pop") . It was officially termed the "evaluation period," which most of us ignored. The original MPOP was to run 90 days from Monday, April 3 through Sunday, July 2, 2000. In June it was extended to 105 days, ending Monday, July 17, 2000. Interpreting the data, writing the report that was to be the vehicle to prove the tank was safe, and enduring the reviews required to satisfy the TAP, added days and weeks to the schedule. The report was finally issued in late November and quoted tank data clear out to October 1, 2000, almost half a year since the pump shut down.

During the MPOP we were supposed to be looking for conditions warning of unacceptable gas retention or other kinds of bad behavior. Gas retention had historically been measured by surface level rise. Nick's official plan for the MPOP issued in May described unacceptable gas retention as 4,600 cubic feet, equivalent to a 6-inch level rise. But we never saw *any* level rise! The level just kept creeping down. Beric Wells tried correcting for evaporation, but there was too much uncertainty to make any solid conclusions. All we could say was that the waste might be going up or down slightly, but most likely staying about level, but no one could claim there was any serious gas retention going on.

Historically we had always been concerned with the gas generation rate in SY-101, because it was about 10 times higher than any other tank whose rate we could estimate. The hydrogen generation rate is simply the product of the headspace hydrogen concentration and the tank ventilation rate. The total gas generation rate is the hydrogen generation rate divided by the hydrogen fraction in the generated gas.

We might have known the measurement would be difficult. By May 1, the hydrogen concentration in the headspace dropped below 5 parts per million. This was really too low for the tired old GCs (the gas chromatographs that had been installed in the tank) to measure accurately. Hydrogen readings oscillated between 0 and 6 parts per million with the average from 3 to 5 parts per million. This implied a gas generation rate in the range of 10 to 15 cubic feet per day. Though this was about an order of magnitude less than the 100 cubic feet per day it had been in December 1999, it was still on a par with the five other burping tanks.

Of course, we had to watch for crust too. Because crust growth had been stated as the cause of the surface level rise problem, we had to prove that a crust would not come back. We still had that rather undefined white foamy layer. Could it even be called a crust? John Connor guessed it might be up to 6 inches thick, but it wasn't very substantial or permanent. Big pools of open liquid appeared in May but they were gone again by late June. During this time the background ammonia concentration rose from around 60 parts per million way up to 220 parts per million because ammonia is released mostly by evaporation from an uncovered surface. A tank video scan showed a pool that seemed to center around the ventilation exhaust riser where moisture tended to condense and drip down onto the waste. When the foamy layer again covered the pool, the ammonia concentration fell back to between 60 and 70 parts per million. We concluded that the foamy scum layer was not showing any signs of converting itself into a thick crust and would not be an issue in the future.

We were also extremely interested in the depth of the slowly compacting sediment layer because it was the key input for the buoyancy ratio model we were using to prove the tank would never burp again. The trouble was, the waste was a lot more "fluffy" than we had thought it would be. The sediment depth was about 120 inches at the end of May, and just settled to 100 inches October 1, as we inferred it from the neutron and gamma logs and the MIT temperature profiles. But we had predicted it would be only 64 inches deep.

The extra depth meant that our "actual" buoyancy ratio was much closer to unity than the original prediction, implying burps were a lot more probable than we wanted. We had claimed that a 500,000-gallon transfer and dilution would lower the buoyancy ratio to 0.3. But the 100-inch sediment layer held it up to 0.8! It was not the clean, crisp result we wanted at all. Fortunately, there was another non-burping tank, AN-107, that showed a buoyancy ratio of 0.9 that convinced us we were on the right track. Besides, we just did not see the kind of gas retention expected of a burping tank. The final report on the evaluation period claimed:

> "The primary requirement for a . . . [burp] is sufficient gas retention to achieve buoyancy. Available data and evaluations

led to the conclusion that since remediation actions have been completed and the mixer pump has been off, there is no indication of significant gas retention. In addition, analysis of the physical parameters that can correlate observed [burp] behavior and non- . . . [burp] *behavior* in other DSTs [double-shell tanks] shows that tank 241-SY-101 meets criteria for non- . . . [burp] behavior. It is, therefore, concluded that . . . [burp] behavior in tank 241-SY-101 has been remediated."

That pesky, fluffy sediment layer is still 90 to 95 inches deep, so SY-101 remains close to the borderline when evaluated with the buoyancy ratio model. It is certainly a very good thing that Rick and Craig agreed to go all the way to 500,000 gallons. We could not have made the case had we stopped after the second campaign at 300,000 gallons, as originally planned back in 1999.

YOU REALLY WANT TO LEAVE IT OFF?

Many were very nervous about leaving the pump off. The unquestioned doctrine was that the pump had to be bumped every few days to prevent it from plugging. Now we were going to leave it off for months or years! If the tank proved our predictions wrong and began building gas again next year, the pump might be irreparably plugged and could not mitigate an imminent burp! On the other hand, we already knew from the old full-scale tests in 1994 that even a bump every three days was a big disturbance. And we had to have a good long time without any disturbance to really prove dilution had succeeded.

Nick compromised in the MPOP plan by providing for a pump bump once every 90 days, but not more than 120 days, "to verify operability." A little earlier, in March, Rick had called on PNNL to thoroughly evaluate the probability that disuse would plug the mixer pump and recommend what maintenance might be possible and prudent during extended shutdown. Carl Enderlin and Guillermo Terrones did some sophisticated calculations to show that the mixer pump could push any potential plug out of the nozzles. Carl found that the double seals that kept waste out of the motor were what would eventually fail the pump. But an extended shutdown would not have an adverse effect on the seals or any other pump component given the waste's post-dilution properties. Besides, there was no way to do maintenance on any of the pump's

414 | **Part Three: Surface Level Rise**

internal components anyway. In short, Carl's work didn't turn up any need to bump the pump at all. But many were not convinced by this finding. Nick's original plan for a bump 90 days after shutdown, meaning June 30, stayed on the schedule.

Towards the end of the MPOP on June 14, 2000, there was conference call with the TAP to get their opinion on a pump bump. One firebrand from the old SubTAP advised, "Do not mix up the tank! If at 120 days [July 30] everyone agrees that tank is OK, then run it. We need to see some data before running it." Craig Groendyke agreed and urged "caution about running the pump too soon." Mujid Kazimi realized he had a consensus and recommended, "Why not wait until 120 days?"

The TRG met June 22 to decide whether to end the MPOP as scheduled or extend it. To give time for last runs of the neutron probe and validation probe in the MITs, the TRG voted to extend the period to July 15. They would do a pump bump on July 19. Then somebody suggested that the bump be extended to a full, 25-minute run. This brought all the strong opinions out in the open and extended the meeting another hour. The science and engineering faction were all for it. They felt one last run was *required* to prove that the tank was not storing gas. If not much gas came out in a final run, we'd stand tall and say, "Look! The tank is not retaining gas. It's now safe!" Of course if a lot of gas came out we'd have to say, "Woops! We have to do another transfer and dilution!" It just seemed like a crime not to collect this essentially "free" but extremely valuable data.

The operations and safety basis faction, led by Fred Schmorde, TRG chairman and SY Farm Facility Coordinator, fought the last pump run, though he favored a bump. His philosophy was that if we don't think we need the pump, let's be consistent and not run the pump. We had sophisticated calculations that said the tank was safe, and a carefully planned observation period to demonstrate it. We should stand tall and stick to the plan. Fred rightly feared that any late changes would be seen as doubt and open us up to all kinds of second guessing. Besides, if the waste were stirred up again, we'd have to start the "evaluation period" clock again, and it would never end.

Getting good, convincing data to prove remediation sounded very attractive. But the possibility of an unconvincing or negative result was worrisome. It might mean extending the project for another costly dilution campaign, waiting through another six-month evaluation period, just like Fred warned, and losing the final performance incentive payment of $625,000 for not getting done by September 30. Faced with the high risk of the test and because the other data were looking pretty good, Craig agreed with Fred, and Rick and CH2M HILL management made the decision not to run the pump again. Not even a bump.

IT'S OVER!

Our final report apparently made a convincing case. More likely, everyone was simply ready for the project to end and saw no reason why it shouldn't. On November 30, 2000, Harry Boston, the new head of DOE's Office of River Protection, sent a letter to CH2M HILL and DOE Headquarters closing the Surface Level Rise USQ on SY-101. Boston's letter also removed the requirement to operate the mixer pump. The SY-101 Surface Level Rise Remediation Project had reached both its goals. The level rise problem was over, and the tank no longer needed regular mixing to prevent burps.

Closure of the USQ was the signal to start shutting things down. The first things to go were those that cost money to keep doing. Similarly, if something broke, nobody fixed it. Jeanne Lechelt stopped collecting and archiving the DACS data. This meant measurements of the tank dome pressure, and the temperature, relative humidity, and flow rate of the exhaust air were not available after October 2000. We also went blind as camera operators stopped going into the video shack to switch on the big camera and look around the tank. The old Fourier transform infrared spectrometer failed on September 17, after over seven years of service, and was not replaced. One bright spot was replacement of the old GCs with new standard hydrogen monitoring system, or SHMS, E+ units. But, on the whole, SY-101 was beginning to look like just another tank by the fall of 2000.

The paper information flow dried up just like the electronic stream did. Weekly, then monthly, and finally quarterly reports were developed from the sum total of monitoring data on SY-101 since 1993. Reports were begun by Rudy Allemann of PNNL and later edited by Frank Panisko and me. Nancy Wilkins, John Conner, and Gerry Koreski from Lockheed did the reports from 1997 to the end. The last SY-101 data report of November 30, 2000, contains this paragraph:

"These data reports have been issued periodically since the mixer pump was installed and began operating in 1993. . . . The main purpose of the reports has been to monitor and document the tank conditions and pump parameters. Following the recent transfers and water dilutions, the surface level rise and flammable gas issues for the tank have been closed. No further pump runs are planned. There is no further need for this data report. Therefore, this will be the final data report of this type for SY-101. . . ."

Closing the USQ was only the first step. SY-101 was still technically subject to the flammable gas safety issue because it once had the potential to release enough gas that, if ignited, could damage equipment or injure workers. SY-101 was also still on the flammable gas watch list of tanks that had "a serious potential for release of high-level waste due to uncontrolled increases in temperature or pressure." But it clearly no longer had such potential. Accordingly, on December 7, 2000, Harry Boston sent an official request to Carol Huntoon at DOE Headquarters to close the flammable gas safety issue on SY-101 and remove the tank from the flammable gas watch list.

On January 11, 2001, Huntoon replied with a letter approving the request. In her letter she stated, "I want to commend you for resolving what has been an extremely complex and challenging issue . . . Congratulations on this outstanding achievement in eliminating one of the most urgent risks at Hanford." It was over at last!

That event was the crowning achievement of all the strife and labor expended on SY-101 since Mike Lawrence's fateful news conference in March 1990. Now the tank was officially just like any other double-shell tank whose space was so precious for staging waste from the ancient single-shell tanks to the future vitrification plant. But it was sad in a way. No more would SY-101 polish or tarnish careers, test the knowledge of high-powered scientists, turn technologically wondrous instruments into junk, and suck Hanford's budget dry. Or will it?

2001–2003

Epilogue

On a fine November day in 2000 there was a celebration in the big open courtyard in the center of 2750E. Everyone who had worked on SY-101 in the last 10 years was invited. There was a catered barbecue with plenty of tables where people could eat and visit. Under the trees along one side was a long row of posters propped on easels with photos, plots, and paragraphs in big print describing various events in the tank's history. We were physically surrounded by history too. On the west end of the courtyard, two small second-floor windows were once the war room for mixer pump installation and mitigation in 1992-93, and where Nick Kirch's committee began puzzling over level rise in 1998. On the south side, several second-floor windows opened to offices of Dan Reynolds, who had evaporated the last waste into SY-101, and Leela Sasaki, who measured the first burp known to go flammable. On the east side, with no windows to mark them, are two large, wood-paneled conference rooms where many of the early TAP meetings were held to discuss SY-101's behavior.

Dale Allen made a speech lauding the project and announcing the promotion of Rick Raymond to senior vice president of CH2M HILL Hanford Group, Inc. in acknowledgement of his leadership of the Surface Level Rise Remediation Project. Then Rick got up and emotionally thanked the project team, individually and collectively, for their efforts. At the end, somebody got up and presented Rick

Part of the SY-101 SLRRP Team at November 2000 recognition lunch. BACK ROW FROM LEFT: Scott Estey, Chuck Stewart, John Conner, Beric Wells, Nick Kirch, Blaine Barton. Center row from left: (unk.), (unk.), Ken Anderson, (unk.), Joe Brothers. FRONT ROW FROM LEFT: Dale Allen, (unk.), (unk.), (unk.), Ron Tucker, Rick Raymond.
Courtesy Rick Raymond.

with a memento of our appreciation. It was a great time of laughter, honest praise, and some tears when memories overflowed.

With all that had been accomplished on SY-101 and the attention that success had garnered, Rick and CH2M HILL wanted to get all the mileage out of it they could for the company, for Hanford, and for the project team. One such opportunity was to nominate the project for an award by Project Management Institute. In March, the Columbia River Basin Chapter of the Project Management Institute chose SY-101 level rise remediation as their project of the year, making it eligible for the international award. Rick Raymond, Roger Bauer, Joe Brothers, Carl Hanson, Joe Buchanan, and several others wrote up a 25-page nomination, endorsed by CH2M HILL management and DOE Richland. This excerpt from the first page sets the tone of the nomination:

> "CH2M HILL Hanford Group is pleased to nominate the SY-101 Surface Level Rise Remediation Project (SLRRP) for the Project Management Institute's consideration as International Project of the Year for 2001. We selected this project as being

our best recent example of effective project management, having achieved and exceeded our client's expectations in resolving urgent safety issues related to the storage of high-level nuclear waste. . . . Our objective in submitting this nomination is twofold-to share the lessons we have learned with other organizations, and to honor the men and women who contributed to this endeavor. It was by their diligent effort that the successes we relate here were accomplished 10 months ahead of schedule and one million dollars below the authorized budget."

After Rick sent the submission off the nomination fell into a sort of black hole from which no information could escape. After some months we heard SY-101 had made the top three! In another month the final result trickled down. We made second place! Not THE project of the year, but close!

Removing SY-101 from the flammable gas watch list started a cascade of events that swept over the tank farms. SY-101 was the reason for the watch lists in the first place. If SY-101 was off the list, there wasn't much further reason for the list. In August 2001 the DOE Office of River Protection sent a letter to DOE Headquarters requesting that the flammable gas safety issue be closed and all tanks be removed from the flammable gas watch list. Steve Stokes, point man for the Defense Nuclear Facilities Safety Board at Hanford, told Craig Groendyke that the board was in unanimous agreement with the Office of River Protection's request to Headquarters. The approving letter arrived August 13, 2001.

The flammable gas safety issue was now closed and all tanks removed from the flammable gas watch list. Senator Ron Wyden, author of the bill creating the watch list in 1990, came to Richland to commemorate the occasion. In his speech he said, "A decade ago, I responded to the dangerous threat posed by certain nuclear waste storage tanks at Hanford by passing a law to protect the people of the Northwest from possible radioactive tank explosions. Today, I'm proud to see the watch list become extinct. The hard work of the Department of Energy and many others has helped protect the people of Hanford and communities downstream from the potentially devastating effects of a radioactive explosion." In Wyden's opinion, closing the flammable gas issue was "the single most important accomplishment at Hanford." Considering the hundreds of millions of dollars and all the national attention it consumed since 1990, he was probably right.

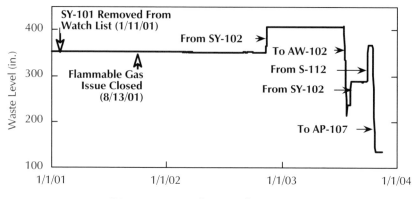

SY-101 Waste Surface Level: *2001–2003*

BACK IN SERVICE

SY-101 was out of the official spotlight, but the tank would not really become "one of the gang" until it received and sent out some waste as a real working tank. According to Craig Groendyke, "When SY-101 actually is in service, it will have been finally remediated and ultimate success can be proclaimed." SY-101's first act of being "in service" happened November 2002, when it received 150,000 gallons of liquid waste from SY-102 to make room for saltwell pumping. It was the tank's first receipt of waste since 1980. It seemed odd, after all the effort we went to transferring SY-101 waste into SY-102 almost 3 years earlier, that we should be reversing the process. But after years of receiving dilute saltwell pumping liquids, receiving waste from SY-102 was almost like a final dilution campaign!

The final rite of passage that would admit SY-101 to all duties and privileges of an operational tank was a cross-site transfer. That would prove that the new generation transfer pump had the power to push brine through several miles of 2-inch pipe. In 2002 DOE gave CH2M HILL a performance incentive to connect up SY-101 for cross-site transfers. It was actually pretty simple to connect the prefabricated pump pit on SY-101 to the SY-A valve pit to make the tank fully capable of sending a cross-site transfer. They met the performance incentive requirements in October 2002.

The cross-site transfer from SY-101 to AW-102 in 200 East began on Thursday, July 17, 2003, exactly 3 years after the end of the MPOP evaluation period. By July 24, 534,000 gallons of waste had been moved through the pipeline with no problems. They didn't even need to use dilution. The next day 64,000 gallons came back in from SY-102, and 2 weeks later on August 6, 2003, it got another 137,000 gallons to make room for saltwell pumping. All this activity had dropped the surface level from 407 inches down to 214 inches and back to 287 inches. SY-101 was now truly in service.

From September 29 to October 2, 2003, there was another transfer of 216,000 gallons of liquid from S-112. This one was special for three reasons. First, this was part of the first full-scale retrieval of a single-shell tank using the saltcake dissolution process. If S-112 can be retrieved successfully, this method of spraying water on and pumping nearly saturated brine out can be applied to a lot more saltcake tanks. Second, Roger Bauer, heir to Rick Raymond's legacy, managed the S-112 project assisted by JR Biggs, Carl Hanson, and many others from SY-101 days. Finally, the transfer reminded us that we had only done just enough to remediate SY-101.

The brine coming from S-112 was much more concentrated than the liquid from saltwell pumping that SY-101 had seen before. Prior to this transfer, the waste compatibility criteria had been revised to include a buoyancy ratio calculation using the estimated post-transfer tank conditions. There was some delay in getting all the information needed to do the calculation, but CH2M HILL decided the transfer could be started anyway. Imagine the embarrassment when the buoyancy ratio came up showing that SY-101 might start burping if the transfer were completed! The transfer had to be stopped, and the calculations were reviewed with great intensity. It turned out that the wrong liquid and sediment densities were used the first time. When corrected, the tank was still predicted benign, as expected.

What makes SY-101 sensitive is the 100-inch sediment layer that we have to assume still has the potential to accumulate gas. Some-day, tank farm operations wants to get rid of that old sediment by running the mixer pump again just before another cross-site transfer. If necessary, they might even pump it through the new slurry line that hasn't yet been used. There are no plans to lower the transfer pump from its current 100 inches to increase transfer capacity. The mixer pump is still operable, though there has been no maintenance since it was shut down. The 480-volt electrical power leads were lifted for safety, but the DACS is still there, though Fred Schmorde thinks "it's probably a mouse hotel by now." The nitro-

422 | Part Three: Surface Level Rise

gen purge system has stayed on to keep the wires and fittings in the pump column dry. All it would take to start the motor would be the will to rewrite the procedures, train a crew, connect the wires, and click "START" on the DACS screen.

As nostalgic and exciting as it would be to act on these plans, they remain just that. The last cross-site transfer to date happened from October 17 to 25, 2003, moving 638,000 gallons of liquid to AP-107 in 200 East Area without the benefit of a mixer pump run.

WHAT DID IT ALL MEAN?

What grand insights and instruction can we derive from the 20-year struggle with SY-101? Are these things inevitable, or can we hope to change society enough to prevent the worst of them? My own opinions went through several transformations in the process of writing this book. I first thought that if DOE and Hanford would read it, they would see past mistakes and avoid repeating them. Yes, there were obvious mistakes of omission, inattention, or expediency. But I also saw examples of valiant, even brilliant, work that caused bad problems later. It is simply not possible to anticipate and plan for every result. I finally saw that human nature will never build a bureaucracy wise enough to set priorities that are consistently correct for the decades required to prevent these problems. Human nature makes people unwilling to admit all problems early enough to fix easily. Nor will it ever supply scientists and engineers smart enough to understand and solve a problem so completely as to never recur. There *will* be future SY-101s as there always have been, and there's not much anyone can do to prevent them.

But there are some things that can be learned from the SY-101 experience besides its inevitability. First, when a big problem gets national notice, the technical-scientific side of it becomes much less important than public perception and political power. Unless you play the game in that arena, you will fail. Second, it is a great advantage to bring active or potential critics into the work and give them responsibility for answering their own criticisms. This turns opponents into allies. Finally, a project gets energy by including as many as possible of its contributors in decision making and giving them the largest possible responsibility. This is what made Rick's team and, to some extent, Jack's team so extraordinarily successful.

But I'm only one person who participated in part of the SY-101 experience, and I can't claim to have all the answers. The rest of this epilogue collects statements of others I have interviewed about what they think made this thing important and how we should view it. Here's what they believe is important. Each paragraph is from a different individual.

"There is a fear that, even after all the effort spent on SY-101 resolution, DOE leadership may not remember it. Wanting to get waste out of single-shell tanks with little double-shell tank space, there is a push to concentrate the waste and maybe repeat history. It is incumbent on us to remind them that we don't want to go down that path again."

"Early safety analyses showed extreme consequences. Predictions were way out in the tails of probability. But SY-101 experience, both early and late, showed that if you avoid hysteria, you can reverse the process and solve the problem. Avoid finger pointing and assigning blame, constantly fearing the worst case scenario."

"An important fact was that, for the first time in a long time, the site really forced a success-oriented team on the client. We put together a team of people that related well with the client's team. We brought all the critics into the team including Los Alamos on the SA [safety analysis] and the TAP for oversight. It brought together the right team with the right resources to get the job done."

"There were two ways to approach the task. Use the available information, consider that it was enough information to make a decision and go ahead, or study the problem forever and not make any decisions. We got the pump designed, built, and installed in less time that it took to study the problem and decide on mixing. It was one of the few times that design, fabrication, installation, and analysis were all going on in parallel."

"It was not a project like putting man on the moon but a brute force job with no fancy system engineering or finely tuned design. It was complex and ugly. The tank was ugly, the pump was ugly, large, and complicated. It was like Red Adair putting out an oil fire, a technical challenge to solve an important safety problem."

"All the players viewed themselves as wanting to do the right thing. But people are imperfect and the scale of the problem was huge, and there were unforeseen causes and effects."

"In the United States there are no long-term government programs. Each new administration runs things differently; there is no continuity. Too much is done in the "panic mode" without looking at future problems. SY-101 was the prime example. The waste concentration to

gain tank space caused burps. Then, in the hurry to fix burps, we created the crust growth problem. What other problems are waiting? Everybody is trying with best ability and intentions to 'save the day,' but in panic to save the day, we create problems in the future."

"SY-101 shouldn't have been that big a deal. DOE and Hanford were over-reacting like crazy. Why should level rise remediation been nominated for best project of the year? However, everyone FELT it was a big deal. We spent a lot of effort in trying to understand a phenomenon we didn't know much about and make sure it didn't occur again. An important issue was that the mixer pump was thought to be the mitigation solution, but it actually created a potentially worse problem. We should have envisioned this earlier. Data was there."

"How could a garbage tank out in the desert be so peculiarly interesting? The enormous technical and political complexity of that tank was what made it interesting. You always ran into situations where you didn't have the information you needed and had to make do with something else."

"There were so many things wrong at the same time: high heat, ferrocyanide, organics, leaking single-shell tanks. It was both the best job and the worst job at the same time. The commitment of the people made all the suffering tolerable."

"The flammable gas safety issue started with a bang and ended with a whimper. It was a much bigger political than actual problem."

"A good analogy for SY-101's effect on folks on site and off site, the community, finance, and society in general might be like Rickover's nuclear navy changing the culture or starting a new, unique one. There are ripple effects that are still going on."

"We had a challenge and we met it though some of our challenges were admittedly contrived. It was always amazing how many people it took to get anything done."

"The human aspects and interactions make the story. The technical record of the Mutiny on the Bounty would be extremely dull. The conflict between Bligh and Christian makes it live. It's the same with SY-101."

"re•sis•ten'tial•ism n. The belief that inanimate objects have a natural antipathy toward human beings and therefore it is not people who control things, but things which increasingly control people."

"The tank was a personality. It was in control and appeared to know it. It generated unreasoning fear. It was as if it thought 'I'm tired of this instrument, so now I'm going to break it,' or 'I'm getting bored and now I'm going to do something unexpected.'"

Chronologies

SY-101 DETAILED CHRONOLOGY

1974–76 . . . . SY double-shell tank farm constructed

April 1977 . . . . First waste (274,000 gallons of double-shell slurry) enters SY-101 raising waste level to 101 inches

October 1977 . . . . Slurry growth in newly added double-shell slurry increases SY-101 volume by 19,000 gallons (6 inches)

November 1977 . . . . 365,000 gallons of complexed concentrate added to SY-101. Waste level now 242 inches

June 1978 . . . . 131,000 gallons of complexed concentrate added from SX-106 makes the waste level 301 inches

August 1978 . . . . 60,000 gallons of complexed concentrate added from U-111. Waste level rises to 318.5 inches

November 1980 . . . . **230,000 gallons of double-shell slurry completes filling of SY-101. Waste level at 408 inches, including 23 inches of slurry growth**

January 1, 1981 . . . . **First level drop, or "burp," identified**

426 | Chronologies

March 15, 1981 . . . . First cycle of level growth and sudden collapse recorded with a level drop of about 3 inches

July 4, 1983 . . . . SY-101 domespace first pressurized above 1 atmosphere during a burp

October 20, 1983 . . . . . . Water lancing of SY-101 crust first performed

July 16, 1985 . . . . Gas monitoring on SY-101 measured hydrogen concentrations over 5 percent in the ventilation duct during a burp

November 25, 1986 . . . . Air lancing first performed. Discontinued in 1989 because it was not only ineffective but was apparently exacerbating burps

December 29, 1989 . . . . Large burp with a 6.5-inch level drop. There were many burps before this. This is the first to be analyzed in detail. Each burp to follow was of high interest

January 1990 . . . . Westinghouse requested Paul d'Entremont of the Savannah River Site to review a white paper on the SY-101 flammable gas issue

February 6, 1990 . . . . . . Paul d'Entremont concluded that that retained gas in SY-101 could be flammable within the waste due to presence of nitrous oxide (oxidant)

March 23, 1990 . . . . Mike Lawrence, DOE Richland, holds press conference announcing new information about flammable gas hazard in SY-101

April 19, 1990 . . . . Burp with a 9-inch level drop and 3.5 percent hydrogen concentration

May 14, 1990 . . . . Mike Lawrence declares that hydrogen and nitrous oxide in waste tanks and hypothetical ignition is an unreviewed safety question

May 1990 . . . . DOE Headquarters forms the Tank Advisory Panel

July 7, 1990 . . . . Mike Lawrence resigns as Manager of DOE Richland Operations. John Wagoner named to replace Lawrence as of July 12

Chronologies | 427

August 5, 1990 . . . . Burp with a 5-inch level drop and 1 percent hydrogen concentration

October 24, 1990 . . . . . . First officially labeled burp, **Event A**, occurred with a 10.2-inch level drop and 4.7 percent hydrogen concentration

February 16, 1991 . . . . . **Event B**, 4.2-inch level drop with low hydrogen concentration

May 17, 1991 . . . . **Event C**, 7-inch level drop with 2.8 percent hydrogen concentration. First core samples taken in Window C

August 27, 1991 . . . . **Event D**, 5.75-inch level drop, low hydrogen concentration. Window D did not open

December 4, 1991 . . . . . **Event E** with a 12-inch level drop and 5.3 percent hydrogen concentration. First video of a major burp showing violent hydraulics and bending in-tank hardware

December 1991 . . . . Paper written by Rudy Allemann for April 15, 1992, Waste Management Conference in Las Vegas formally documents the "gob" model explaining how burps happen in SY-101

February 4, 1992 . . . . . . **Fiasco in Pasco workshop at Pasco Red Lion Inn to prepare the project plan to mitigate SY-101 burps. Project team formed of Westinghouse Hanford Company, Los Alamos National Laboratory, and Pacific Northwest Laboratory**

April 20, 1992 . . . . **Event F**, 5-inch level drop and 1 percent hydrogen. Window F did not open

September 3, 1992 . . . . . **Event G**, 9.75-inch level drop and 5.1 percent hydrogen. Bent lances and thermocouple tree removed in Window G. No time for mixer pump installation

February 1, 1993 . . . . . . **Event H**, 9-inch level drop, 2.7 percent hydrogen concentration, but incomplete turnover of waste on the bottom. Did not allow mixer pump installation in Window H

428 | **Chronologies**

June 26, 1993 . . . . **Event I, 9.75-inch level drop with 3.4 percent hydrogen. Window I open for mixer pump installation June 30**

July 3, 1993 . . . . **MIXER PUMP INSTALLED!**

July 5, 1993 . . . . **FIRST MIXER PUMP RUN AT 6:43 AM for 2 seconds!**

July 15, 1993 . . . . The first of three phases of pump testing begins. The first phase, called Phase A, a low-speed testing, begins

July 19, 1993 . . . . **Defense Nuclear Facilities Safety Board issues Recommendation 93-5 for improved waste tank sampling studies to understand waste behavior**

July 20, 1993 . . . . Phase A testing is completed

July 25-26, 1993 . . . . Pump is plugged. This forced the pump run to be increased to 5 minutes at 1,000 revolutions per minute for "bumps"

August 4, 1993 . . . . Unplanned mixer pump start in SY-101 during DACS testing

August 10, 1993 . . . . **Rock-on-a-rope contamination incident on Tank C-106. This incident plus the unplanned pump start in SY-101 forced the great 1993 stand down that lasted for weeks and delayed the second phase, Phase B, of mixer pump testing**

August 27, 1993 . . . . Moderate gas release that pressurized the tank and released about 2,000 cubic feet of gas. This was the first of three "mini-burps" of gassy waste left over from Event I in June

September 19, 1993 . . . . Second "mini-burp," small but noticeable

October 15, 1993 . . . . . . Last small "mini-burp"

October 21, 1993 . . . . . . **Phase B, high-speed testing, begins. Testing was completed December 17, resulting in mixing and de-gassing most of the tank**

February 4, 1994 . . . . . . Full-scale testing begins; completed April 13. Jet penetration testing leads to excavation runs, where the pump jet excavated the waste off the bottom of the tank, through spring and summer of 1994

September 13, 1994	Three excavation runs per week of 25 minutes at 1,000 revolutions per minute is set as the normal mixer pump operation schedule
December 15, 1994	New Enraf-Nonnius waste level gauge installed in Riser 1A to complement the old FIC gauge in Riser 1C. The difference in the readings was the first major indication of level growth
June 21, 1996	**DOE Richland closes flammable gas unreviewed safety question on SY-101 because the mixer pump eliminated spontaneous gas releases**
November 19, 1996	**SY-101 level growth noted in Flammable Gas Program Manager Jerry Johnson's weekly hydrogen meeting. First official note of level growth**
December 18, 1996	FIC in Riser 1C replaced with Enraf buoyancy gauge
October 19, 1997	Mixer pump schedule accelerated to four runs per week in attempt to slow level growth
December 8, 1997	Return to three mixer pump runs per week because four runs per week was found to accelerate level growth. This behavior was contrary to the mixer pump safety controls and initiated the SY-101 level rise unreviewed safety question
December 29, 1997	A group, led by Jerry Johnson and process engineer and manager Nick Kirch, begins studying SY-101 surface level rise
February 18, 1998	Standing Order 98-5 for running the pump deletes requirement for increased pump operations at 402 and 406 inch level, requires monthly Test Review Group review of tank data.
February 26, 1998	**DOE Richland declares a new unreviewed safety question related to waste surface level changes in SY-101. Became known as the "Surface Level Rise USQ"**

430 | Chronologies

May 5, 1998 . . . . Level rise formally verified as a real, tank-wide phenomenon

June 29, 1998 . . . . First run of the void fraction instrument in SY-101 with two more on July 22 and September 11. Results confirmed gas accumulation in crust layer was cause of level rise

October 1, 1998 . . . . **SY-101 Surface Level Rise Remediation Project authorized**

November 7, 1998 . . . . . SY-101 surface level exceeds maximum operating level of 422 inches

November 10–12, 1998 . **SY-101 level rise workshop at Hanford Square Conference Center. Developed the plan for remediating SY-101 surface level rise**

December 18, 1998 . . . . First core sampling campaign in SY-101 complete. Retained gas samples showed gas fractions up to 50% in a "bubble slurry layer." Second core sample January 5–18, 1999, also showed a bubble slurry layer

January 27, 1999 . . . . . . 400 gallons of water poured on the crust through Riser 5B to make a hole for the in-tank camera

April 16, 1999 . . . . Crust base drop event shown in neutron logs. Growth rate slows

April 27, 1999 . . . . **The "Wagoner Letter,"** which was actually from Dick French, Manager of DOE Office of River Protection, suspends controls and directs Lockheed to proceed with "all diligent speed" using "prudent controls necessary to safely conduct such operations" to mitigate level growth in SY-101

May 20, 1999 . . . . SY-101 waste surface hits its **maximum level of 434.28 inches** and begins to drop slowly

May 20, 1999 . . . . **Mechanical Mitigation Arm** operated in Riser 11B

May 26, 1999 . . . . Second run of Mechanical Mitigation Arm, again in Riser 11B

August 9, 1999	Deployment of 42-inch "cruciform" spray lance to bore a hole for the transfer pump. This is the same lance used to burrow a hole for the mixer pump in July 1993
November 15, 1999	**Defense Nuclear Facilities Safety Board closed Recommendation 93-5** based on improved waste characterization data and better understanding of the flammable gas hazard
December 18, 1999	**First remediation campaign.** Waste transfer completed 5:30 AM on December 19, 1999. Back dilution completed December 20, 1999
January 25, 2000	**Second remediation campaign.** Waste transfer completed January 27, 2000. Top back dilution completed January 28, 2000. Most of the crust dissolved. Bottom back dilution February 21–23, followed by a small top dilution
February 29, 2000	**Third remediation campaign.** Waste transfer completed March 2, 2000. Top and bottom dilution series completed March 13–15
April 3, 2000	**SY-101 mixer pump permanently shut down.** Mixer pump observation period begins
November 30, 2000	Harry Boston, DOE Richland, **closes the Surface Level Rise Unreviewed Safety Question on SY-101**. Closure letter also removes the requirement to operate the mixer pump
January 11, 2001	Carol Huntoon, DOE Headquarters, **removes SY-101 from the flammable gas watch list**
August 13, 2001	**Flammable gas safety issue is closed, and all tanks removed from the flammable gas watch list**
November 2002	SY-101 receives its first waste since 1980, a 150,000 gallon waste transfer from SY-102
July 15, 2003	First cross-site transfer from SY-101 to AW-102

CHRONOLOGY OF GOVERNMENT OVERSIGHT

1943 .... Manhattan Engineer District, Corps of Engineers

1946 .... Atomic Energy Commission

January 29, 1975 ...... Energy Research and Development Administration

October 1, 1977 .... U.S. Department of Energy

September 29, 1988 .... Congress passes Public Law 100-456, Section 1441, establishing the Defense Nuclear Facilities Safety Board

May 1989 .... Tri-Party Agreement (officially known as *Hanford Federal Facility Agreement and Consent Order*) established among U.S. Department of Energy, Washington State Department of Ecology, and U.S. Environmental Protection Agency, establishing a 30-year timetable and milestones for Hanford cleanup

November 5, 1990 ..... Congress enacts Public Law 101-510, Section 3137, "Safety Measures for Waste Tanks at Hanford Nuclear Reservation," also known as the "Wyden Amendment," after its author, Senator Ron Wyden of Oregon. Initiated the watch list of dangerous tanks

February 8, 1991 ...... First official watch lists published for 23 flammable gas, 23 ferrocyanide, 8 organic, and 1 high-heat tank

1998 .... Office of River Protection separated from DOE Richland Operations office to manage specific parts of Hanford cleanup

CHRONOLOGY OF HANFORD CONTRACTORS

Until 1967, a single contractor managed Hanford work. With the entry of Atlantic Richfield, several companies shared the contract. Those listed here were in charge of the tank farms and managed SY-101.

1943	E. I. du Pont de Nemours (now simply called DuPont)
1946	General Electric Corporation
1967	Atlantic Richfield Hanford Company
July 1, 1977	Rockwell Hanford Operations
June 29, 1987	Westinghouse Hanford Company
August 6, 1996	Project Hanford Management Contract (PHMC). Lockheed Martin Hanford Company managed the tank farms
December 1999	CH2M HILL Hanford Group, Inc.

Hanford Production Reactors

Hanford Reactors	B	D	F	DR	H	C	KW	KE	N
Construction start	8/43	11/43	12/43	12/47	3/48	6/51	11/52	1/53	5/59
Start up	9/44	12/44	2/45	10/50	10/49	11/52	12/54	2/55	3/64
Shut down	2/68	1/67	6/65	12/64	4/65	4/69	2/70	1/71	1/87
Years of operation	22	12	20	14	15	17	15	16	22
Design power (MW)	250	250	250	250	250	750	1,800	1,800	4,000
Peak power (MW)	1,940	2,005	1,935	1,925	1,955	2,310	4,400	4,400	3,950
Total plutonium created (tons)	6.1	6.0	4.9	4.2	4.5	7.1	14.1	14.8	12.1
Fuel grade (tons)	0.7	0.7	0.06	0.06	0.1	0.6	2.0	1.1	9.0
Weapon grade (tons)	5.4	5.3	4.8	4.1	4.4	6.5	12.1	13.7	3.1

Plutonium Separation Production

Process	Fuel Processed (tons)	Waste Produced (million of gallons)	Average Waste Rate (gallons per ton fuel)
Bismuth phosphate	8,900 (8%)	98	10,000
REDOX	24,600 (23%)	42	1,600
PUREX	73,100 (69%)	39	500
Total	106,600 (100%)	179	

Hanford Processing Plants

Plant	Function	Process	Start Up	Shut Down
T Plant	Pu separation	Bismuth phosphate precipitation	Dec. 1944	Mar. 1956
	Equipment decontamination, repair, and storage		1957	1983
B Plant	Pu separation	Bismuth phosphate precipitation	Apr. 1945	Nov. 1956
	Cs, Sr recovery	Cs: precipitation & ion exchange Sr: solvent extraction TBP, D2EHP, complexants)	1968	Cs - Sept. 1983 Sr - Feb. 1985
U Plant	U recovery	Solvent extraction (TBP + kerosene), nickel ferrocyanide to precipitate Cs	Jul. 1952	Mar. 1957
S Plant	Pu, U, Np separation	REDOX solvent extraction (hexone)	Jan. 1952	Jul. 1967
A Plant	Pu, U, Np	PUREX solvent extraction	Jan. 1956	June 1972

Cs	cesium	REDOX	reduction-oxidation process	
Hexone	a commercial organic solvent like kerosene	Sr	strontium	
Np	neptunium	TBP	tributyl phosphate, an organic solvent	
Pu	plutonium	U	uranium	
PUREX	plutonium-uranium extraction			

Underground Waste Storage Tanks

Tank Farm	Number of Tanks	Capacity (1000 gallons)	Years Built	Note
Single-Shell Tanks				
A	6	1,000	1954–55	Self boiling
AX	4	1,000	1963–64	Self boiling
B	4	55	1943–44	
	12	530		
BX	12	530	1946–47	
BY	12	750	1948–49	In-tank solidification
C	4	55	1943–44	
	12	530		
S	12	750	1950–51	Self boiling, DSSF
SX	15	1,000	1953–54	Self boiling, DSSF
T	4	55	1943–44	
	12	530		
TX	18	750	1947–48	
TY	6	750	1951–52	
U	4	55	1943–44	DSSF
	12	530		
ALL SST	149	93,680	1943–64	
Double-Shell Tanks				
AN	7	1,160	1980–81	DSS, DSSF
AP	8	1,160	1983–86	
AW	6	1,160	1974–76	DSSF
AY	2	1,160	1968–70	High temperature
AZ	2	1,160	1971–77	High temperature
SY	3	1,160	1974–77	DSS
ALL DST	28	34,280	1968–86	
HANFORD	177	126,160	1943–86	

DSS   Double-shell slurry
DSSF Double-shell slurry feed
DST   Double-shell tank
SST   Single-shell tank

Glossary

Air lancing: Pressurized air is directed into the waste through pipes, "lances," to stir it up and release gas. It was discontinued in 1989 because it was not only ineffective but was apparently causing the waste level to rise.

ASSD: Anti-Siphoning Slurry Distributor, a piping system installed in tank SY-102 for the 1999-2000 transfer and dilution campaigns to distribute the slurry pumped out of SY-101 below the waste surface while an air gap prevented siphoning.

Atomic Energy Commission: This commission took over management of the national defense nuclear production complex after World War II and served until 1975. In 1975, the organization was split into the Nuclear Regulatory Commission and the Energy Research and Development Administration.

Barometric pressure effect: A calculation procedure for estimating the volume of gas retained in a tank from surface level changes in response to barometric pressure fluctuations.

Bismuth phosphate process: A process for separating plutonium from irradiated reactor fuel where plutonium is precipitated with bismuth phosphate.

Blush Report: *Report on the Handling of Safety Information Concerning Flammable Gases and Ferrocyanide at the Hanford Waste Tanks,* written by Steve Blush and others, and published in July 1990. The report was very critical of Hanford's handling of tank safety issues.

438 | **Glossary**

Bump: A short run, seconds to a few minutes, of the mixer pump in Tank SY-101 to prevent nozzle plugging without mixing very much of the waste.

Burp: A sudden release of flammable gas that has accumulated in the sediment layer in Tank SY-101 and several other Hanford waste storage tanks. Also called a gas release event or GRE. In SY-101, burps consisting of a few thousand to 10,000 cubic feet of flammable gas occurred very 100 to 150 days between 1980 and 1993.

CIUALG (pronounced by loosely following the Irish Gaelic "K^{y}OO-l^{i}g"): Committee for Insight, Understanding and Alleviation of Level Growth in SY-101 that helped investigate and suggest solutions for the surface level rise issue from 1998-1999.

Crust: A floating layer of solid particles with trapped gas bubbles from 1 to 3 feet thick in very concentrated radioactive waste tanks. SY-101's crust eventually grew to about 10 feet thick before remediation in December 1999.

DACS: Data acquisition and control system, which ran the mixer pump as well as collected data from various in-tank sensors. The surplus underground nuclear test control trailer that housed the data acquisition and control system from 1992 on was also called the DACS.

Decay heat: The energy generated in a material as the particles and rays emitted by radioactive decay deposit their energy on other atoms.

Defense Nuclear Facilities Safety Board: Oversight board established by Congress in 1989 to improve safety at U.S. Department of Energy nuclear production sites.

DOE: U.S. Department of Energy, created in 1977 from the Energy Research and Development Administration.

DOE Richland: U.S. Department of Energy Richland Operations Office. The local Energy Department's monitor of the Hanford Site.

Domespace: The volume of air next to the top, or dome, of the tank that is not occupied with waste. Also called headspace or vapor space.

Double-shell tanks: These tanks store radioactive waste in an inner steel tank housed within a secondary steel and reinforced concrete shell. A total of 28 double-shell tanks were built at Hanford between 1968 and 1986. Double-shell SY-101 was built around 1975. Each double-shell tank has a capacity of 1,160,000 gallons.

Electron: A small, negatively charged atomic particle. In an element, the cloud of electrons around the atomic nucleus balances the charge of the protons, one electron for each proton. The number and arrangement of electrons determine the chemical behavior of an atom.

Glossary | 439

Enraf: A waste surface level gauge produced by the Enraf-Nonius Corporation. Enrafs sensed a buoyancy or reduction in the weight of a bob when it touched the waste.

Fast Flux Test Facility (FFTF): A liquid sodium cooled nuclear test reactor operated in the 1980s to test fuel for liquid-metal fast breeder reactors. The reactor and associated facilities were shut down in the late 1980s, making many engineers and scientists available to work on Tank SY-101 mitigation.

FIC: Food Instrument Corporation, which made a waste surface level gauge that lowered a metal bob until it made electrical contact with the waste; it gave its name to the instrument. Most FICs were replaced with Enraf buoyancy gauges in the mid-1990s.

Fission: The reaction that happens when the already neutron-heavy nucleus of a fissile atom absorbs a neutron and splits apart, releasing one or more additional neutrons and a lot of heat.

Fission products: The fragment atoms left over after a nuclear fission. Fission products are usually radioactive and include such isotopes as iodine-131, cesium-137, and strontium-90.

Flammable gas watch list: A list of tanks required by the "Wyden amendment," Public Law 101-510, Section 3137, that had the potential to release large volumes of flammable hydrogen gas that could burn, fail the tank, and release radioactivity.

FTIR: Fourier transform infrared gas monitoring system, which was installed on Tank SY-101 in early 1993 to measure the concentration of ammonia and nitrous oxide in the tank dome.

Gas release event: See Burp.

GC: Gas chromatograph, a sophisticated gas concentration measuring instrument, several of which were installed on Tank SY-101 to monitor hydrogen and methane in the tank domespace.

High-efficiency particulate air (HEPA) filters: These filters were usually installed on both the ventilation inlet and exhaust ducting of radioactive waste tanks to prevent airborne radioactive contamination from escaping.

Isotope: An element with a specific number of neutrons plus protons in its nucleus. While an atom of an element has a fixed number of protons, it can have different numbers of neutrons, each identifying a different isotope of that element. For example, two isotopes of uranium are uranium-235 and uranium-238. The number is the sum of the number of protons and the number of neutrons in the isotope's nucleus.

440 | **Glossary**

Joint Test Group (JTG): Established to organize and monitor work during the 20- and 30-day windows where in-tank work was allowed after large burps. The group was established in early spring of 1991, after Window B.

LTOP: Long-term operating plan, which prescribed how the mixer pump was to be run for long-term hydrogen mitigation. The type of pump run depended on the waste surface level, assuming that running the pump more would lower the level and running the pump less would allow the level to rise.

Manual tape: A measuring tape with a "bob," or weight, that was lowered manually until an ohm-meter indicated waste contact to the operator. It could also be run in the "slack tape" mode, where the bob was simply lowered until the tape went slack.

MASF: Maintenance and Storage Facility (pronounced "massif"), a massive, climate-controlled warehouse that is part of the Fast Flux Test Facility complex. The SY-101 mixer pump was tested and stored at MASF prior to installation in July 1993.

Mechanical Mitigation Arm: A 6-foot length of pipe hinged to a mast that was rotated in the horizontal plane by workers walking on a circular catwalk pushing on handles while a crane lifted the arm up against the base of the crust layer. Promoted by engineer JR Biggs, it was also called "Biggs' Lift." It was run twice in SY-101 in May 1999 in an attempt to slow crust growth until the dilution and transfer system could be installed.

MIT: Multifunction instrument tree, which mainly measured the waste temperature with 22 thermocouples welded to the inside of a 3-inch pipe. It also had ports for gas monitoring and strain gauges to measure bending forces.

Mitigation: Preventing or reducing the size of large flammable gas releases in SY-101 by periodic actions, as opposed to remediation which would permanently prevent burps with no further action. The Hydrogen Mitigation Project installed a mixer pump in the tank in 1993 that prevented burps with three 25-minute runs per week until April 2000.

MMA: See Mechanical Mitigation Arm.

MPOP: Mixer pump off period, from April through July 2000 while SY-101 was being evaluated for potential gas retention without mixer pump operation.

NAI: Numerical Applications, Inc. A small local company started by former Pacific Northwest National Laboratory staff who developed the FATHOMS code used to model SY-101 burps.

Neutron: An un-charged or neutral particle, which, with protons, makes up the nucleus of an atom. A neutron is slightly more massive than a proton. Neutrons are freed when atoms fission or split and can be absorbed by other atoms, causing further fission, or can transmute the atom into a new element, such as converting uranium-238 to plutonium-239.

Office of River Protection (ORP): Established by the U.S. Department of Energy in 1998 to oversee tank waste cleanup at Hanford.

Parts per million (ppm): A unit used to express dilute gas concentrations (1% = 10,000 ppm).

PHMC: Project Hanford Management Contract. A team of contractors brought on to manage Hanford as part of the Clinton Administration's idea to privatize work done at the Site.

PIAB: Potential inadequacy in the safety authorization basis. In a PIAB, operation controls might not cover all important aspects of a hazard. A PIAB may be a precursor to an unreviewed safety question (USQ).

PNNL: Pacific Northwest National Laboratory operated for DOE by Battelle of Columbus, Ohio. Established in 1965 as Pacific Northwest Laboratory (PNL), it became a DOE multi-program research laboratory in 1986. Renamed Pacific Northwest *National* Laboratory (PNNL) by DOE in 1995.

PPP: Prefabricated Pump Pit. A new, clean pump "pit" system designed to sit above ground attached to a tank riser to avoid having to install equipment in the radioactive contamination typical of old existing pump pit.

Proton: A positively charged particle, which, with neutrons, makes up the nucleus of an atom. The number of protons is the atomic number that identifies an element. The number of neutrons plus protons is the atomic weight that identifies an isotope of the element.

PUREX: Plutonium-uranium extraction. A solvent extraction process used to separate plutonium from irradiated reactor fuel slugs in A Plant from 1956 to 1972 (and later restarted from 1983–1992). The A Plant was also called the PUREX Plant.

QA: Quality assurance.

Radioactive decay: The process by which atoms of unstable isotopes emit particles and rays to become stable.

Radioactivity: The stream of particles and rays emitted by unstable isotopes as they decay into stable ones. Typical emissions include alpha particles, beta particles, neutrons, and gamma rays.

442 | **Glossary**

REDOX: Reduction-oxidation. A solvent extraction process, used to separate plutonium from irradiated reactor fuel slugs in the S Plant from 1952 to 1967. The S Plant was also called the REDOX Plant.

Remediation: Permanently preventing large flammable gas releases from SY-101 in such a way that no future action is needed, as opposed to mitigation that requires continuous or periodic actions. The Surface Level Rise Remediation Project diluted the waste in Tank SY-101 in 2000 such that large gas releases or burps ceased.

Retained gas sampler (RGS): The RGS collected a sample of waste along with its retained gas in a tightly sealed chamber that was extracted and analyzed in a hot cell. The RGS gave the gas concentration as well as the volume fraction in the sample.

Revolutions per minute (rpm): The measure of the rotational speed of a shaft or motor. The second hand on a clock turns at one revolution per minute.

SA: Safety analysis, or safety assessment, which documented all the calculations, assumptions, conclusions, and controls that allowed the mixer pump to be installed in, operated in, and removed from Tank SY-101. The SA was written and maintained by Los Alamos National Laboratory.

Safety and Environment Advisory Council (SEAC): A Westinghouse internal senior advisory group whose approval was required for major activities in the tank farms.

Sediment: Ten- to fifteen-foot layer of settled solids on the bottom of the tank with roughly the consistency of mud or clay. The sediment had sufficient strength to trap gas bubbles generated by chemical reactions and radioactive decay.

SHMS: Standard hydrogen monitoring system, developed to be installed on tanks that were on the flammable gas watch list. The early ones monitored only hydrogen, but later versions also monitored ammonia, nitrous oxide, and other gases.

Single-shell tanks: These Hanford tanks store radioactive waste in a single steel tank within a reinforced concrete shell. At total of 149 single-shell tanks were built from 1944 through 1964. These tanks have capacities from 530,000 to 1,000,000 gallons each. Leaks in these older tanks drove construction of the newer double-shell tanks.

SLRRP (pronounced "slurp"): SY-101 Surface Level Rise Remediation Project.

Slurry growth: The phenomenon caused by gas accumulation in the sediment that was first noticed in the late 1970s as a rise in waste level in tanks containing concentrated waste.

SubTAP: A subset of the full Tank Advisory Panel (TAP) with specific areas of cognizance. The Chemical Reactions SubTAP was set up in 1993 to review the major tank safety issues, including flammable gas, and was the primary oversight group involved with SY-101 mitigation. Another was the Worker Health and Safety SubTAP.

Supernatant liquid: A layer of liquid above the sediment layer in a radioactive waste tank. Also called supernate.

Tank farm: A group of 2 to 18 underground waste tanks, including interconnecting underground pipes and equipment, built together to serve a specific plant or process. A total of 18 tank farms were built in the 200 Areas at Hanford from 1944 to 1980.

Tank Riser Characterization Unit (TRCU): An automated robotic inspection device originally designed to measure the big risers in SY-101 for the test chamber but was used to measure the riser that was to take the mixer pump. The data from the TRCU showed the pump was slightly too big and had to be trimmed to fit the riser.

TAP: Tank Advisory Panel, set up by DOE Headquarters in 1990 to review and make recommendations on mitigation of SY-101 flammable gas releases and the other major safety issues.

TCR: Tank Characterization Report, issued to document results of tank waste core sample analysis on a tank.

Thermocouple (TC): A temperature-measuring device consisting of two wires of different material welded together. A thermocouple uses the principle that a junction of dissimilar metals produces an electric voltage proportional to its temperature.

TRC: Technical Review Committee, the new name of the CIUALG from early 1999.

TRG: Test Review Group, established in 1992 to oversee operation of the mixer pump in SY-101. The TRG met monthly from 1993 through the fall of 2000.

Tri-Party Agreement: *Hanford Federal Facility Agreement and Consent Order* whereby the U.S. Department of Energy, Environmental Protection Agency, and the Washington State Department of Ecology, agreed to specific objectives and schedules for Hanford cleanup.

TWRS: Tank Waste Remediation System, an organization established in late 1991 that was responsible for safely managing the waste in the tanks and preparing it for disposal.

444 | Glossary

USQ: Unreviewed safety question, which was declared when a situation was discovered that was not analyzed in the current safety analysis or had worse consequences than those analyzed. A USQ was closed by further analysis, modification of the safety analysis documentation, and, depending on the question, additional safety controls.

VDTT: Velocity-density-temperature tree, which was designed to measure the velocity, density, and temperature of waste flows induced by burps or by the mixer pump. It also measured pressure at several elevations. Two VDTTs were installed in Tank SY-101 in late 1992, but both had failed by inleakage of waste by early 1993.

Void fraction instrument (VFI): Designed to measure the local gas volume fraction in a sediment layer by pressurizing a sample collected in a small test chamber. The instrument was deployed in Tank SY-101 in 1994 and in five other burping tanks the next year.

Watch list: A list of underground radioactive waste storage tanks at the Hanford Site required by the "Wyden amendment," Public Law 101-510, Section 3137, that had a "serious potential to release radioactive waste." Lists were developed for each major safety issue (flammable gas, organic nitrate, ferrocyanide, and high heat). SY-101 was on the flammable gas and organic nitrate watch lists.

Water lancing: Water was injected into the waste through a 2-inch pipe or lance to break up the sediment and release trapped gas to prevent burps. Water lancing was performed regularly in SY-101 until about 1987. It was abandoned because added water was raising the waste level.

Window: A window was a 20- to 30-day period immediately following a large burp in Tank SY-101 during which it was considered safe to perform work in the tank.

Window criteria: A document specifying the conditions required to open a window during which work could be performed in a tank after a large burp. The window criteria included a minimum waste level drop, temperature change, and hydrogen concentration during a burp.

Bibliography

Though much of the information presented in this book is not publicly available, there are many books and reports available that describe Hanford and the people that worked there, as well as the SY-101 story. Some of the best at describing Hanford, the tank, and the various reactions to it from my experience are described here.

HANFORD IMPACT, CONTAMINATION AND DOWNWINDERS

- *Atomic Farmgirl*, Teri Hein, Fulcrum Publishing, Golden, Colorado, 2000. Describes how Hein grew up in the wheat farming community of Fairfield, south of Spokane. It's a delightful, non-sensational homespun story of the Hanford "downwinders" experience.
- *Atomic Harvest*, Michael D'Antonio, Crown Publishers, New York, 1993. D'Antonio writes more of a sensational story that purports to describe "Hanford and the lethal toll of America's nuclear arsenal." The main value of this book is to identify the personalities of the Hanford stakeholders and downwinders, and describe their attitudes about the health problems ascribed to Hanford emissions.
- *On the Home Front: The Cold War Legacy of the Hanford Nuclear Site,* Michelle Gerber, University of Nebraska Press, Lincoln, Nebraska, 1997. This is an exhaustive historical study of Hanford's impact on the local communities. It also gives details on what kind of contamination was released from Hanford, where it went, how the

445

446 | **Bibliography**

Manhattan Project and the Atomic Energy Commission set limits on releases, and the health studies of the mid-1990s.

HANFORD HISTORY, PLANT OPERATIONS, AND WASTE CLEANUP

- *Linking Legacies: Connecting the Cold War Nuclear Weapons Production Processes to Their Environmental Consequences,* U.S. Department of Energy report to Congress, DOE/EM-0319, January 1997. This report gives the history of the entire Energy Department's weapons complex and operations at each.

- *The Making of the Atomic Bomb,* Richard Rhodes, Simon & Schuster, 1986. Describes the workings of politics, personalities, and technology that evolved into the Manhattan Project. Describes how the entire U.S. nuclear weapons complex developed and how the Hanford Works related to it during Manhattan Project days.

- *Hanford Site Historic District: History of the Plutonium Production Facilities, 1943-1990,* by a large group of Hanford authors, Battelle Press, June 2002. A huge collection of details and historical information from pre-history to near present. This is also available on the internet at http://www.hanford.gov/docs/rl-97-1047/index. htm.

- *Hanford Tank Cleanup: A Guide to Understanding the Technical Issues,* by Roy Gephart and Regina Lundgren, Battelle Press, September 1998. This book gives an excellent and non-technical introduction to the basic issues and has several concise appendixes on chemistry, physics, plutonium processing, and waste tanks.

- *Hanford: A Conversation About Nuclear Waste and Cleanup,* by Roy Gephart, Battelle Press, March 2003. Gephart's book deals with both the technical issues and adds the political and emotional dimension with some ideas on how to accommodate both in meaningful cleanup.

- *Hanford and the Bomb: An Oral History of World War II,* SL Sanger, Living History Press, Seattle. As the title indicates, this book presents a series of interviews with people who worked at Hanford, mainly from the Manhattan Project and early post-war era. Some of the physicists and chemists who developed the bomb and separations processes tell their stories, as do Colonel Matthias and General Groves.

- *Historical Tank Content Estimate for the Southwest Quadrant of the Hanford 200 West Area,* by Chris Brevick and JL Stroup, HNF-SD-

WM-ER-352, February 1997. This report presents the details of processes creating the waste that actually entered SY-101. Similar reports by Brevick are available for all the single-shell waste tanks. A good knowledge of chemistry is required to understand this report.

EARLY HANFORD SAFETY ISSUES

- *Report on the Handling of Safety Information Concerning Flammable Gases and Ferrocyanide at the Hanford Waste Tanks,* Steve Blush and others, DOE/NS-0001P, July 1990. Written by a team led by Steve Blush, this report was nicknamed the Blush Report. Though it does a lot of finger pointing, it is an excellent resource to the history of these two safety issues.

HYDROGEN MITIGATION PROJECT

- *A Safety Assessment for Proposed Pump Mixing Operations to Mitigate Episodic Gas Releases in Tank 241-SY-101, Hanford Site, Richland, Washington,* LA-UR-92-3196, Revision 14, March 31, 1995. This is THE Los Alamos safety assessment that changed the way Hanford works as much as any other factor surrounding SY-101. This two-volume document clearly depicts the depth and breadth of thought about what could possibly go wrong during mixer pump insertion, operation, or removal; how bad it might be; and what efforts could justifiably be applied to prevent them.
- *Evaluation of the Generation and Release of Flammable Gases in Tank 241-SY-101,* WHC-EP-0517, November 1991, by Harry Babad, Jerry Johnson, Jeanne Lechelt, and Dan Reynolds. This is the first comprehensive summary document describing SY-101 burp behavior. The theory is generally correct, and there is a good description of tank history and the big burps of 1990.
- *Mitigation/Remediation Concepts for Hanford Site Flammable Gas Generating Waste Tanks,* WHC-EP-0516, April 1992, by Harry Babad, John Deichman, Ben Johnson, Doug Lemon, and Denis Strachan. Actually completed in October 1991, this report details the brainstorming and evaluation process from which the mixer pump was chosen along with the concepts that were to go into the ill-fated test chamber.
- *Tank 101-SY Flammable Gas Mitigation Test Project Plan,* WHC-EP-550, May 1992, by Jack Lentsch. This is the bible for the mitigation project. It described the basic schedule and milestones of the project without

448 | Bibliography

modification through and beyond the installation of the mixer pump.

THE PROCESS AND EVALUATION OF MITIGATION

- *Mitigation of Tank 241-SY-101 by Pump Mixing: Results of Testing Phases A and B.* PNL-9423, 1994, by a host of Pacific Northwest Laboratory and Westinghouse authors. Describes the high-speed mixing tests that actually accomplished mitigation of SY-101's burps.
- *Mitigation of Tank 241-SY-101 by Pump Mixing: Results of Full-Scale Testing.* PNL-9959, 1994, by a team of Pacific Northwest Laboratory and Westinghouse authors. Provides the details of the full-scale testing program that began the process of long-term mitigation.
- *The Behavior, Quantity, and Location of Undissolved Gas in Tank 241-SY-101.* PNL-10681, 1995, by Mary Brewster, Neil Gallagher, John Hudson, and Chuck Stewart. This report is the final word on SY-101 mitigation. The actual retained gas volume in the waste is calculated with fair accuracy for the first time, based on measurements by the void fraction instrument and the ball rheometer. This report essentially proves the tank has been mitigated.

FLAMMABLE GAS SAFETY ISSUE

- *Gas Retention and Release Behavior in Hanford Single-Shell Waste Tanks.* PNNL-11391, 1996, by a team of Pacific Northwest National Laboratory authors. Describes and evaluates the potential gas release mechanisms in single-shell tanks. This is the first report to state that huge SY-101-style gas releases were not possible in the single-shell tanks.
- *Gas Retention and Release Behavior in Hanford Double Shell Waste Tanks.* PNNL-11536, 1997, by a team of Pacific Northwest National Laboratory authors. A follow-on to the single-shell tank study, this report describes the important bubble migration theory that is currently used in the tank farms safety basis to predict whether a tank will burp or not.
- *Data and Observations on Double-Shell Flammable Gas Watch List Tank Behavior.* RPP-6655, 2000, by a team of CH2M HILL and Pacific Northwest National Laboratory authors. An exhaustive discussion of tank data, observations, and measurements in the double-shell tanks that were on the flammable gas watch list. The report provides the background data for removing the tanks from the watch list and resolving the flammable gas safety issue.

Bibliography | 449

- *Data and Observations of Single-Shell Flammable Gas Watch List Tank Behavior.* RPP-7249, 2001, by CH2M HILL and Pacific Northwest National Laboratory authors. This is the sequel to RPP-6655, for single-shell tanks.
- *Flammable Gas Safety Issue Resolution.* RPP-7771, 2001, by Jerry Johnson, Dave Hedengren, Mike Grigsby, Chuck Stewart, Jim Zach, and Leon Stock. This report gives the basis for closing the flammable gas safety issue originally established in 1991.

SURFACE LEVEL RISE ISSUE

- *Tank 241-SY-101 Level Confirmation Report,* HNF-2772, 1998, by Cheryl Benar. Summarizes all the evidence that the waste surface level in SY-101 was, indeed, rising. Contains the level confirmation studies of Bauer and Antoniak and the attempt at hydrostatic pressure measurement by Jim Bates.
- *In Situ Void Fraction and Gas Volume in Hanford Tank 241-SY-101 as Measured with the Void Fraction Instrument,* PNNL-12033, by Chuck Stewart, Jim Alzheimer, Guang Chen, and Perry Meyer, 1998. This report details the three deployments of the void fraction instrument in SY-101 in June, July, and September 1998. These data allowed us to quantify the gas volume in the crust fairly accurately for the first time.
- *Tank 241-SY-101 Surface Level Rise Remediation Options and Evaluation,* HNF-3705, 1998, by Rick Raymond. Presents the recommendations of the Despair at the Hanford Square value engineering workshop.
- *Tank 241-SY-101 Surface Level Rise Remediation Project Plan,* HNF-3834, 1999, by Rick Raymond. This is the actual plan for the SY-101 Surface Level Rise Remediation Project resulting from the Despair at Hanford Square workshop.
- *A Discussion of SY-101 Crust Gas Retention and Release Mechanisms.* PNNL-12092, by Scott Rassat, Phil Gauglitz, Stacey Caley, Lenna Mahoney, and Donny Mendoza, 1999. Scott and his team summarize all the gas retention testing experience that could have revealed why SY-101's crust was growing. This was the starting point for our understanding of the crust growth phenomenon.

SURFACE LEVEL RISE REMEDIATION

- *Process Control Plan for Tank 241-SY-101 Surface Level Rise Remediation,* HNF-4264, by Scott Estey, 2000. Describes the actual process of waste

450 | Bibliography

transfer and back dilution that was followed to remediate gas retention in SY-101.

- *Results of Waste Transfer and Back-Dilution in Tanks 241-SY-101 and 241-SY-102*. PNNL-13267, by Lenna Mahoney and a team of CH2M HILL and Pacific Northwest National Laboratory co-authors. This is THE report giving all the detailed data and observations during the three campaigns of transfer and dilution from December 1999 through April 2000.
- *Evaluation of Hanford High-Level Waste Tank 241-SY-101*. RPP-6517, by Jerry Johnson with a team of CH2M HILL and Pacific Northwest National Laboratory co-authors, 2000. Presents the results of the evaluation period after SY-101 remediation, concludes that gas is no longer being retained, and recommends that SY-101 be removed from the watch list.

REACTION TO A PROBLEM

- *Isaac's Storm*, Erik Larson, Crown Publishers, New York, 1999. Tells the story of the 1900 Galveston hurricane following the career of Isaac Cline, U.S. Weather Bureau station chief, and how he tried but failed to forecast the path and severity of the storm. This story is an interesting parallel to the approach and impact of the SY-101 safety issue.

WEBSITES

- http://www.hanford.gov/doe/culres/historic/info.html. The primary resource for Hanford history documents.
- http://picturethis.pnl.gov. A large collection of high-quality photos of Pacific Northwest Laboratory and Pacific Northwest National Laboratory research as well as historic Hanford photos and Hanford scenery.
- http://www2.hanford.gov/DDRS/index.cfm. Hanford declassified document retrieval system including access to 77,000 historic photos.

Index

Pages containing figures are denoted by *f*, pages containing maps are denoted by *m*, pages containing photographs are denoted by *p* and pages containing tables are denoted by *t*.

242-A, B, S and T evaporators, 23, 31–32, 102
300 Area, 5*m*, 10, 186
1100 Area, 5*m*
3000 Area, 5*m*
100 Areas, 5*m*, 10, 10*p*, 11*t*, 132
200 Areas
    B, T and U Plants, 13*p*
    Cold Test Facility, 176
    double-shell tanks, 20, 21*p*, 24*t*
    evaporators, 23
    map, 5*m*
    plutonium separation plants, 11–14, 17
    single-shell tanks, 19, 21*p*, 24*t*
222-S laboratory, 16*p*, 94, 127, 129*p*
271-SY instrument building, 102, 102*p*, 103

A
A-105 tank, 23
A Plant (PUREX), 16*p*
    double-shell tank farms, 20
    functions and processes, 17*t*
    plutonium separation production, 15*t*
    proposed new tanks, 118
    solvent extraction process, 14
Abrams, Charles, 69

accident prevention, 168, 182–184
accomplishments
    hose-in-hose system, 375
    mitigation and remediation, 291, 299, 416, 417–419
    mixer pump, 246, 256–258
    Surface Level Rise Remediation Project, 369, 422
    waste transfer and dilution readiness review, 380–381
    window A work, 105
    window C work, 125
    window G work, 214–215
accounting plane, 194
acronyms, 370
Advisory Committee on Nuclear Facility Safety, 68, 71–73, 79, 144
Agnew, Steve
    chemistry model, 404
    crust growth, 386
    ultrasonic mitigation, 144, 151, 154, 294–295
    wastes chemical composition, 383
Ahearne, John, 71
Ahearne Committee, 68, 71–73, 79, 144
air emissions of radioactive contamination, 4, 12, 22–23

| 451

452 | Index

air lances
broken during window G, 208–209
contamination after window G, 218–219
crust sampling, 99
definition, 437
use in tank, 202
Allemann, Rudy, 90*p*
accomplishments, 280–281
bubble slurry flow experiment, 153
gob theory, 89–91, 186
mini-burps, 271
mitigation project, 147–148
mixer pump installation, 249
SY-101 data reports, 415–416
tank data management, 242
TEMPEST computer program use, 212, 226
Test Review Group (TRG), 196
wasteberg simulation, 351–352
Allen, Dale, 321, 328, 360, 417, 418*p*
aluminum, 12, 26, 168, 172, 383
Alumkal, Bill, 257*p*, 259*p*
C-106 tank contamination incident, 268
employee retraining, 269
Hazel O'Leary visit, 259*p*
mixer pump installation, 249
recognition dinner, 257, 257*p*
unplanned pump start, 377
Westinghouse reorganization, 238
Alvarez, Bob, 385
Alzheimer, Jim, 298, 322
ammonia
concentrations during releases, 246, 271, 405, 412
concern raised by Event I, 245
explosion hazards, 55–56, 238
Fourier transform infrared spectrometer, 185, 186, 245
mixer pump installation, 252
release rates, 308
safety concerns, 245–246, 273–276
solvent extraction process, 26
vapors, 375–376
waste transfer and dilution, 397, 400
zone of death, 247
ammonium nitrate and fuel oil explosive (ANFO), 56
ammonium nitrate explosion potential, 56, 238
AN-104 and 105 tanks, 42
AN-103 tank, 42, 338
AN-107 tank, 412
Anderson, Ken, 418*p*
Anderson, Tom
management style, 165
mixer pump test plan review, 232

Anderson, Tom *(cont.)*
Pasco mitigation workshop briefing, 155–156
Westinghouse reorganization, 107–108, 148
window work, 215, 223, 248
annulus, double-shell tank, 221, 286
anti-siphoning slurry distributor (ASSD), 226, 230, 370, 375, 437
Antoniak, Zen, 249, 317–318, 321
Anttonen, John, 214
AP-107 tank, 422
AQ tank farm (proposed), 118
Arcaro, Ralph, 354
Archimedes, 345, 365, 366, 381
Argonne National Laboratory, 85, 145
Army Corps of Engineers, 3, 6, 9
Atlantic Richfield Hanford Company, 29, 57
atomic bombs, 5–6, 9
Atomic Energy Act, 92
Atomic Energy Commission
cesium and strontium capsules, 16
description, 437
double-shell tank farms, 20
Hanford management, 30
nuclear reactors, 10, 14
radioactive wastes, 4
authorization basis, 194, 198, 313, 337, 388, 410
authorization letter, Surface Level Rise Remediation Project, 354–355, 378, 382
AW-102 tank, 420
AW tank farm, 21*p*
awards
Accountability in Government, 74
Most Important Person, 330, 331*p*
project of the year, 340, 418–419, 424

B
B Plant, 11–12, 13, 17*t*, 22, 26
Babad, Harry, 226
Babcock & Wilcox Hanford, 305
badges, radiation, 190
ball rheometer, 297, 298–299, 298*p*, 337
Bamberger, Judith, 376–377
Barnes, Dave, 348, 349
barometric pressure effect, 185, 187–189, 320–321, 437
Barrett, Haentjens and Company, 157, 158–159, 167, 180
See also Haentjens, Walter
Barton, Blaine, 326–327, 392, 418*p*
Barton Company, 177
Basalt Waste Isolation Project, 52
Bates, Jim, 314, 319, 326–327, 360
bathtub experiment for gas retention and release, 33–34

Battelle, 29–30
  See also Pacific Northwest National
    Laboratory
Bauer, Roger
  Committee for Insight, Understanding,
    and Alleviation of Level Growth
    (CIUALG), 326–327
  mixer pump test plan, 242
  Standard Hydrogen Monitoring
    Systems (SHMS), 185
  Surface Level Rise Remediation
    Project, 329
  surface level rise remediation task
    force, 314, 316–318, 321
  waste transfer and dilution, 392
Bear, Guy, 249
Bellingham Herald, 268
Benegas, Tony, 161*p*, 170*p*
  critical lift procedures, 182
  drop leg, 375
  employee morale, 165
  flexible hose tests, 374
  mitigation project, 147, 157
  mixer pump, 160, 166–167, 174, 175,
    231–232
  nozzle clogs, 264–265
  Surface Level Rise Remediation
    Project, 329
  transfer pump selection, 370
Benton County Sheriff's Dept., 248
benzene flammability, 60
Biggs, J.R., 331*p*, 357*p*
  Biggs Lift, 334–335, 346, 356, 357*p*
  Committee for Insight, Understanding,
    and Alleviation of Level Growth
    (CIUALG), 326–327
  densimeter, 376
  Hanford Square value engineering
    session, 332
  mixer pump, 249, 250, 253, 264
  Most Important Person Award, 330,
    331*p*
  nozzle clogs, 264–265
  readiness review, 380
  siphon break, 375
  Surface Level Rise Remediation
    Project, 329, 339, 357, 357*p*
bismuth phosphate process
  B and T Plants, 13, 22
  definition, 437
  plutonium separation, 15*t*
  tank wastes, 26
  U Plant, 14
blow-off riser cover (Giant Nickel), 130,
    183, 193, 219–220, 225–226
Blush, Steve
  back-up ventilation, 129
  onsite review, 73–75

Blush, Steve *(cont.)*
  safety documentation criticism, 150
  ventilation recommendations, 120
  waste sampling, 81, 82–84, 104
Blush Report, 73–74, 106, 150, 437
boiling radioactive wastes, 22
bombs, atomic, 5–6, 9
Boston, Harry, 415, 416
Brager, Howard, 89
brake shoes, mixer pump, 173, 179*p*, 230,
    232, 235, 256
Brennan, Chris, 363
brine, liquid waste, 365–366, 420, 421
Brookhaven National Laboratory, 156
Brothers, Joe, 242–243, 329, 339, 392, 418*p*
Bryan, Sam, 186, 238
bubble migration theory, 337, 384
bubble slurry, 323*p*, 401*p*
  crust base drop, 363
  dilution, 367
  effect on mixer pump operation, 364
  existence of, 361, 362
  flow hazard, 398, 401–402
  gas content, 347–348
  gas release mechanisms, 351–354
  safety hazard, 349
bubbles
  ammonia concentrations, 273
  barometric pressure effect, 187
  explosion hazards, 34
  flammable gas hazards, 45, 65
  ignition hazards, 50, 65, 71–72
  migration, 337, 381, 384
  in sediment, 89–90, 337–338
  semi-liquid slurry, 323*p*
  slurry growth, 33, 48
  waste samples, 128
  window criteria, 80–81
  See also gas release events
bump, mixer pump
  definition, 438
  hydrogen suppression, 285
  mini-burps during, 270–271
  mixer pump operation, 282
  mixer pump shut down, 413–414
  pump performance concerns, 244
  safety, 195, 292
  during stand down of 1993, 269–270
  tests, 235, 263
  unplanned pump start, 266–267
  waste mixing, 265
buoyancy, 50, 89–90, 345, 364–366
buoyancy ratio
  calculations for dilution, 381, 384, 404,
    421
  model, 337, 383
Burger, Lee, 57
Burger report, 57, 58

454 | Index

Burke, Tom, 147, 196
burn hazards. *See* flammable gas
  hazards; ignition hazards
burp, definition, 438
burps. *See* gas release events
burrowing ring, mixer pump, 255
Bush, George H. W., 64
Bush, Spence, 222
BY-105 tank, 26

C

C-106 tank, 23, 266, 288, 380
C Plant, 11–12
calibration of instruments, 129, 277, 308,
  319
cameras, 44, 48, 64, 65*p*
  See also video cameras
Campbell, Dave, 138, 200
Cannon, Scott, 348
canyon buildings, 12
carbon dioxide release from Lake Nyos,
  Cameroon, 275
Carlson, Fred, 69
Carothers, Kelly, 46
caustic wastes, 26
Central Waste Complex, 218
cesium and cesium-137
  gamma radiation, 348–349
  reactor byproducts, 9, 14–17
  remediation methods, 147
  removal from tank waste, 145
  steam bumps, 22–23
  SY-101 tank, 32–33
Chang, Shih-Chih, 134–136, 157
Change Control Board, 203
chemical mitigation, 145–146
chemical reactions
  crust formation, 49
  gas generation, 85
  hazard definition, 79
  hydrogen-nitrous oxide mixtures, 41,
   299–300
  mixer pump test plan, 283
  runaway, 57
  safety recommendations, 93
  tank wastes, 40
  waste transfer and dilution, 397
chemical wastes, 26
Chen, Guang, 320–321
CH2M Hill Hanford Group, Inc.
  Hanford contractors, 29
  leadership, 417
  performance incentives, 409–410, 420
  Surface Level Rise Remediation
   Project, 389, 415
Christensen, Roger, 211, 217, 229
Christmas Chrust message, 346*f*, 356

Clark, William, 3
Clinton Engineering Works, 6
Clover Island Inn, Kennewick,
  Washington, 155
coal-fired power plant, 13*p*
cobalt-60 slugs, 26
Cogema Engineering, 305, 358
Cold Test Facility, 176, 178*p*
cold tests
  Cold Test Facility (mixer pump), 176,
   178*p*, 215
  MASF (mixer pump), 173–174,
   177–180, 215, 217
  transfer and dilution, 378–379
Cold War, 4, 13–14, 57
colloids, 385–386
Colson, Jim, 147–148
Columbia River, 7, 14, 94
Columbia River Basin, 3
command center for remediation project,
  328–329, 339, 342
Committee for Insight, Understanding,
  and Alleviation of Level Growth
  (CIUALG), 326–328, 347, 438
complexed wastes, 17, 26, 31–32, 41–42,
  145
computer models. *See* models
conduct of operations, 113, 379–380
Conner, John, 392, 408, 412, 415–416, 418*p*
contamination, radioactive
  airborne, 12
  C-106 tank, 266, 267–268
  data acquisition and control system
   (DACS) trailer, 192
  prevention, 206
  pump pits, 371
  in soil, 4, 218–219
  from waste transfer pipes, 374
contractors, 29–30, 305–306
convection currents, 87, 275
Cook, Bob, 58, 78, 84, 238
core sampling. *See* waste sampling
Corn, Morton, 387
corn meal mush experiment, 135–136
Corps of Discovery, 3
corrective actions, 267, 268–270, 380–381
corrosion, 25–26, 221–222
cost cutting measures, 203–204
cranes
  critical lift procedures, 182
  mechanical mitigation arm, 358–359
  mixer pump installation, 250, 253–255,
   253*p*, 255
  operators, 166, 176
  pump drop hazard, 182
  safety, 198, 238
  SY- 101 tank farm, 102*p*

cranes *(cont.)*
    transfer pump installation, 371
    video monitoring of, 249
    waste sampling, 297–298
    water lancing, 362
cribs, 22
crisis management, 175–176, 228–229,
        312–313
critical lift procedures, 182
criticism
    Blush report, 67, 73–75
    Los Alamos safety analysis and
        TEMPEST code, 228–229
    of remediation plan by DOE
        Headquarters, 385–389
    sampling plan, 83–84
    summary of 1990, 107–108
    Westinghouse safety documents,
        150–151
    window G work, 215–218
cross-site waste transfers, 420, 421, 422
cruciform lance. *See* water lance
crush plate on mixer pump, 172, 173
crust
    base drop, 356, 363
    buoyancy, 345, 382
    definition, 438
    gas buildup, 304, 308–309, 326, 345
    gas fraction measurement, 323,
        325–326
    gas releases through, 48–50, 90,
        130–131, 138, 273, 350–354
    lancing, 45, 298–299, 322, 361–362
    layers, 32, 347, 349
    neutron logs, 349–350, 350*f*
    sampling, 84, 99–111, 137, 345
    solubility, 336, 344–345, 348, 367, 381,
        395–396
    tank-adhering ring, 395, 395*p*, 397, 401
    temperature measurements, 87–88,
        88*f*
    temporary mitigation, 79–81, 358–361
    thickness, 342, 347, 363, 412
    waste transfer and dilution, 366,
        394–395, 398, 401, 403
crust growth
    acceleration, 38, 195, 309, 333
    gas retention, 302
    lessons learned, 424
    mitigation methods, 334–335, 400
    mixer pump off period (MPOP), 412
    prediction models, 341, 346
    waste temperature effect on, 344
culture
    Cold War, 57
    Fast Flux Test Facility (FFTF), 173–174
    Hanford, 157, 378–380, 424

culture *(cont.)*
    safety, 66, 72, 81, 113, 269, 392
    Westinghouse Hanford Company,
        148, 165–166

D
data acquisition and control system
        (DACS)
    control room, 191*p*
    definition, 438
    densimeter, 376–377
    installation, 158, 189–192
    mitigation project, 204
    mixer pump, 249
    relay calibration, 277
    safety measures, 220
    software, 243
    SY-101 in-service operation, 421
    tests, 177, 262, 266–267
    trailer contamination, 192
    video camera operation, 402
    waste transfer and dilution, 392, 395
Davis-Bacon Act, 167
death
    falling into waste tank scenarios,
        28–29
    Texas City explosion, 56
    zone of, 247, 249, 262, 273
decay heat, 22, 32, 275, 344, 438
declassified documents, 52–53
decontamination, 192, 252–253
Defense Nuclear Facilities Safety Board
    description, 58, 92–94, 438
    gas flammability hazard decision, 61
    mitigation authorization letter, 354
    recommendations, 96, 119, 151,
        296–297
    safety analysis and assessments,
        150–151, 197
    Surface Level Rise Remediation
        Project, 331
    waste level measurements, 303
    waste transfer and dilution, 388
    Westinghouse reports to, 75, 116
defense waste environmental impact
        statement, 92
DeFigh-Price, Cherri, 224
Deichman, John
    cost cutting measures, 203
    external publication policy, 108
    Hanford House public meeting, 83
    ignition hazards, 138
    leadership change, 205
    mitigation plan summary, 158
    submersible pump, 160
    Tank Advisory Panel (TAP) meeting,
        156–157

456 | Index

Deichman, John (*cont.*)
    Tank Safety Task Team, 78
    waste tank safety program support, 103
    Westinghouse reorganization, 106
Delegard, Cal, 41, 48
Delozier, Fran, 387
densimeter, 376–377
d'Entremont, Paul, 60, 64, 70, 74
Despair at Hanford Square, 332–336
dilution of waste
    flammable gas remediation, 72, 143, 145, 147, 293, 295
    studies, 366–368
    surface level rise remediation, 335, 338, 365
dipsticks, 187, 240, 317–318, 317*p*, 394
discharge legs, mixer pump, 171
DOE. *See* U.S. Dept. of Energy
dome or domespace of waste tank
    blow-off riser cover, 183
    collapse, 61, 193
    computer model, 193–194
    convection in, 41
    definition, 438
    equipment installation, 121
    flammable gas hazards, 34, 46, 80, 350
    gas release events, 90, 98*t*
    hydrogen concentrations, 43–44, 365, 398, 411
    ignition hazards, 138, 182
    load, 220, 233, 382
    mixer pump installation pressure changes, 252
    pressure
        gas release events, 44, 58–59, 121, 225, 245
        instrument problems, 46
        measurement of, 51, 86–87
    risers, 100*f*, 216*f*, 258*f*
    ventilation, 47, 117, 120
double-shell tanks
    200 Areas, 24*t*, 78
    definition, 438
    diagram, 50*f*
    gas volume measurement, 188
    history of, 20–22
    shortage of, 118–119
    slurry, 31–32, 42, 47
    SY-101, 100*f*
    tank farms, 20, 24*t*
    void fraction instrument data, 326
    wastes, 27–29, 39*f*, 423
down winders, 12, 53
downcomer legs or pipes, 166–167, 171, 262, 264
drop leg, 375, 397

Duffy, Leo, 95*p*
    confidence in Steve Marchetti, 124–125
    crust growth, 386
    five-year program, 119
    Hanford review team, 64, 65
    Hanford visits, 95*p*, 218, 223
    Tank Advisory Panel, 67–68
    waste sampling, 69, 81, 84, 116
    weapons complex cleanup, 148
    window extension request, 212
    window work, 104, 121, 216
Duke Engineering Services Hanford, 305
Dunford, Gary, 43, 196
DuPont, 29, 58
dust devils, 255

E

earthquake protection, 220
EBASCO Services, 155, 157
Edwards, Jack
    ammonia concentrations, 246, 247
    mixer pump installation, 226, 249
    safety, 180, 182, 192, 197, 227
    Test Review Group (TRG), 196
    window event G, 208
Efferding, Larry, 244, 249, 276, 287
Eisenhawer, Steve, 89, 188
electricians, 261, 262
electron, definition, 438
Elmer Staats Award, 74
emergency response plan, 93
employees
    morale, 107, 122–123, 268
    overtime, 236
    recalibration, 166, 174, 190, 207
    retraining, 268, 269, 380
    safety, 207, 214, 225–226, 247
    team building, 330, 333
    working conditions, 206, 210, 211–212, 241–242, 250, 399
    See also teamwork
Enderlin, Carl, 413–414
Energy Daily, 256–257
Energy Research and Development Administration, 30, 52
Energy Technology Center, 165
Enraf-Nonius, 240
Enrafs
    crust measurements, 344–345
    definition, 439
    free liquid level system, 318
    instrument of record, 307, 325
    level history, 307*f*
    reliability, 306, 308
    surface level rise, 316–317, 323
    use of, 240

Enrafs *(cont.)*
  waste levels, 302–304
  waste surface currents, 408
  waste transfer and dilution, 403
environmental assessment, 130
environmental impact statements, 52, 57, 92
environmental risks, 12, 29, 55
Erhart, Mike, 243, 262
Estey, Scott, 384, 404, 418*p*
Estey criterion, 384, 404
ethylenediaminetetra-acetate (EDTA), 17
evaluation period, mixer pump, 410–413
evaporation from waste, 411, 412
evaporators, 21*p*, 22–24, 25, 39, 102
excavation runs, mixer pump, 292, 293, 301
experiments
  bubble slurry flow, 353
  corn meal mush, 135–136
  gas mixtures, 299–300
  mitigation methods, 294–295
  physics of buoyancy, 366
  slurry growth, 33–34
  See also simulations
explosion hazards
  ammonium nitrate fertilizer, 56
  chemical reactions, 41, 57
  emergency response plan, 93
  ferrocyanide, 14–15
  hydrogen-nitrous oxide mixture, 60
  mixer pump, 158
  public opinion, 84, 94–95
  risks, 34, 42
  solid wastes, 55–56
  static electricity, 123

F
F reactor, 10*p*
facial hair survey, 110
falling ball viscometer, 297, 298–299, 298*p*, 337
FAST diagram, 333, 334
Fast Flux Test Facility (FFTF), 52, 76, 173–174, 439
fast tracking, project, 339
Fat Man atomic bomb, 9
FATHOMS computer program, 133–134
Federal Aviation Administration, 248
ferrocyanide
  explosion hazards, 14, 55–57
  liquid waste cleanup, 22, 26
  multifunction instrument trees (MITs), 151–152
  safety concerns, 127
  safety recommendations, 93
  watch list, 96
Fiasco in Pasco, 152–156

FIC. *See* Food Instrument Corporation (FIC)
fire, Hanford Site, 132
fire hazards. *See* flammable gas hazards; ignition hazards
fire protection, 220
First, Mel, 69
Fisher, Al, 236
fission
  atomic bomb, 6, 9
  definition, 9–10, 439
  fragments, 9, 12
  products, 439
  reactors, 9, 10*t*, 11*t*
flammable gas hazards
  ammonia, 247
  crust, 364–365
  description, 34
  in domespace, 117
  gas release events, 103–104, 207, 221
  gas volume calculations, 189
  hydrogen gases, 73
  mitigation, 68, 79–81, 119, 143–148, 282, 334
  problem identification, 37–53, 326
  public opinion, 61, 83–84
  risks, 146
  safety, 77, 93, 127, 192–193, 198, 355
  in solid wastes, 55–56
  task force, 49
  unreviewed safety question (USQ), 288
  ventilation, 120–121
  white paper, 60
  See also Flammable Gas Program; flammable gas watch list
Flammable Gas Program, 76
flammable gas watch list
  change in focus, 299
  closing of, 419
  definition, 439
  development, 75, 96
  proposed additions to, 188
  saltwell pumping, 355
  SY-101 removal from, 196, 384, 416
  Wyden amendment establishes, 94–96
flexible hose, 372, 374
flexible receiver, 176, 178*p*
float, project, 150
flow meters, 176–177, 244, 249, 262–263
Flowers, Ken, 255, 256
Fluor Daniel Hanford, 305
Fluor Hanford, Inc., 29
foaming of waste
  flow, 362, 398, 400*p*, 401–402
  gas volume, 347–348
  during mixer pump operation, 309, 313

458 | **Index**

foaming of waste (*cont.*)
  simulation, 352
  See also bubble slurry
Food Instrument Corporation (FIC)
  crust samples, 99, 104–105
  definition, 439
  level gauge, 51, 130, 136, 202, 207, 277
  malfunctions, 124, 280
  mixer pump tests, 293
  replacement of, 307
  waste level measurements, 240,
    244–245, 271, 302–304
Fourier transform infrared gas
    monitoring system (FTIR)
  ammonia and nitrous oxide
    concentrations, 185–186, 245
  ammonia monitoring, 271, 273
  definition, 439
  failure, 415
free liquid level system, 318
freeboard, crust layer, 345–346, 382
Freedom of Information Act, 52–53
French, Dick, 268, 342, 354
Fry, Herb, 185
fuel. *See* nuclear fuel
Fulton, John, 78, 205, 212, 226, 246
function analysis system technique
    (FAST), 333, 334
funding, 4, 119, 127, 163, 203–205,
    235–236

G
Gable Butte and Gable Mountain, 11,
    15*m*
Galbraith, John, 230
Gale, Larry, 41–42
gamma logs, 403, 412
gamma probes, 348–349
gamma radiation, 109
Gardner, Booth, 62, 73, 95*p*
gas buildup
  barometric pressure effect, 320–321
  buoyancy, 270–271
  crust, 304, 308–309, 326, 345
  mitigation, 72, 412–413
  mixer pump off period (MPOP), 411,
    412–413, 414
  slurry growth, 33–34, 42, 48–50
  surface level rise, 315
  void fraction instrument, 322
gas chromatographs (GCs)
  data corruption, 239
  definition, 439
  gas concentration measurements, 185
  mixer pump off period, 411
  operation problems, 123, 224, 241

gas generation
  in bubble migration model, 337–338
  calculations, 308, 411
  experimental work, 131, 145
  influences, 383–384
  mitigation and remediation, 147, 299,
    409
  organic complexants, 42, 144
  radiochemistry of, 85
  waste levels, 302, 304
gas monitoring
  barometric pressure effect method,
    187–189
  instruments, 185, 186, 187, 239
  mixer pump installation, 252–253
  probes, 109, 113
  safety authorization basis, 198
  slurry flows, 398
  standard hydrogen monitoring
    system (SHMS), 185, 186, 415,
    442
gas release events (burps), 98*t*
  causes, 17
  explosion risks, 42, 120–121
  gas buildup, 33–34, 48–50, 187,
    270–271
  large, 138, 143, 207–208
  lessons learned, 423–424
  measurements, 71*f*, 87–88, 138–139
  mitigation and remediation, 79–80,
    143–148
  mixer pump tests, 276, 280, 286
  models, 132–135, 135–136
  prediction, 140*t*, 223, 226, 337
  prevention, 78, 334
  rollover, 88, 89
  safety analysis, 193–198
  tank damage, 65, 90
  videotapes of, 130–131
gas release windows, 98*t*
  case study, 82
  criteria, 101, 114, 200–202
  descriptions, 97–98, 98*t*
  instruments, 186, 208–210
  pump installation, 235–256
  waste sampling, 82
  work planning, 136–137
gas releases (non-burp), 98*t*
  ammonia, 405
  bubble slurry, 348, 401–402
  calculations, 306
  carbon dioxide from Lake Nyos,
    Cameroon, 275
  crust-base drop, 356
  deficit relative to gas generation, 308
  mechanical mitigation arm, 359, 361
  mixer pump, 262, 263, 272, 278, 292

Index | 459

gas releases (non-burp) *(cont.)*
    volumes predicted during mixer
        pump installation, 245, 249,
        299, 304
    waste transfer and dilution, 382–383,
        400
gas volume
    barometric pressure effects, 185,
        187–189
    calculations, 33–34, 86–87
    crust, 346, 346*f*
    fractions, 322–323, 347, 384
    instruments, 297–298, 315–316
    measurement, 117–118, 272
Gauglitz, Phil, 90, 294, 326–327, 332, 333,
    337–338
General Electric, 29, 57
Georgia Institute of Technology, 85
Gerton, Ron, 71–72
    hydrogen concentration prediction, 249
    mitigation project funding, 203
    mixer pump, 158, 177
    tank waste priorities, 120
    window work, 125, 209, 213, 223–224,
        246
Giant Nickel (blow-off riser), 130, 183,
    193, 219–220, 225–226
Glenn, John (Senator), 58, 92
glycolate, 17
go-no-go gauge, 230, 248
gob theory, 90–91, 131, 134, 138, 186, 281
Grand Coulee Dam, 7
graphite in reactor cores, 10
Gray, John, 196, 215, 240, 249
Green Run, 12
Gregory, Barry, 243, 249
Grelecki, Chet, 75, 138
Grigsby, Mike, 329
Groendyke, Craig, 331*p*
    blanket authorization letter, 354–355
    Hanford Square value engineering
        session, 332, 334
    mixer pump, 410, 415
    mixer pump off period (MPOP), 414
    press conference, 392
    surface level rise remediation, 385,
        386, 387
    Surface Level Rise Remediation
        Project, 330, 331–332, 331*p*
    SY-101 in-service operation, 420
    Test Review Group (TRG), 196
ground disposal of liquid wastes, 14, 16,
    22
grout, 149
Groves, Leslie (General), 3, 6–7
Grumbly, Thomas, 64, 256–257, 268
Guthrie, Mike, 384

H

Haentjens, Walter, 158, 160, 166, 363–364
    See also Barrett, Haentjens, and
        Company
Hall, Butch, 249, 360
Hall, James, 354
Hall, Mark, 153–154, 155, 236, 294
halon fire suppression system, 220
Hamrick, Doug, 196
Hamrick, Phil, 95*p*
Hanford, Washington, 7–8
Hanford Federal Facility Agreement and
    Consent Order, 30, 52, 119–120,
    375, 443
Hanford Grout Program, 159, 160
Hanford High School, 7, 8*p*
Hanford House, 83
Hanford Meteorological Station, 166, 187
Hanford Patrol, 247, 248
Hanford Reach, 10
Hanford Site
    contractors, 29–30, 305–306
    culture, 57, 157, 378–380, 424
    dignitary visits, 95*p*, 103, 257–258,
        259*p*
    fire, 132
    maps, 5*m*, 15*m*
    mission change, 52
    operations, 266–268, 311–312
    processing plants, 17*t*
    production reactors, 11*t*
    safety culture, 66, 72, 81, 113, 269, 392
    selection and development, 6–7, 9
    stand down of 1993, 261, 266, 269–270
    See also tank farms; individual
        facilities by name
Hanford Square value engineering
    session, 332–336
Hanford Waste Management Technology
    Plan (HWMTP), 57
Hanford Works, 3–4, 9–10, 11
Hanson, Carl, 109*p*
    Committee for Insight, Understanding,
        and Alleviation of Level Growth
        (CIUALG), 326–327
    data acquisition and control system
        (DACS), 189–190
    Hanford Square value engineering
        session, 332
    instruments, 109–110, 109*p*, 118,
        241–242, 318
    mixer pump, 157, 173, 248, 249
    Pasco mitigation workshop, 152
    PHMC contract changes, 306
    safety analysis and assessment, 237
    surface level rise remediation, 314,
        329, 339, 367, 369, 387

460 | Index

Hanson, Carl *(cont.)*
  Test Review Group (TRG), 196
  water lancing plan, 227
  window G work, 217
Harmon, Harry, 107*p*
  fiscal responsibility warning, 120
  Hazel O'Leary visit, 259*p*
  management style, 165–166
  mitigation project funding, 203–204
  mixer pump, 157, 160, 226, 228, 232,
    249, 256
  Pasco mitigation workshop, 154, 155–156
  recognition dinner, 257, 257*p*
  remediation, 143–144, 392
  Westinghouse criticisms, 151
  Westinghouse reorganization,
    106–107, 108, 148–149, 238
  window work, 122, 125, 211, 213–214,
    215, 223–224
Harrington, Rich, 332
Hatcher, Charlie, 190
Hauck, Marshall, 370
headspace. *See* dome or domespace of
    waste tank
heat transfer characteristic of
    temperature profiles, 87–88
helicopters, 132
Herting, Dan, 332, 366
high-efficiency particulate air filters
    (HEPA), 102, 120, 129, 439
HiLine Engineering, 306, 322
Hiroshima, Japan, 9
H&N Electric, 174
Hoida, Hiroshi, 190
Honeyman, Jim, 39
hose-in-hose system, 374, 375
hot cells, 105, 109, 127–128, 129*p*, 315
Hudson, Billy, 237, 385, 388
Hudson, John, 249
HULL computer program, 222–223
humidity, 109, 113, 123
Huntoon, Carol, 416
hurricane analogy, 37–38, 288, 301
hydrodynamic oscillation, 219
hydrogen concentrations
  controls, 195, 201, 237–238
  crust-base drop, 356
  domespace, 194, 365
  gas release events, 207, 245
  instruments, 224
  Mechanical Mitigation Arm effects
    on, 358–359, 361
  mixer pump off period, 411
  mixer pump testing, 271, 279, 280,
    285, 286, 292
  predictions during mixer pump
    installation, 249

hydrogen concentrations *(cont.)*
  suppression, 285
  waste transfer and dilution, 397, 398,
    400, 401, 407
  water lancing, 362
hydrogen gas
  from concentrated wastes, 33
  flammability, 34, 73
  mitigation success criterion, 146–147
  nitrous oxide mixtures, 41, 45, 75,
    299–300
  waste simulation, 40
  See also flammable gas hazards; gas
    releases; hydrogen
    concentrations
Hydrogen Gas Mitigation Project.
    See SY-101 Hydrogen Gas
    Mitigation Project
hydrogen suppression, 285
hydrostatic pressure, 32, 117, 318–319,
    395
hydroxyethylenediaminetetra-acetate
    (HEDTA), 17

I
Idaho National Engineering Laboratory,
    85
Ignition Control Set, 194
ignition hazards
  earthquake, 220
  during gas release events, 138, 221
  hydrogen-nitrous oxide, 40, 42, 75
  identification, 34
  measurements, 47
  mixer pump, 159, 184
  public opinion, 55–56
  safety measures, 44, 194
  sludge weight samplers, 110–111
  waste sampling, 79–80
impact limiters, mixer pump, 168, 183
impeller, mixer pump, 170*p*, 171
inside the fence contractors, 305
instruments, measurement
  ammonia gas, 274, 275
  calibration, 277, 308, 319
  densimeter, 376
  gas monitoring, 184–192
  gas release events, 207
  gas retention, 297
  gas volume, 299
  installation, 101, 108–109
  neutron and gamma probes, 348–349
  performance issues, 175–178, 224,
    239–242, 319–320
  removal from tanks, 209, 212–213
  surface level rise, 315–317
  waste gases, 296–297

Index | 461

instruments, measurement *(cont.)*
    waste levels, 303–304, 317, 344–345
    See also Enrafs; Food Instrument
        Corporation (FIC); Fourier
        transform infrared gas
        monitoring system (FTIR);
        radar gauge; Standard
        Hydrogen Monitoring Systems
        (SHMS)
iodine-131, 9, 12
Islam, Mohammed, 196
isotopes, 22, 439

J
Jackson, Henry (Senator), 52
Jacobs, Roy, 60
Japan, 4
Jefferson, Thomas, 3
Johnson, Ben, 147–148, 152, 157–158
Johnson, Jerry, 77*p*
    ammonia gas issue, 273
    Committee for Insight, Understanding,
        and Alleviation of Level Growth
        (CIUALG), 326–327
    computer models, 133, 228
    data management, 126
    Flammable Gas Program, 76, 77, 77*p*
    Hanford Square value engineering
        session, 332
    ignition hazards, 138
    mitigation and remediation projects,
        144, 147–148, 314, 321
    mixer pump removal, 410
    Pasco mitigation workshop, 152
    PHMC contract changes, 306
    recognition dinner, 257
    Science Panel, 86, 131
    waste levels, 240, 307
    work coordination, 134–135
Joint Test Group (JTG)
    description, 440
    event predictions, 223
    formation, 113–114
    lessons learned, 125
    meeting discipline, 136
    TEMPEST computer program, 133
    window criteria, 201
    window work, 196, 207, 225
Jordan, Ken, 329, 331*p*

K
Kaiser Engineers
    C-106 contamination incident,
        267–268
    mixer pump design and construction,
        163, 167, 173, 175–176, 262
    velocity-density-temperature trees
        (VDTTs), 186

Kazimi, Mujid, 68, 69, 200, 330–331, 404,
    414
Kinzer, Jackson, 354
Kirch, Nick, 59*p*, 331*p*
    Committee for Insight,
        Understanding, and
        Alleviation of Level Growth
        (CIUALG), 326–327, 417
    crust measurements, 326, 342
    data management, 126, 224, 240
    Hanford Square value engineering
        session, 332
    mitigation project, 86, 103, 147–148
    mixer pump, 249, 281
    mixer pump off period (MPOP),
        410–411, 413–414
    Most Important Person Award, 331*p*
    Science Panel, 131
    surface level rise remediation, 196,
        314, 329, 339, 345–346, 418*p*
    tank safety, 59–60, 59*p*
    Waste Tank Safety Task Team, 76
    waste transfer and dilution, 335–336,
        392
    window criteria, 79–80, 200
    window work, 136, 225, 245–246
Klein, Keith, 67
Knecht, Walter, 244
Knotek, Mike, 226
Koreski, Gerry, 308, 408, 415–416
Kostelnick, Al, 358
Kouts, Herb, 388
Kreiter, Max, 135, 147–148, 196, 243, 280
Krogsrud, Steve, 249
Kryzstofski, John, 321, 324–325
Kubic, Bill, 274, 332
Kuhn, Bill, 366, 385–386
Kyshtym, Soviet Union waste tank
    explosion, 58

L
Lake Nyos, Cameroon carbon dioxide
    release, 275
Lampson Company, 182, 233
Lane, Tom, 40, 45
Larsen, Doug
    Committee for Insight,
        Understanding, and
        Alleviation of Level Growth
        (CIUALG), 326–327
    Hanford Square value engineering
        session, 332
    mixer pump tests, 174–175, 177–178,
        262
    surface level rise task force, 314
    Test Review Group (TRG), 196
Lawrence, Mike
    declassified documents, 52

462 | Index

Lawrence, Mike *(cont.)*
  DOE Headquarters Review Team
    briefing, 65
  openness policy, 63–64
  press conferences, 34, 38, 61, 416
  unreviewed safety question (USQ), 66
leaking tanks, 4, 20–22, 26, 43, 372, 374
Lechelt, Jeanne, 126, 239–240, 242, 415
Lee, Jim, 249
Leggett, Bill, 76, 77–80, 143
Lenkersorfer, Doug, 230
Lentsch, Jack, 149*p*
  corrosion concerns, 222
  data acquisition and control system
    (DACS), 189
  explosion hazards, 41
  management style, 164–166, 328
  mitigation project, 149–159, 149*p*,
    235–236
  mixer pump, 156, 174, 226, 228, 249,
    269
  mixer pump tests, 277, 279
  Pasco mitigation workshop, 152–153,
    154–155
  PHMC contract changes, 306
  safety analysis, 180, 183, 219–221
  tank data, 298
  Test Review Group, 196
  ultrasonic mitigation, 295
  waste level measurements, 304
  window work, 209, 212–213, 215, 217,
    245, 246, 247
lessons learned, 113, 125, 136–137,
  217–218, 299, 422–424
level cycling, 42–43, 76
Lewis, Meriwether, 3
lightning hazards, 238
liquid level in waste, 318, 345, 347, 348,
  349
liquid wastes
  brine, 365–366, 420, 421
  concentration, 22–24
  density, 117, 383–384
  ground disposal of, 14, 16, 22
  measurement, 118, 318
  processing, 14–17
  sources of, 19
  tank layers, 32, 48, 322
  videotape, 138
Little Boy atomic bomb, 9
load frame, mixer pump, 168
lock and tag procedures, 267
Lockheed Martin Hanford Company
  Hanford contract, 29, 303, 305
  performance incentives, 328, 388
  surface level rise remediation, 342
  waste level measurement, 303

lollipops, waste buildup, 220, 382, 394,
  399
long term operating plan, 284, 293, 303,
  312, 440
Lopez, Tom, 167, 174, 175
Los Alamos, New Mexico, 6
Los Alamos National Laboratory
  data acquisition and control system
    (DACS), 189–190
  gas volume calculation, 187
  multifunction instrument trees
    (MITs), 185
  radiation badges, 190
  safety analysis and assessment, 116,
    155, 163, 180–181, 197–198, 237
  SY-101 Hydrogen Gas Mitigation
    Project collaboration, 151–152
  ultrasonic vibration, 144
  window criteria, 200–201
low dilution *vs.* top dilution, 366–367,
  381, 396–397, 398, 403, 406–407

M
Magnuson, Warren, 52
Mahoney, Lenna, 326–327, 330
maintenance and storage facility (MASF),
  170*p*, 173–174, 177, 179*p*, 217,
  296, 440
man-lift, crane, 253*p*, 254
Manhattan Project, 3–4, 6, 10, 131
manual tape
  definition, 440
  methods, 51, 99
  mixer pump testing, 277
  waste level measurements, 240, 271
  waste levels, 104, 111, 224, 244–245
  window events, 130, 136, 202
maps, 5*m*, 15*m*
March, Lawrence, 74, 86
Marchetti, Steve, 124*p*
  leadership change, 205
  management style, 124, 124*p*, 165
  mitigation project funding, 203
  mixer pump, 157–158, 160, 174
  Pasco mitigation workshop, 152–153,
    154, 155–156
  Westinghouse reorganization,
    106–107, 148, 149
  window criteria, 200
  window work, 113, 119–120, 121, 122,
    136
Martin, Earl, 57
Marusich, Bob, 110–111, 181, 225–226, 354
Massachusetts Institute of Technology, 68
Matthias, Frank (Colonel), 3
maximum burps, 193, 226
McDuffie, Norton, 147–148, 273

Index | 463

McElroy, Mike, 249
mechanical mitigation arm (MMA),
    358–361, 440
Mellinger, George, 146
Mendoza, Donny, 351–352
Merriman, Ray, 167, 174
methane gas, 185
Meyer, Perry, 89
    bubble migration, 381
    bubble slurry flow experiment, 353
    Committee for Insight,
        Understanding, and
        Alleviation of Level Growth
        (CIUALG), 326–327
    Hanford Square value engineering
        session, 332
    Mechanical Mitigation Arm trials, 360
    mixer pump operation, 363
    Surface Level Rise Remediation
        Project, 330
    SY-101 level forecaster, 342
    transfer pump, 371
    waste transfer and dilution, 335–336,
        375
mice, 192
Michener, Tom, 132–133, 194, 222, 226, 272
Mishko, George, 217, 224, 249, 257
mitigation
    accomplishments, 299, 416, 417–419
    definition, 440
    errors, 423–424
    flammable gas hazards, 68, 119, 206,
        282
    gas buildup, 72, 412–413
    methods, 143–148, 151, 293–296,
        334–335, 400
    mixer pump, 156–157, 158, 280
    Pasco, Washington workshop,
        152–156
    project funding, 203–204, 235–236
    temporary measures, 79–81, 358–361
    test acceptance, 291
    waste transfer and dilution, 391
    See also SY-101 Hydrogen Gas
        Mitigation Project
Mitigation Project. *See* SY-101 Hydrogen
    Gas Mitigation Project
mixer pump off period (MPOP), 410–411,
    413, 440
mixer pump operations
    aggressive schedule, 309
    backup, 296
    excavation runs, 292, 293, 301
    gas releases, 270–271
    hazards, 38, 364–365
    long term operating plan, 284, 293,
        303, 312, 440

mixer pump operations *(cont.)*
    maintenance, 413–414
    photos, 159*p*, 161*p*, 170*p*, 171*p*, 179*p*,
        399*p*
    power sweeps, 407
    problems, 230–233, 266–267
    shut down, 407, 410*f*, 415–416
    surface level rise issues, 343
    waste transfer and dilution, 382, 395,
        398–399, 403, 406–407
    wiring, 167, 172, 174–175
    See also bump
mixer pump safety analysis (SA)
    ammonia, 246–247, 273–274
    approval process, 129
    backup pumps, 296
    criticisms of, 150–151, 228–229
    description, 180–184
    emergency amendment, 254
    instrument of record, 303
    Los Alamos National Laboratory, 116,
        155
    requirements, 168, 177–178, 237, 238,
        255, 279
    review, 156
    revisions, 219–223, 237, 273–274
    surface level rise issues, 312, 314
    TEMPEST computer program, 133,
        272
    tests, 291–292
    waste de-gassing, 295
    waste level limits, 281
mixer pump strategies
    budget and funding, 156, 224, 300
    design and modification, 109,
        163–169, 169–173, 172*f*, 182
    installation, 45, 208, 214, 226–227, 228,
        246–258, 254*p*
    planning, 78, 131, 154–156, 156–161,
        205, 282–284
    safety, 145, 168, 182–184, 237–238, 273,
        421
    success, 91, 295–296
    temperature limits, 180
    tests, 173–180, 242–244, 261–265, 272,
        276–288, 292–293
models
    Archimedes, 381
    buoyancy ratio, 337, 366, 381, 383, 384,
        404, 412
    chemistry, 383–384, 404
    computer, 131–135, 193–194, 222–223,
        226, 228–229, 272
    crust buoyancy, 367, 382
    crust growth prediction, 341
    gob theory, 90–91, 131, 134, 138, 281
    mixer pump test, 176

464 | **Index**

models *(cont.)*
  sediment viscosity, 132
  statistical, 229
Monte Carlo simulations, 339, 404
Morgan, Larry, 76
Morrison, Bruce, 164, 164*p*
Most Important Person Award, 330, 331*p*
Muhlstein, Lewis, 80
Mulhan, Rajiv, 249
multifunction instrument trees (MITs)
  calibration, 241, 277
  definition, 440
  design and installation, 151–152, 185,
     214, 248, 262
  excavation runs, 292
  funding for, 203
  measurements, 225, 345, 396, 412
  mixer pump off period (MPOP), 414
  mixer pump tests, 276, 284
  neutron and gamma probes inside,
     348–349
  prototypes, 148

N
N Reactor, 13, 52
Nagasaki, Japan, 9
National Defense Authorization Act, 94–95
National Explosion Prevention
     Association, 123
National Weather Service, 37
nature preservation of Hanford area, 4
neutron, 441
neutron logs
  bubble slurry flow, 362
  crust base drop, 355–356
  crust liquid level, 349–350, 350*f*
  dissolved crust, 403
  sediment depth, 412
  waste transfer and dilution, 396
  wasteberg capsize event, 401
neutron probes, 348–349, 414
New Generation Transfer Pump, 370, 372,
    373*p*
  See also transfer pump
New York Times, 58, 257, 386
news media
  Hanford House public meeting, 83
  mixer pump installation, 248, 255,
    256–257
  press conferences, 34, 38, 61, 259*p*,
    392, 416
  zone of death, 247
  See also individual newspapers by
    name
Nichols, Roger, 106, 107
nickel ferrocyanide, 14, 22
  See also ferrocyanide

Niebuhr, Dan
  air lances, 219
  blow-off plate, 183
  measurement instrument
    management, 241–242
  Mechanical Mitigation Arm trials, 360
  mixer pump installation, 249
  Surface Level Rise Remediation
    Project, 329
  ventilation restart, 102
  working conditions, 206
Nike missile batteries, 131–132
nitrate salts, 32, 105
nitric acid, 12, 25
nitrogen inerting, 72
nitrogen purge, 172, 246, 277, 421–422
nitrous oxide, 34, 40, 245
  concentrations, 356
  flammability, 60
  gas, 185
  hydrogen gas mixtures, 41, 45, 75,
    299–300
  release rates, 308
nominal pump operation, 292
non-waste intrusive windows, 200, 203
nozzles, mixer pump
  clogs, 195, 263–264, 265, 413
  flow meters, 177
  flush line plumbing, 254
  modification, 166–167
  operation parameters, 169–170
  photos, 159*p*, 256*p*
  pressure gauges, 318–319
  tests, 276, 278*f*
nozzles, water lance, 45, 250, 362
nuclear fuel, 5*m*, 10–11, 12–13, 26
nuclear reactors
  building of, 9–10
  computer models, 133
  map, 5*m*
  plutonium production, 6–7, 11*t*, 52
  reactor cores, 10
  shut down of, 52
  water use, 10–11
Nuclear Safety and Licensing, 312
Numatec, 305
Numerical Applications, Inc., 133–134,
    440
nylon skid pads, mixer pump, 231–232

O
Oak Ridge Site, 4
Oakley, Don, 68–69, 386
Office of Environment, Safety, and
    Health, 81
Office of Environmental Restoration and
    Waste Management, 30, 64, 148

Office of Nuclear Safety, 75, 81, 82
Office of River Protection. *See* U.S. Dept. of Energy, Office of River Protection
Ogden, Don, 133–134
oil, pump motor, 171, 178, 180, 284, 296, 344
O'Leary, Hazel (U.S. Secretary of Energy), 75, 257–258, 259*p*
Onishi, Yasuo, 133
Oregon Public Broadcasting, 190
Oregonian, 74
organic compounds
  carbon, 383
  in complexed waste, 16–17, 26
  explosion hazards, 55
  removal, 144, 145
  safety recommendations, 93
  slurry growth, 41–42
Ostrom, Mike, 155, 231–232
outside the fence contractors, 305
overpacks, 218
Owendoff, Jim, 385, 388
ox team, 358–359
oxidants, 60

P

Pacific Northwest National Laboratory
  competition with Westinghouse, 133–134
  description, 29–30, 441
  mitigation project roles, 147–148, 151, 294
  mixer pump test plan, 155, 243–244
  research, 57, 85
  slurry growth report, 49
  waste analysis, 127
  wasteberg simulation, 352
Panisko, Frank, 196, 415–416
partial neutralization of waste, 31
Pasamehmetoglu, Kemal, 227*p*
  ammonia concentrations, 246
  bubble migration, 381
  buoyancy ratio model, 337–338
  mixer pump, 181, 222, 226, 227–228, 227*p*, 230
  tank de-gassing, 281
  tank waste models, 194
  TEMPEST computer program, 272
  Test Review Group (TRG), 196
  waste viscosity tests, 297
  window event G, 208
Pasco, Washington mitigation workshop, 152–156
Payne, Mike
  hydrogen concentration prediction, 249, 252

Payne, Mike *(cont.)*
  pump installation, 226
  surface level rise issue, 321, 328
  tank farm management, 149
  window G work, 217
  worker safety, 387
Pearl Harbor, attack on, 5
Pederson, Larry, 238
Pepson, Dave, 215, 232, 237, 248, 249
performance incentives
  agreement, 339
  Surface Level Rise Remediation Project, 328, 404, 409–410, 415
  SY-101 cross-site transfers, 420
  SY-101 Safety Mitigation, 388
personality of SY-101 tank, 308, 356, 424
physical mitigation, 146
pitot tube, 51
Plant Review Committee, 246, 274, 312–313
plastic fiber hose, 374
plutonium and plutonium-239
  fuel production, 4, 6, 9–13
  production, 52
  separation, 11–12, 15*t*
  waste products, 12
plutonium-uranium extraction (PUREX), 22, 441
plywood pump pit cover, 250, 251*p*, 255
political pressure
  Blush Report, 74
  Defense Nuclear Facilities Safety Board, 92–93
  flammable gas safety issue, 424
  lessons learned, 422
  letter from Booth Gardner, 61–62
  for safety improvements, 164
  Wyden Amendment, 94–96, 127
potential inadequacy in the safety authorization basis (PIAB), 313, 314, 441
prefabricated pump pit (PPP), 371–372, 373*p*, 420, 441
Project Hanford Management Contract (PHMC), 29, 305–306, 441
Project Management Institute, 340, 418
project of the year award, 340, 418–419, 424
Propson, John, 200–201
protective clothing, 103, 105*p*, 206
proton, definition, 441
prototypes, 148, 168
public meetings, 83–84, 107
public opinion
  Berger report, 58
  Blush report, 74
  criticisms, 107–108

466 | Index

public opinion (cont.)
    crust growth, 386
    explosion hazards, 61–62
    flammable gas hazard, 38, 55–56
    Hanford House public meeting, 83–84
    lessons learned, 422
    radioactive waste, 52–53
    waste sampling safety, 84
    Wyden Amendment, 94–95
pump column, 246
pump drop accident, 168, 182
pump pit
    mixer pump installation, 254–255
    plywood cover, 250, 251p, 255
    prefabricated, 371–372, 373p, 420
    radiation, 169, 172–173, 231
    Tank C-106, 267
    transfer pump, 371
punch list, 130, 378

Q

quality assurance (QA), 230–231, 236, 441

R

radar gauge
    calibration, 129
    gas release events, 130, 202
    installation, 105p, 109, 121, 124
    performance issues, 117, 239, 240
    waste levels, 111, 122f, 136, 224
radiation
    badges, 190
    dose, 172–173, 190–191, 218, 226, 231,
        375
    shine, 206, 250, 255, 371
    warning alarms, 107
radioactive activation products, 10, 12
radioactive contamination. See
        contamination
radioactive decay, 344, 441
radioactive fission fragments, 12
radioactivity, 32, 71, 73, 441
radiochemical analysis, 127–128
radiochemistry of gas generation, 85
radiolysis, 40
Rassat, Scot, 351–352, 366
Rayleigh, Lord, 89
Rayleigh-Taylor instability, 89
Raymond, Rick, 328p
    flammable gas watch list, 96
    Hanford Square value engineering
        session, 332, 334
    management style, 328–329, 328p
    mixer pump removal, 410
    political backing, 388–389
    readiness review, 378–379, 380
    success factors, 422

Raymond, Rick (cont.)
    surface level rise remediation, 346,
        385, 386, 415
    Surface Level Rise Remediation
        Project, 339, 417–418, 418p
    tank leak criteria, 43
    Waste Tank Safety Task Team, 76
    waste transfer and dilution, 336, 369,
        388, 392, 403
    wasteberg simulation, 352
reactors, nuclear. See nuclear reactors
readiness reviews, 129–130, 211, 217,
        377–381
Reberger, Dave, 196
Recknagle, Kurt, 353
Red Lion Hotel, Pasco, Washington, 152
reduction-oxidation (REDOX), 15t, 20, 22,
        442
    See also S Plant
remediation
    definition, 442
    final steps, 405–407
    management concern, 404
    methods, 147, 335–336, 369–370
    report, 412–413
    SY-101 in-service operation, 420–421
    technical basis, 336–338
    top dilution, 367
    unreviewed safety questions (USQ),
        312–314, 385–387, 415
    waste transfer and dilution, 381–383,
        391–396, 398–403, 404–408
    white paper, 404
    See also Surface Level Rise
        Remediation Project (SLRRP)
research laboratories, 5m
resistentialism, 424
Respond and Pump in Days (RAPID),
        370
retained gas sampler (RGS), 275, 315–316,
        344, 347, 442
retraining, 268, 269, 380
revetment, shielded, 253
Reynolds, Dan, 40p
    calculations, 87–88, 187
    Committee for Insight, Understanding,
        and Alleviation of Level Growth
        (CIUALG), 326–327
    double-shell slurry, 39, 40, 40p
    evaporators, 417
    gob model, 131
    predictions, 129, 249
    Senior Chemists Panel, 79
    temperature profiles, 89
    TEMPEST computer program, 228
    waste density, 384
    waste levels, 212

Reynolds, Dan *(cont.)*
    waste temperature changes, 138–139
    Weibull distribution, 202
    window criteria, 80–81, 101, 122,
        200–201, 203
rheometer, 297, 298–299, 298*p*, 337
Richland, Washington, 7, 83
riggers, crane, 182, 212, 253, 253*p*, 255
risers
    C-106 contamination incident,
        267–268
    identification, 100–101, 100*f*
    maps, 100*f*, 216*f*, 258*f*
    mixer pump fit problem, 230–233
    mixer pump installation, 254
    pump ejection accident, 182–183
    sealing plates, 168–169
    shield plug, 250, 251*p*
    shine radiation, 206, 250
    simulated, 176
robotic inspection, 231
rock-on-a-rope incident, 266, 267–268,
    378
Rockwell Hanford Operations, 29, 31, 34,
    44, 56–57
Rodriguez, Ed, 193, 222
rollover, sediment, 88, 89, 102
Roosevelt, Franklin Delano, 3–4, 5
Rosenwald, Gary, 249
Ross, Bill, 380
Ruud, Casey, 222

S
S-112 tank, 421
S Plant (REDOX), 14, 16*p*, 17*t*, 20, 48
    See also reduction-oxidation (REDOX)
S tank farm, 21*p*
Saddle Mountain, 132
safety analysis or assessment (SA)
    crust gas release hazard, 350
    crust wetting system, 79–80
    definition, 442
    sludge weight samplers, 109, 110
    TEMPEST computer program, 133,
        272
    unresolved safety question, 67
    waste behavior predictions, 423
    waste sampling, 69, 82–83
    See also mixer pump safety analysis
Safety and Environment Advisory
        Council (SEAC)
    definition, 442
    safety analysis approval, 196–197
    safety recommendations, 112–113, 238
    waste sampling approval, 104
    window criteria, 114
    window extension, 212

safety authorization basis, 194, 198, 313,
    337, 388, 410
    See also potential inadequacy in the
        safety authorization basis
        (PIAB)
safety concerns
    ammonia, 238, 246–247, 273–276, 400
    employee, 207, 225–226
    explosion hazards, 14–15, 56–57, 58
    flammable gas hazards, 47–48, 72,
        83–84, 146, 221–222
    gas release, 79–80, 193–194, 195,
        364–365
    mishandling of, 55, 67, 73, 107
    mixer pump
        bubble slurry layer, 348
        flammable gas hazards, 181–184
        installation, 168–169, 214, 254
        nozzle clogs, 264–265
        operations, 219–221, 236–238, 283,
        309, 311–312, 353–354
    overdue inspections, 212
    priorities, 127, 148–149
    radioactivity releases, 71, 73
    steam bumps, 23
    waste sampling, 82–83
    waste transfer and dilution, 375–376,
        381–385, 382
    whistleblowers, 53
    white papers, 59–60
    worst case scenarios, 48, 57, 61, 110, 423
safety improvement plan, 77, 143
safety measures
    ammonia concentrations, 247
    confidence in, 78, 84
    criticism of, 67
    flammability, 61, 80–81, 194
    gas release events, 103
    Hanford Works, 10, 11
    mixer pump, 175, 219–221, 228–229,
        248, 421
    prudent controls, 355
    stand down of 1993, 269–270
    waste transfer and dilution, 374–375,
        376
    Westinghouse Hanford Company,
        75–76
    window criteria, 200–201
    Wyden Amendment, 95–96
Safety Measures for Waste Tanks at
        Hanford Nuclear Reservation,
        94–96, 127, 151
safety recommendations
    Defense Nuclear Facilities Safety
        Board, 92–93
    Safety and Environment Advisory
        Council (SEAC), 112–113

468 | Index

safety recommendations (cont.)
  tank waste, 72–73
  task force, 45–46, 82–83
  water lancing, 44
safety reviews, 184
safety window. *See* window
salmon, 14
saltcake, 255
  description of, 27–28
  on electrical probes, 51
  on measurement probes, 99
  soluble, 386
  waste analysis, 128
  waste sampling, 84, 99
saltwell, 255
  dissolution process, 421
  pumping, 22, 266, 349, 355, 375, 420,
    421
Sandia National Laboratories, 64
Sasaki, Leela
  calculations, 44, 131
  flammable gas hazards, 417
  hydrogen gas monitoring, 46–47, 49
  waste temperature profiles, 87, 105
Savannah River Site, 58, 59–60, 85, 106,
  119, 188
scallops, in sludge layer during mixer
  pump testing, 293
Schmorde, Fred, 379
  data acquisition and control system
    (DACS), 421
  mixer pump off period, 414–415
  Surface Level Rise Remediation
    Project, 329
  Test Review Group (TRG), 196
  transfer pump installation, 371
Schneider, Tom, 185, 242
Schultz, Wally, 86
Schutte, William, 69
Science Panel
  description, 85–86
  funding, 203
  gob model theory, 131
  mitigation and remediation report to,
    147
  recommendations, 144
  working group studies, 145
seal plates, 168–169, 179*p*, 183–184, 255,
  296
Seattle Post-Intelligencer, 74, 95
Seattle Times, 84, 257
sediment
  ammonia gas in, 275
  buoyancy, 226, 265, 270–271
  definition, 442
  depth, 383–384, 384, 412, 421
  after gas release events, 48–50, 117

sediment *(cont.)*
  measurement, 87–88, 88*f,* 99, 319
  mitigation and remediation, 145, 294,
    336
  mixer pump tests, 279, 283, 292
  models, 132, 337
  scallops in, 293
  Science Panel, 85
  steam bump, 22–23
  strength, 89–90
  tank waste layers, 32, 322
  temperature, 140*f,* 343
  trenches in, 222, 265, 270, 276
seismic shutdown switch, 220
Senate Armed Services Committee, 58
Senior Chemists Panel, 79
separation plants, plutonium, 11–12, 15*t*
Shaw, Craig, 157, 166–167
Shepard, Chet, 297
Shepherd, Joe, 299–300
shield plug, 250, 251*p*
Shofield, John, 196
shrouded test assembly, 158
Siano, Mike, 205
Silver Cloud Inn, Kennewick,
  Washington, 282
simulations, 193, 339, 351–352, 352–353,
  404
  See also models
single-shell tanks
  definition, 442
  gas volume, 189
  leaks, 20–22
  prefabricated pump pits (PPP), 372
  saltwell pumping, 266
  sizes and capacities, 19–20, 24*t*
  wastes, 27–29, 56–57, 375, 423
Slezak, Scott, 64, 69, 86, 145–146
SLRRP. *See* Surface Level Rise
  Remediation Project (SLRRP)
sludge. *See* sediment
sludge weights
  removal, 104–105
  samplers, 106, 121, 123, 124, 125*p*
  waste sampling, 98*t,* 99
slurry
  densimeter, 376–377
  description, 169–170
  dilution, 367, 370
  double-shell, 31–32, 42, 47
  gas fraction analysis, 323, 324, 325
  production, 32, 42
  removal, 370
  simulation, 40, 41, 48–50
  solid-liquid, 322
  tank transfers, 335
  temperature, 318

slurry *(cont.)*
    waste transfer and dilution
        measurements, 396–397
    See also bubble slurry
slurry distributor, 226, 230, 370, 375, 437
slurry growth
    bathtub experiment, 33–34
    definition, 33, 443
    mitigation, 83
    problem identification, 39–42
    Waste Tank Safety Task Team, 76
slurry line, 421
sodium, 383
sodium aluminate, 397
sodium hydroxide, 12, 25, 294
sodium nitrate, 25, 55–56
sodium salts, 409
sodium sulfate, 26
soil contamination, 4, 218–219
    See also ground disposal of liquid
        wastes
solvent extraction processes, 14, 17, 26
sonic agitation, 153–154, 333
    See also ultrasonic agitation
sonic probe, 155, 236, 293–294, 295
Soviet Union, 4, 6, 13
spark hazards. *See* ignition hazards
spark-proof ventilation, 129
specific gravity, 384, 404
Spokesman Review, 53
stalactites, 111, 130, 302
stand down of 1993, 261, 266, 269–270, 277
standard hydrogen monitoring system
        (SHMS), 185, 186, 415, 442
statistical models, 229
steam bumps, 22–23
Stewart, Burton, 7–8
Stewart, Chuck, 8, 418*p*
    Committee for Insight,
        Understanding, and
        Alleviation of Level Growth
        (CIUALG), 326–327
    crust gas volume projections, 346
    gas fraction data analysis, 323–326
    mixer pump success, 281–282
    Surface Level Rise Remediation
        Project, 330, 418*p*
    surface level rise task force, 314
    Test Review Group (TRG), 196
    waste level instruments, 303–304
    waste transfer and dilution, 392, 394,
        402
Stewart, Horace Burton, 7–8
Stokes, Steve, 388
Stokes, Troy, 250, 297, 306, 322
stop-leak experiments, 26

stop work, 261, 266, 269–270, 277
Straalsund, Eric, 185, 239, 242
Strachan, Denis, 86, 145
strain gauges, 184, 265, 276, 277
Strehlow, John, 167
strontium and strontium-90, 9, 14–17,
        22–23, 32–33
SubTAP
    ammonia gas hazard, 274
    Chemical Reactions, 86, 223, 295, 388,
        404
    description, 69, 443
    gas fraction analysis, 324
    level rise remediation issue, 330–331
    mitigation success, 299
    pump installation safety, 227, 237
    recommendations, 272, 280, 287–288,
        318, 319
    waste level measurements, 303
    Worker Safety & Health, 387, 388
success factors, 422, 423, 424
sucker (retrieval tool), 212–213, 213, 214*p*
Sullivan, Harold, 68, 152, 154–155, 196
supernatant liquid
    definition, 28, 443
    mixer pump testing, 244
    remediation, 147
    SY-101 Tank, 32
    temperature, 87–88, 88*f*
Surface Level Rise Remediation Project
        (SLRRP)
    accomplishments, 409–416, 417–419
    acronyms, 370
    budget, 340
    description, 328–332, 442
    goals, 334
    mixer pump off period (MPOP),
        410–413
    official authorization, 327
    performance incentives, 415
    planning, 339–340
    readiness review, 377–381
    team photo, 418*p*
    temporary mitigation measures,
        356–362
    Test Review Group (TRG), 196
    urgency of, 342–344, 354–355
    U.S. Dept. of Energy. Headquarters
        concerns, 385–389
    waste transfer and dilution, 381–385,
        391–396, 404–408
    See also remediation
surface level rise task force, 314, 316, 318
surveillance cameras, 249
SX-106 Tank, 31–32

470 | Index

SY-101 Hydrogen Gas Mitigation Project, 143–159
  achievements, 190
  closing of, 300
  conduct of work, 164
  culture change, 165–166
  full scale test meeting, 282–283
  funding, 163, 203–205
  team, 76
  test acceptance, 291
SY-101 level forecaster, 342
SY-101 Safety Mitigation performance incentive, 388
SY-102 tank
  cross site transfers, 397, 420, 421
  domespace ammonia concentration, 405
  waste transfer and dilution, 78, 400
SY-103 tank, 42, 78, 102*p*
SY-101 tank personality, 308, 356, 424
SY-A valve pit, 420
SY tank farm, 102*p*
  dignitary visits, 257–258
  double-shell tanks, 20
  Gardner, Booth (Governor) visit, 95*p*
  Lockheed Martin operation of, 308
  Secretary of Energy visit, 103
  ventilation, 119, 120, 254*p*, 400, 405
  zone of death, 247

T

T Plant, 11–12, 13*p*, 17*t*, 19–20, 22
Tank Advisory Panel (TAP)
  crust sampling, 106
  definition, 443
  free liquid level measurement, 118
  gob theory, 90–91
  Hanford Site contributions, 70
  meetings, 155–156, 332, 417
  members, 67–69
  mitigation and remediation, 151, 156–157, 330–331, 368, 404
  mixer pump, 222–223, 227, 232
  mixer pump off period (MPOP), 414
  recommendations, 144, 154, 296–297
  reports and studies, 116, 145–147
  safety analysis, 80, 192
  waste transfer and dilution, 338, 387
  window criteria, 200–201
  See also SubTAP
Tank Characterization Database, 242
Tank Characterization Report (TCR), 128, 443
tank damage
  corrosion concerns, 221–222
  crust movement, 101
  gas release events, 146

tank damage *(cont.)*
  during gas release events, 207–208
  maximum burps, 193
  photographs of, 64, 65*p*
  pipes, 34, 90
  wastebergs, 169
tank data, 89–90
  analysis and interpretation, 113, 128, 195–196, 200, 201
  instruments, 111, 177, 239–242, 297, 298, 298*p*, 309
  management, 126, 242–243
  measurements, 185, 322–323, 325–326
  mixer pump, 286, 408, 414
  monitoring, 101, 103
  need for, 83, 84, 86, 94, 314–315, 424
  preponderance of evidence, 114–115
  reliability, 306, 404
  reports, 224, 416
  safety applications, 114
  temperature measurement changes, 343
  trends, 241, 272, 300, 302, 308–309, 319
  waste level predictions, 342
  waste transfer and dilution, 401
  window criteria, 101–102
  window G evaluation, 210
Tank Farm Project, 106
tank farms
  capacities, 24*t*
  culture, 66, 72
  dedicated, 11–12
  definition, 443
  dignitary visits, 95*p*, 103
  double-shell, 20–22
  evaporators, 23
  history of, 19–20
  operational discipline, 378, 392, 393
  safety, 47–48, 72, 410
  stand down of 1993, 261, 266, 269–270
  supervisors, 269
  temperature instruments, 320
tank pressure, 44, 47, 102, 193–194
Tank Riser Characterization Unit (TRCU), 231, 443
Tank Safety Program, 144
Tank Surveillance Program, 43
tank waste. *See* wastes
Tank Waste Information Network System (TWINS), 70
Tank Waste Remediation System (TWRS), 118, 148–149, 214, 238, 443
tanks, double-shell. *See* double-shell tanks
tanks, single-shell. *See* single-shell tanks
tape measure. *See* manual tape
task force, SY-101 tank, 44, 45–46

Taylor, Geoffrey, Sir, 89
team building, 330, 333
teamwork
    among companies, 175, 177
    mechanical mitigation arm (MMA),
        358–359
    siphon break, 375
    success factors, 423
    Surface Level Rise Remediation
        Project, 329
    window G work, 217
technical basis document, 381
Technical Basis Review (TBR), 339
Technical Review Committee (TRC)
    definition, 443
    densimeter, 376
    Mechanical Mitigation Arm, 359, 361
    mixer pump off period (MPOP), 414
    origins, 327
    waste transfer and dilution, 382, 383,
        404
temperature profiles
    changes in, 58–59, 71–72, 71f
    mixer pump off period (MPOP), 412
    models, 132–133
    related to heat transfer mechanisms,
        87–89, 88f
    waste behavior, 104
    waste transfer and dilution
        campaigns, 401
TEMPEST computer program
    crust base drop, 363
    description, 131–135
    gas release model, 193, 212, 226, 272
    mixer pump models, 244, 353
    nozzle clogs, 264
    safety analysis and assessment,
        222–223, 228–230
Terrones, Guillermo, 413–414
test chamber, 155, 156, 158, 204–205, 236
test plan, mixer pump. *See* tests, mixer
        pump
Test Review Group (TRG)
    description, 195–196, 443
    Enraf gauges, 307
    excavations runs, 292–293
    hydrogen concentrations, 285
    mixer pump off period, 414
    mixer pump tests, 263, 276–277, 278,
        279, 293
    surface level rise, 312–314
    tank data review, 300
    unplanned pump start, 267
    void fraction instrument test, 325
    waste level rise, 309
tests, mixer pump
    excavation runs, 292
    full-scale, 282–288

tests, mixer pump *(cont.)*
    jet penetration, 286–287, 291
    knockdown phase, 286–287, 291
    phase A, 242–243, 262–265
    phase B, 277–282
    regrowth phase, 285
    safety analysis and assessment (SA),
        291–292
tests, void fraction instrument, 322, 325,
        327
Texas City, Texas, 56
thermal convection currents, 41
thermocouples (TCs)
    calibration, 277
    definition, 443
    Event B, 112
    gas release event data, 201
    mixer pump tests, 265, 284, 286–287,
        292
    multifunction instrument trees
        (MITs), 241
    temperature measurement, 44, 51,
        138, 319–320, 322, 343
    tree removal, 209, 211, 213–214, 214p
    window D event, 130
Thompson, Jim, 152
Thurgood, Marv, 133–134
top dilution *vs.* low dilution, 366–367,
        381, 395–396, 395p, 397
    to dissolve crust, 398
    waste levels, 403
    waste transfer and dilution, 401, 407
total organic carbon, 383
    see also organic compounds
toxicity hazards, 247
training, 269, 380
transfer and back dilution campaigns,
        393t, 410f
transfer pump
    campaigns, 394, 399
    capacity, 421
    creating a crust hole for, 362
    flow rate, 371
    inlet dilution, 366–367, 372, 373p
    selection, 370
    startup procedure, 393
    video camera surveillance, 376
Travis, Jack, 193
trenches, waste disposal, 22
trenches in sediment, 222, 265, 270, 276
Trent, Don, 131p
    gas release projections, 226
    safety analysis and assessment, 193
    safety analysis and assessment, 222
    TEMPEST computer program,
        131–133, 194, 272
    waste viscosity testing, 297

472 | Index

Tri-City Herald, 61, 64, 127, 135, 192, 230, 256
Tri-Party Agreement, 30, 52, 119–120, 375, 443
Trojan Nuclear Power Plant, 149
Tseng, John
    data analysis, 127
    Headquarters task force, 70
    lessons learned meeting, 217–218
    mitigation methods, 146
    mixer pump design, 158
    Pasco mitigation workshop, 152–155, 155–156
    progress report, 116
    public opinion, 84
    SY-101 Hydrogen Gas Mitigation Project, 147–148
    Tank Advisory Panel (TAP), 68
    tank waste priorities, 120
    ultrasonic mitigation system, 151
    waste sampling safety analysis, 80
    Waste Tank Safety Program, 118
    Westinghouse criticisms, 150
    windows, 136, 199–200, 209, 210, 224
Tucker, Ron, 418*p*

U
U-101 tank, 26
U-111 Tank, 31–32
U Plant
    under construction, 13*p*
    description, 12
    functions and processes, 17*t*
    liquid waste disposal, 22
    SY-101 tank wastes, 14
    tank wastes, 26
ultrasonic
    agitation, 144, 146, 153–154, 158, 293, 295
    probe, 236
    waves, 376
uncertainty, 229, 339, 404, 411
underground tank farms. *See* tank farms
University of Chicago, 9
unreviewed safety questions (USQ)
    ammonia release, 274
    definition, 444
    flammable gas safety issue, 288, 295, 300, 416
    hydrogen-nitrous oxide ignition issue, 66–67
    Surface Level Rise, 410*f,* 415
    surface level rise issue, 312–314, 328, 340, 341, 385
unusual occurrence reports, 43, 267
uranium and uranium isotopes, 4, 6, 9–11, 22

U.S. Dept. of Energy, Headquarters
    description, 438
    explosion risks reaction, 34
    Hanford Site contractors, 29–30
    Headquarters task force, 70–71, 80
    mixer pump safety review, 248
    performance incentives, 404, 409–410
    public opinion, 53
    remediation white paper, 404
    safety culture, 81
    surface level rise remediation, 368
    Surface Level Rise Remediation Project, 331, 385–389
    SY-101 Hydrogen Gas Mitigation Project, 164
    tank data reports, 224
    weapons complex, 92
    work approvals process, 125, 129–130, 136
    See also Office of Environmental Restoration and Waste Management
U.S. Dept. of Energy, Office of River Protection (ORP), 342, 354, 415, 419, 441
U.S. Dept. of Energy, Richland Operations Office
    description, 30, 438
    explosion risks announcement, 34
    leadership changes, 67
    performance incentives, 328, 388
    progress report, 116
    Surface Level Rise Remediation Project, 340
    SY-101 Hydrogen Gas Mitigation Project, 164, 204
    Tank Waste Remediation System (TWRS), 148–149
    waste sampling approval, 104
    Westinghouse criticisms, 151
    window G data evaluation, 210
U.S. Environmental Protection Agency, 266

V
value engineering, 333
valve pit, 420
Van Tuyl, Harold, 57, 76
vapor space. *See* dome or domespace of waste tank
Vargo, George, 157, 160, 161*p*, 189–190, 243
variable frequency drive, 393
Veith, Don, 188–189
velocity-density-temperature trees (VDTTs)
    definition, 444

velocity-density-temperature trees *(cont.)*
  as dipsticks, 317*p*
  gas monitoring, 185
  installation, 98*t*, 213, 214
  mixer pump testing, 276
  performance issues, 186–187, 239, 240
  readiness reviews, 211
ventilation
  backup, 120–121, 129, 136
  exhaust, 400, 405, 412
  findings, 117
  flow, 113, 194, 306
  gas release events, 102–103, 104, 121
  measurement, 51
  mixer pump tests, 286
  spark-proof, 83, 120, 129
  systems, 83–84, 119
  tank, 46, 83, 110
  vacuum, 168
video cameras
  data collection and management, 126,
    245, 271, 309, 317
  full-tank scans, 318, 398
  gas release events, 130–131, 207
  hazard identification, 221–222
  in-tank, 146, 169, 348, 395*p*
  infrared, 286
  installation, 121, 123–124, 123*p*
  mixer pump, 255, 399*p*
  mixer pump off period (MPOP), 412
  portable, 402–403
  repairs and replacement, 129, 136, 248
  shut down, 415
  surveillance, 249
  waste transfer and dilution, 376, 394,
    400*p*, 401
video shack, 376, 392, 402, 415
videotapes, 128, 138
viscosity, 127, 132
vitrification, 4, 118, 146, 147, 410, 416
void fraction instrument (VFI), 298*p*,
    322–326
  crust analysis, 346–347
  definition, 444
  gas retention and release, 297–298
  sediment gas distribution, 337–338
  surface level rise, 315, 321
  tests, 322, 325, 327
volute, mixer pump
  exposed after transfer, 406
  flush, 264–265
  photos, 161*p*, 171*p*
  pressure, 319, 395

W

Wagoner, John, 95*p*, 259*p*
  fiscal responsibility warning, 120
  Hazel O'Leary visit, 259*p*

Wagoner, John *(cont.)*
  Headquarters work approval, 121
  project team recognition, 257
  safety concerns, 246
  SY tank farm visit, 95*p*
  tank C-106 contamination incident, 268
  Westinghouse reorganization, 67
  window H work approval, 226
  window I work, 248
Wagoner letter, 354, 382
Wald, Matt, 58, 257, 386
Wallis, Graham, 135
war room, 249, 262, 314, 417
Washington State Dept. of Ecology
  ammonia releases, 405
  Burger Report, 58
  Hanford Square value engineering
    workshop, 332, 334
  proposed new tanks, 119
  remediation work, 104, 266
  safety concerns, 84
Waste Encapsulation and Storage Facility, 16
waste layers
  caked on walls, 407
  foamy crust, 412
  mixer pump off period (MPOP), 412
  temperature profile measurements,
    87–88
  transfer and dilution, 406
  void fraction instrument
    measurements, 322–324
waste levels
  changes, 117, 303–304, 309
  cross-site transfers, 421
  after gas release events, 58–59, 66,
    103–104, 112, 122, 270–271
  gas retention, 307–308
  history
    1977-1981, 39*f*
    1989-1990, 59*f*, 66*f*, 81*f*
    1990-1991, 104*f*, 111*f*, 122*f*
    1991-1992, 130*f*, 202*f*
    1992-1993, 207*f*, 225*f*, 245*f*, 270*f*, 279*f*
    1994-1995, 284*f*, 303*f*
    1995-1996, 306*f*
    2001-2003, 420*f*
    Enraf gauges, 307*f*, 313*f*, 342*f*
    safety window, 80
    transfer and dilution, 391*f*
  limits, 194, 341
  measurement
    crust, 344–345
    dipstick, 187, 240, 317–318, 317*p*, 394
    gas release event, 136
    methods, 51, 99, 244–245, 301–302,
      317
    radar gauge, 111

474 | Index

waste levels (cont.)
Mechanical Mitigation Arm effects
on, 361
mixer pump off period (MPOP), 411
mixer pump tests, 279, 280, 283–284,
285, 292–293
projections, 342, 343f
safety analysis limits, 281
surface level rise project, 311
temperature correlation, 44
transfer and dilution effect on,
396–398, 399, 406, 409
windows, 130, 210
Waste Management and Environmental
Advisory Committee
(WMEAC), 151
Waste Management Federal Services, 305
waste samples, 79–84
approval process, 104
crust, 345, 347
data, 70
errors, 137
full-length core, 124
need for, 72, 73, 77–78, 79
non-waste intrusive, 113
processes, 128
recommendations, 116–117
retained gas samplers, 315–316
sludge weight samplers, 106, 109–110
techniques, 121
waste-penetrating, 61, 81, 112
waste tank safety, 72, 77, 106, 118, 127
Waste Tank Safety Program
budget, 119
change in focus, 139
change recommendations, 118
credibility, 108
fiscal responsibility requirements, 120
formation of, 106
Los Alamos National Laboratory, 116
progress report, 116
support for, 103
Waste Tank Safety Task Team, 76–78, 91,
106, 144
waste temperature profiles. see
temperature profiles
waste temperatures
changes, 138–139, 343
event I, 245
history, 140f
instrument readings, 319–320
limits, 195
measurement, 119
mixer pump tests, 244, 285, 286
monitoring, 46, 151
tank level correlation, 44
thermocouple measurement, 319–320,
396–397

waste temperatures (cont.)
transfer and dilution effect on, 401,
409
See also waste temperature profiles
waste transfer and dilution
campaigns, 393–407, 393t, 410f
cross-site transfers, 420
equipment, 369–377
level history, 391f
mixer pump data reports, 416
pipes, 372, 374
planning, 72, 78, 341
remediation, 145, 147, 293, 295
safety concerns, 381–385
safety improvement plan, 143
studies, 366–368
surface level rise remediation,
335–336, 338, 365
system design, 109
waste to water ratio, 409
waste water, 14
wastebergs
breaking, 212
bumpers, 169, 183
capsize event, 401, 402
crust gas release mechanisms,
351–352
potential pump obstruction, 232
waste transfer and dilution safety
concerns, 383
wastecicles, 111, 130, 302
See also saltcake
wastes, 12
behavior, 144, 281–282, 285, 343, 408
characterization studies, 94
chemical composition, 25–29, 137, 347,
348–349, 383, 409
chemical reactions, 40
cleanup, 52, 106, 148
compatibility criteria, 421
complexed, 17, 26, 31–32, 41–42, 145
concentration, 4, 22–24, 32–33, 118,
423–424
dilution, 147, 236, 413
gas retention, 33–34, 42–43, 306, 411
generated 1944-1998, 25f
heating, 22, 136, 294, 295, 333
intrusive windows, 200, 203, 214
models, 115, 194, 337
public opinion, 52–53, 107
solubility, 367
sources of, 14–17, 22–24, 25
storage, 56–57, 94
temperature profiles, 88f
See also liquid wastes; waste levels;
waste temperatures
watch list
criteria, 76

Index | 475

watch list *(cont.)*
definition, 444
development, 75, 76–77
flammable gas, 188–189, 299, 384
Tank Waste Remediation System
(TWRS), 149
temperature monitoring
requirements, 151
Wyden Amendment requirement for,
95–96
water lance
crust penetration, 298–299
definition, 444
gas release, 44–45
mixer pump installation, 226–227, 249,
250–253, 251*p*
transfer pump hole, 361–362
window I work, 248
water skid, 372
water use
to dissolve crust, 395–396
lancing, 361–362
mixer pump installation, 252
nuclear reactors, 10–11, 14
plutonium production, 11
transfer pump, 370
waste transfer and dilution, 336, 403,
406, 407, 409
waste water solutions, 14
water wands, 237, 248
Watkins, James (U.S. Secretary of Energy)
Ahearne Committee, 71
Booth Gardner letter, 62
Mike Lawrence rebuke, 63
report to congress, 127
SY tank farm visit, 95*p*, 103
tank safety emphasis, 118
waste sampling approval, 112
Weibull distribution, 202, 203
Wells, Beric, 330, 366, 392, 404, 411, 418*p*
Westinghouse Corporate Science and
Technology Center, 103
Westinghouse Hanford Company
Change Control Board, 203
criticisms of, 150–151
culture, 165–166
FATHOMS computer program,
133–134
flammable gas watch list, 188
mixer pump, 155, 163, 173
prime contractor, 29, 57–58
progress report, 116
public relations, 62
reorganization, 106, 148, 205
Safety and Environment Advisory
Council, 196–197, 212, 238, 263
safety measures, 75–76

Westinghouse Hanford Company *(cont.)*
Science Panel, 85
SY-101 Hydrogen Gas Mitigation
Project, 147–148, 151
Westinghouse New Generation Transfer
Pump, 370, 372, 373*p*
See also transfer pump
Westinghouse Nuclear Safety and
Licensing, 274
whistleblowers, 53, 222
Whitaker electrochemical cells, 185, 224,
245
White, Bob, 181*p*
ammonia hazards, 246–247, 273, 275
mixer pump, 195, 226
safety analysis and assessment,
180–181, 185, 197, 219–221, 237
surface level rise task force, 314
tank de-gassing, 281, 295
Test Review Group (TRG), 196
waste level measurement, 303
window I work, 245
White, Mike, 175
White, Mitsy, 309
White Bluffs, Washington, 7
white papers, 59–60, 160, 404
whites (protective clothing), 103, 105*p*,
206
Whitney, Paul, 188
Wiegman, Steve, 354
Wilkins, Nancy, 223, 308, 415–416
window criteria
definition, 444
formalized, 101
revision request, 199–201
revisions, 113–114, 122, 202–203
tank work, 97
waste level drop, 80, 82, 112
windows, 97–140
definition, 444
development, 75
extending, 209–210, 210
waste sampling, 79–80, 82
window A, 98*t*, 99–108, 104
window B, 98*t*, 108–119
window C, 98*t*, 119–129
window D, 98*t*, 129–136
window E, 98*t*, 136–140, 169
window F, 98*t*, 140*t*
window G, 205–219
data acquisition and control
system (DACS), 191*p*
pump installation plans, 163, 180,
196
schedule, 140*t*, 158, 160
work accomplished, 98*t*, 186–187
work costs, 203–204

windows *(cont.)*
   window H, 223–233
      blow-off plate installation, 183
      embarrassment of, 180
      numerical data, 114
      schedule, 140*t*, 158
      work accomplished, 98*t*
   window I, 98*t*, 140*t*, 158, 246–258
   windows J and K, 140*t*
wiring
   mixer pump, 167, 172, 174–175
   nitrogen purge system, 422
   strain gauges, 184
Wise, Barry, 187–188, 249
Wodrich, Don, 59, 60–61, 75, 76, 79–81

work packages, 254, 269–270
working conditions, 206, 211–212, 250, 358–359, 399
World War II, 4, 5–6, 9
Wyden, Ron (Senator), 94, 419
Wyden Amendment, 94–96, 127, 151

Y
Yakama Nation, 238, 246

Z
Zaman, A., 196
zip cord measurements, 129
zirconium, 26
zone of death, 247, 249, 262, 273